Recent Advances in Robotics

Recent Advances in Robotics

Editor: Lauren Barrett

NY RESEARCH PRESS
New York

Published by NY Research Press
118-35 Queens Blvd., Suite 400,
Forest Hills, NY 11375, USA
www.nyresearchpress.com

Recent Advances in Robotics
Edited by Lauren Barrett

International Standard Book Number: 978-1-63238-723-3 (Hardback)

Cataloging-in-Publication Data

Recent advances in robotics / edited by Lauren Barrett.
p. cm.
Includes bibliographical references and index.
ISBN 978-1-63238-723-3
1. Robotics. 2. Automation. 3. Machine theory. 4. Robots. I. Barrett, Lauren.
TJ211 .R43 2020
629.892--dc23

Contents

Preface VII

Chapter 1 **A Single-Way Ranging Localization of AUVs based on PSO of Outliers Elimination** 1
Xinnan Fan, Zhongjian Wu, Jianjun Ni and Chengming Luo

Chapter 2 **Kinematics and Transmission Performance Analyses of a 2T2R Type 4-DOF Spatial Parallel Manipulator** 15
Haiyan An, Bin Li, Shoujun Wang and Weimin Ge

Chapter 3 **Kinematics and Dynamics of a Tensegrity-Based Water Wave Energy Harvester** 26
Min Lin, Tuanjie Li and Zhifei Ji

Chapter 4 **Adaptive Neuro-Fuzzy Inference System based Path Planning for Excavator Arm** 39
Nga Thi-Thuy Vu, Nam Phuong Tran and Nam Hoai Nguyen

Chapter 5 **Service Arms with Unconventional Robotic Parameters for Intricate Workstations: Optimal Number and Dimensional Synthesis** 46
Satwinder Singh and Ekta Singla

Chapter 6 **Method for Effectively Utilizing Node Energy of WSN for Coal Mine Robot** 57
Xiliang Ma and Ruiqing Mao

Chapter 7 **A Low-Cost Model Vehicle Testbed with Accurate Positioning for Autonomous Driving** 65
Benjamin Vedder, Jonny Vinter and Magnus Jonsson

Chapter 8 **A Bioinspired Gait Transition Model for a Hexapod Robot** 75
Qing Chang and Fanghua Mei

Chapter 9 **Error Analysis and Adaptive-Robust Control of a 6-DoF Parallel Robot with Ball-Screw Drive Actuators** 86
Navid Negahbani, Hermes Giberti and Enrico Fiore

Chapter 10 **On the Direct Kinematics Problem of Parallel Mechanisms** 101
Arthur Seibel, Stefan Schulz and Josef Schlattmann

Chapter 11 **Allocating Multiple Types of Tasks to Heterogeneous Agents based on the Theory of Comparative Advantage** 110
Toma Morisawa, Kotaro Hayashi and Ikuo Mizuuchi

Chapter 12 **MapFuse: Complete and Realistic 3D Modelling** 128
Michiel Aernouts, Ben Bellekens and Maarten Weyn

Chapter 13 **Decentralized Cooperative Localization Approach for Autonomous Multirobot Systems** 141
Thumeera R. Wanasinghe, George K. I. Mann and Raymond G. Gosine

Chapter 14 **Theoretical Design and First Test in Laboratory of a Composite Visual Servo-Based Target Spray Robotic System** 159
Dongjie Zhao, Ying Zhao, Xuelei Wang and Bin Zhang

Chapter 15 **Dynamic Surface Adaptive Robust Control of Unmanned Marine Vehicles with Disturbance Observer** 170
Pengchao Zhang

Chapter 16 **Long Short-Term Memory Projection Recurrent Neural Network Architectures for Piano's Continuous Note Recognition** 176
YuKang Jia, Zhicheng Wu, Yanyan Xu, Dengfeng Ke and Kaile Su

Chapter 17 **Modular Self-Reconfigurable Robotic Systems: A Survey on Hardware Architectures** 183
S. Sankhar Reddy Chennareddy, Anita Agrawal and Anupama Karuppiah

Permissions

List of Contributors

Index

Preface

It is often said that books are a boon to mankind. They document every progress and pass on the knowledge from one generation to the other. They play a crucial role in our lives. Thus I was both excited and nervous while editing this book. I was pleased by the thought of being able to make a mark but I was also nervous to do it right because the future of students depends upon it. Hence, I took a few months to research further into the discipline, revise my knowledge and also explore some more aspects. Post this process, I begun with the editing of this book.

Robots are designed to perform a complex series of tasks automatically. The field which deals with design, operation, construction and use of robots is referred to as robotics. It is an interdisciplinary field of science and engineering which includes mechanical engineering, information engineering, computer science and electronic engineering. Robotics utilizes computer systems for sensory feedback, information processing and to control robots. They are especially used in the situations that are dangerous for humans such as in manufacturing processes, bomb detection and deactivation, high heat, space and clean-up of hazardous materials and radiations. Industrial robots, mobile robots, modular robots, collaborative robots, service robots and educational robots are the examples of the most commonly used modern robots. This book presents the complex subject of robotics in the most comprehensible and easy to understand language. It strives to provide a fair idea about this discipline and to help develop a better understanding of the latest advances within this field. Through this book, we attempt to further enlighten the readers about the new concepts in this field.

I thank my publisher with all my heart for considering me worthy of this unparalleled opportunity and for showing unwavering faith in my skills. I would also like to thank the editorial team who worked closely with me at every step and contributed immensely towards the successful completion of this book. Last but not the least, I wish to thank my friends and colleagues for their support.

Editor

1

A Single-Way Ranging Localization of AUVs based on PSO of Outliers Elimination

Xinnan Fan,[1,2] Zhongjian Wu,[1] Jianjun Ni,[1,2] and Chengming Luo[1]

[1]*College of IOT Engineering, Hohai University, Changzhou 213022, China*
[2]*Jiangsu Universities and Colleges Key Laboratory of Special Robot Technology, Hohai University, Changzhou 213022, China*

Correspondence should be addressed to Jianjun Ni; njjhhuc@gmail.com

Academic Editor: L. Fortuna

Localization of autonomous underwater vehicles (AUVs) is a very important and challenging task for the AUVs applications. In long baseline underwater acoustic localization networks, the accuracy of single-way range measurements is the key factor for the precision of localization of AUVs, whether it is based on the way of time of arrival (TOA), time difference of arrival (TDOA), or angle of arrival (AOA). The single-way range measurements do not depend on water quality and can be taken from long distances; however, there are some limitations which exist in these measurements, such as the disturbance of the unknown current velocity and the outliers caused by sensors and errors of algorithm. To deal with these problems, an AUV self-localization algorithm based on particle swarm optimization (PSO) of outliers elimination is proposed, which improves the performance of angle of arrival (AOA) localization algorithm by taking account of effects of the current on the positioning accuracy and eliminating possible outliers during the localization process. Some simulation experiments are carried out to illustrate the performance of the proposed method compared with another localization algorithm.

1. Introduction

In a three-dimensional underwater environment, the problem of localization of autonomous underwater vehicles (AUVs) has been widely investigated in recent years, as they make it accessible to those untouchable areas for human beings and assist with complex and arduous underwater tasks, which has important theory and application value in various robotic applications, such as the underwater target detection [1], the underwater target tracking [2, 3], and underwater search and rescue missions [4, 5]. Localization is one of the key components that enable the autonomy of AUVs. AUVs need accurate localization for the accuracy of the gathered data. For a mobile robot to be truly autonomous, it needs to be able to operate and navigate without human intervention and in a nonspecially engineered environment [6, 7]. And thanks to the uncertainty and complexity of the underwater environment and technical restrictions, it is still challenging and needs more complementary researches that localize a moving AUV. Indeed, the research of robotic navigation in two-dimensional space has achieved great results and extended to the study of multirobot systems [8], while during underwater navigation, AUVs cannot rely on Global Navigation Satellite Systems (GNSSs) due to the attenuation of electromagnetic radiation in the water domain, and in the absence of specific positioning systems, they can exclusively relay on dead-reckoning techniques. As the latter approaches integrate noisy and biased measurements from Inertial Measurement Units (IMUs) and velocity sensors, they suffer from numerical drift that makes them usable only for relatively short periods [9]. And the inertial navigation system (INS) is also mostly applied on AUVs, but INS usually suffers from error accumulation.

Commercially available underwater positioning solutions are mainly based on acoustic devices that, through the measurement of the time of flight of acoustic signals, allow us to measures the ranges from source to receiving nodes. Acoustic localization system for AUVs mainly includes long baseline (LBL), short baseline (SBL), and ultra-short baseline (USBL). More recently, research efforts focused on position systems based on the use of range measurements to a single node with the aim of developing the solutions which is simple, cheap, and easy to operate. Such approaches, known in the literature as single beacon localization [9], single range localization [10], or range-only localization [11], are based on the fusion of range measurements to the single source with information from AUV's onboard sensors as Inertial Measurement Unit (IMU), Doppler Velocity Logger (DVL), and depth sensors.

However, time-synchronization is hard to achieve in underwater environment, which baffles the accuracy of ranging and leads to inaccuracy of localization. Sound speed uncertainty enlarges the inaccuracy of distance estimation. Long propagation delay of acoustic signals, the impact of AUV mobility, and the drift term caused by the water current are the urgent problems to be solved, which will affect the accuracy of underwater acoustic positioning. Moreover, the severe multipath property in underwater environment caused by the scattering from the seabed and water surface is also needed to be overcome. In LBL, AUV communicates with one or more transponders fixed on the seabed or the surface of the water to measure distance and achieve underwater acoustic localization; it can also be classified as time of arrival (TOA), time difference of arrival (TDOA), and angle of arrival (AOA) according to the specific positioning mode. TOA needs strict time-synchronization, which is energy-intensive and hard to achieve. Multiple-node cooperation in TDOA [12, 13] increases cost and is computationally complex. Moreover, the LBL multiple-array acoustic localization [14] based on multiple-sensor data fusion mostly adopts the above two or more methods, so as to acquire more accurate location estimation. However, the positioning algorithms based on data fusion are relatively complicated and need to take much time to calculate and fuse the results of different positioning methods; thus, practical application is not desirable in view of low-cost requirements.

As mentioned above, various improvements have been proposed to deal with the problems about underwater localization of AUVs. For example, Tan and Li [15] proposed a centralized algorithm to overcome the severe multipath property of the underwater environment. Cheng et al. [16] presented a TDOA-based localization scheme for stationary underwater acoustic sensor networks, which does not require time-synchronization among network nodes. Gao et al. [17] proposed the round-trip ranging (RTR) technology frustrated by the long propagation delay, the drift term of the water current, multipath property of the underwater environment, and AUVs' mobility. And Li et al. [18] presented a self-localization algorithm with accurate sound travel time solution (SL-STTS) which solved the problem of time-synchronization and sound speed uncertainty; however, the drift term of the water current was ignored and the measurements of multiple sensors are not fully utilized.

In this paper, an improved single-way ranging localization approach based on PSO of outliers elimination is proposed. In the proposed approach, the problem of vehicle autonomous localization under complex underwater environment is fully considered and some solutions are presented in the following paragraphs, including the drift term caused by the current and the outliers from multiple-array acoustic localization caused by sensors. Finally, various experiments

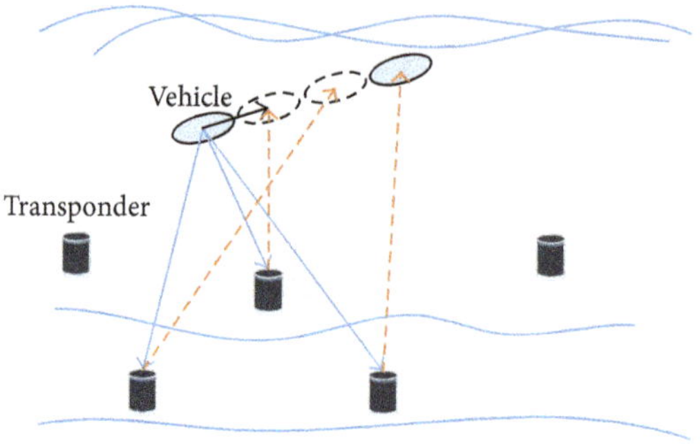

FIGURE 1: Network model of the underwater localization.

are conducted under different noise conditions for the vehicle autonomous localization. The results show the efficiency and affectivity of the proposed approach. The main contributions of this paper are summarized as follows. (1) The effect of the water current on the localization for AUV is fully taken into account and the relative analysis is carried out. (2) An outlier detection approach of multiple-array acoustic localization is proposed to eliminate the abnormal results of multiple localization combination.

This paper is organized as follows. Section 2 presents the improved SL-STTS-based method for the vehicle autonomous localization method. Section 3 presents a method for detecting and removing outliers from sensor in multiple-array acoustic localization. Performance evaluation of the proposed algorithms is presented in Section 4 through several simulations. Section 5 gives out some discussions on the robustness and fault tolerance of the proposed approach. Finally, the conclusions are given in Section 6.

2. Single-Way Ranging Localization

Considering the 3D AOA localization problem of the underwater vehicle in Figure 1, where the vehicle at an undetermined location $R_t : (x_{R_t}, y_{R_t}, z_{R_t})$ is to be located by communicating with acoustic transponders at the fixed inertial location $S_i : (x_{s_i}, y_{s_i}, z_{s_i})$, $i = 1, 2, \ldots, N$ (where N is the number of the acoustic transponders), using the azimuth and elevation angles. Here, at least three transponders are required for 3D AOA vehicle localization. The orientation of the vehicle includes yaw and pitch angle (ignore the posture of the vehicle, so skip the roll angle). With no loss of generality, the vehicle and transponders are assumed to be at the Cartesian coordinate system. The vehicle velocity is given by a superposition of a drift term v_f and a controlled input term v_r. For underwater vehicle, the velocity term v_r can be generated through a guidance controller exploiting an onboard navigation sensor as a Doppler velocity logger (DVL). The drift term v_f models a constant unknown water current. The vehicle is also equipped with axis gyro and inclinometer for measuring its yaw angle and pitch angle and AOA antennas for measuring AOA of received acoustic signals. Similarly, the transponders are equipped with AOA antennas for measuring AOA of received acoustic signals. Conductivity-temperature-depth (CTD) instruments are preloaded onto the vehicle and transponders to help

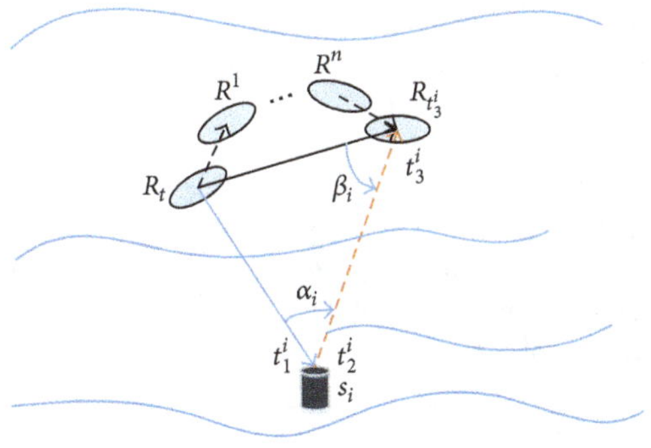

FIGURE 2: Model of TWTT between the AUV and transponder s_i.

estimate sound speed by measuring salinity, temperature, and depth [19].

2.1. Problem Formulation. As the vehicle transmits a sound signal at time instance t, meanwhile, a timer starts. Assume that there are N ($N \geq 3$) transponders detecting the sound signal from the vehicle. For transponder s_i, after s_i receiving the sound signal at its local time t_1^i, it puts its number, the location, the sound speed measured by CTD [20], and the AOA of received acoustic signal into a localization packet Φ_i. At its local time t_2^i, it adds the leaving time t_2^i into Φ_i as soon as it responds to the AUV with Φ_i. The vehicle receives Φ_i at its local time t_3^i. When the timer is out, the vehicle stops receiving replies from transponders and processes the received localization packets; then the vehicle at time instance t is located according to the location packet. In order to overcome the severe multipath property in underwater environment, the AUV will no longer receive acoustic signals from the same transponder before it completes one localization.

It can be observed that the vehicle may have moved to the next location when it receives Φ_i at t_3^i due to the vehicle's motion and the drift term of the current as shown in Figure 2. Thus, the one-way travel time (OWTT) of an acoustic signal traveling from the AUV to a transponder may not equal the opposite case. The problem of the time is not synchronized between transponders and AUV is considered in [18], which measures the two-way travel time (TWTT) Δt_i based on transponder s_i with

$$\Delta t_i = \left(t_3^i - t\right) - \left(t_2^i - t_1^i\right). \quad (1)$$

Assume the vehicle has moved from R_t to $R_{t_3^i} : (x_{R_{t_3^i}}, y_{R_{t_3^i}}, z_{R_{t_3^i}})$ when the vehicle receives Φ_i as shown in Figure 2. For simplicity we assume v_r to be a constant in a short time, but v_f is unknown but bounded; thus, the traveling distance of the vehicle is calculated by

$$D_{if} = \left(t_3^i - t\right) \cdot \left(v_r + v_f\right). \quad (2)$$

Define the OWTT of a signal traveling from the vehicle (at R_t) to transponder s_i as t_{if}^{VT}, from transponder s_i to the vehicle as t_{if}^{TV}. Because t_{if}^{VT} and t_{if}^{TV} may not be equal; thus, the TWTT model is redefined as follows:

$$\Delta t_i = t_{if}^{VT} + t_{if}^{TV} = 2t_{if}^{VT} + \varepsilon_i, \quad (3)$$

where $\varepsilon_i = t_{if}^{TV} - t_{if}^{VT}$ denotes the difference between t_{if}^{VT} and t_{if}^{TV}.

In Figure 2, let α_i and β_i, respectively, denote the angle between vectors $\overrightarrow{R_t S_i}$ and $\overrightarrow{R_{t_3^i} S_i}$ and the angle between vectors $\overrightarrow{R_t R_{t_3^i}}$ and $\overrightarrow{S_i R_{t_3^i}}$. Define the elevation AOA and azimuth AOA of the received acoustic signal from vehicle as χ_i and γ_i and the elevation AOA and azimuth AOA of the received acoustic signal from transponder s_i as ϕ_i and φ_i, which are measured from the positive z-axis and x-axis by AOA antennas, respectively. Since in shallow water, the ray trace of sound can be treated as straight line [21], the values of direction cosine for $\overrightarrow{R_{t_3^i} S_i}$ along x-axis, y-axis, and z-axis are $\sin\chi_i \cos\gamma_i$, $\sin\chi_i \sin\gamma_i$, and $\cos\chi_i$, respectively. And the values of direction cosine for $\overrightarrow{S_i R_{t_3^i}}$ along x-axis, y-axis, and z-axis are $\sin\phi_i \cos\varphi_i$, $\sin\phi_i \sin\varphi_i$, and $\cos\phi_i$, respectively [22].

The pitch angle θ_i and yaw angle ψ_i of the vehicle can be measured by inclinometer and z-axis gyro. The only difference between the vehicle's local coordinates and the global Cartesian coordinates is that z_l-axis is in the opposite direction of z-axis. Thus, in the global Cartesian coordinates, the pitch angle of vehicle is $(\pi - \theta_i)$ and the yaw angle is still ψ_i. The values of direction cosine for $\overrightarrow{R_t R_{t_3^i}}$ along x-axis, y-axis, and z-axis are $\sin(\pi - \theta_i)\cos\psi_i$, $\sin(\pi - \theta_i)\sin\psi_i$, and $\cos(\pi - \theta_i)$, respectively [18].

Define the unit vectors of $\overrightarrow{R_t S_i}$, $\overrightarrow{S_i R_{t_3^i}}$, and $\overrightarrow{R_t R_{t_3^i}}$ as μ, λ, and κ, respectively. Then it can easily get their expression according to the angle measurements, resulting in the fact that the cosine value of angles α_i and β_i can be solved by using their unit vectors as follows.

$$\mu = \left(\sin\chi_i \cos\gamma_i, \sin\chi_i \sin\gamma_i, \cos\chi_i\right), \quad (4)$$

$$\lambda = \left(\sin\phi_i \cos\varphi_i, \sin\phi_i \sin\varphi_i, \cos\phi_i\right), \quad (5)$$

$$\kappa = \left(\sin\left(\pi - \theta_i\right)\cos\psi_i, \sin\left(\pi - \theta_i\right)\sin\psi_i, \cos\left(\pi - \theta_i\right)\right), \quad (6)$$

$$\cos\alpha_i = \frac{\overrightarrow{R_t S_i} \cdot \overrightarrow{R_{t_3^i} S_i}}{\left|\overrightarrow{R_t S_i}\right| \cdot \left|\overrightarrow{R_{t_3^i} S_i}\right|} = -\lambda \cdot \mu, \quad (7)$$

$$\cos\beta_i = \frac{\overrightarrow{S_i R_{t_3^i}} \cdot \overrightarrow{R_t R_{t_3^i}}}{\left|\overrightarrow{S_i R_{t_3^i}}\right| \cdot \left|\overrightarrow{R_t R_{t_3^i}}\right|} = \lambda \cdot k. \quad (8)$$

It can be observed that $\alpha_i = \angle R_t S_i R_{t_3^i}$ and $\beta_i = \angle S_i R_{t_3^i} R_t$ in Figure 2; thus, cosine value of α_i and β_i can also be denoted as follows:

$$\cos\alpha_i = \frac{R_t S_i^2 + S_i R_{t_3^i}^2 - R_t R_{t_3^i}^2}{2 \cdot R_t S_i \cdot S_i R_{t_3^i}} = \frac{\Delta t_i^2 + \varepsilon_i^2 - 2D_{if}^2/C^2}{\Delta t_i^2 - \varepsilon_i^2}, \quad (9)$$

$$\cos\beta_i = \frac{R_t R_{t_3^i}^2 + S_i R_{t_3^i}^2 - S_i R_{t_3^i}^2}{2 \cdot R_t R_{t_3^i} \cdot S_i R_{t_3^i}} = \frac{D_{if} + \Delta t_i \varepsilon_i C^2}{D_{if}\left(\Delta t_i + \varepsilon_i\right)C}, \quad (10)$$

where C is the sound speed. Thus, the following system of equations can be derived by substituting (7), (8) into (9), (10), respectively,

$$\frac{\Delta t_i^2 + \varepsilon_i^2 - 2D_{if}^2/C^2}{\Delta t_i^2 - \varepsilon_i^2} = -\lambda \cdot \mu,$$
$$\frac{D_{if} + \Delta t_i \varepsilon_i C^2}{D_{if}(\Delta t_i + \varepsilon_i)C} = \lambda \cdot k. \tag{11}$$

It easily leads to ε_i and v_f by substituting (1), (2) into (11); then the OWTT t_{if}^{VT} of a signal traveling from the vehicle (at R_t) to transponder s_i and the OWTT t_{if}^{TV} from transponder s_i to the vehicle (at $R_{t_3^i}$) can be further solved by combining (3).

Note that the vehicle orientation may have changed when the vehicle receives the reply from transponder s_i, as shown in Figure 2. Thus, in this case, we should reconsider the combined displacement $\overrightarrow{R_tR_{t_3^i}}$ according to displacement in each vector velocity. The specific strategies are as follows: the vehicle changes direction every time; its direction can be expressed as $\overrightarrow{R_tR^1}, \overrightarrow{R^1R^2}, \ldots$, and $\overrightarrow{R^nR_{t_3^i}}$ separately. Define the unit vectors of them as $\lambda_1, \lambda_2, \ldots$, and λ_n, respectively. Then it can be known that

$$\overrightarrow{R_tR_{t_3^i}} = \overrightarrow{R_tR^1} + \overrightarrow{R^1R^2} + \cdots + \overrightarrow{R^nR_{t_3^i}},$$
$$\lambda_{\text{sum}} = \lambda_1 + \lambda_2 + \cdots + \lambda_n, \tag{12}$$

where $\lambda_1, \lambda_2, \ldots$, and λ_n can be calculated with the same form as λ described in (5).

2.2. Multiple-Array Underwater Acoustic Localization. In underwater environment, the sound speed is not constant because of the different temperature, pressure (depth), and salinity. But with varying salinity, temperature, and depth, thus the sound speed C is computed by exploiting the empirical formula [12]: namely,

$$\begin{aligned} C(T,s,h) &= 1492.9 + 3(T-10) - 6 \cdot 10^{-3}(T-10)^2 \\ &\quad - 4 \cdot 10^{-2} \cdot (T-18)^2 + 1.2(s-35) \\ &\quad - 10^{-2}(T-18)(s-35) + \frac{h}{61}, \end{aligned} \tag{13}$$

where C is the temperature (in Celsius), h is the depth (in meters), and s is the salinity (in parts per thousand). Define the sound speed for the vehicle and transponder s_i as C_V and C_T, respectively, which can be calculated after measuring the corresponding parameters.

The single-way distance measurements are obtained by the distance estimation method [18]; it is called accurate single-way distance measurement (ASDM) owing to the fact that the current is taken into account in this paper: namely,

$$d_i' \approx t_i^{VT} E_i(C), \tag{14}$$

where $E_i(C)$ denotes mathematical expectation of sound speed acquired by calculating the sound speed of different depths.

In a three-dimensional underwater environment, at least three transponders are required for 3D AOA target localization. Thus, N ($N \geq 3$) transponders will have C_N^3 kinds of positioning combinations, which can solve the multiple-group target position according to single-way distance measurements as follows:

$$\sqrt{\left(x_{R_t} - x_{s_l}\right)^2 + \left(y_{R_t} - y_{s_l}\right)^2 + \left(z_{R_t} - z_{s_l}\right)^2} = d_l',$$
$$\sqrt{\left(x_{R_t} - x_{s_m}\right)^2 + \left(y_{R_t} - y_{s_m}\right)^2 + \left(z_{R_t} - z_{s_m}\right)^2} = d_m', \tag{15}$$
$$\sqrt{\left(x_{R_t} - x_{s_n}\right)^2 + \left(y_{R_t} - y_{s_n}\right)^2 + \left(z_{R_t} - z_{s_n}\right)^2} = d_n',$$

where $(x_{R_t}, y_{R_t}, z_{R_t})$ is the undetermined position coordinate of the vehicle, $(x_{s_i}, y_{s_i}, z_{s_i})$ denote the determined position coordinates of the corresponding transponders, and $i = 1, 2, \ldots, N$; $l, m, n \in i$ and $l \neq m \neq n$, d_l', d_m', and d_n' denote the distance measurements from the vehicle to the corresponding transponder, respectively.

Although multiple measurements of the undetermined vehicle location can be obtained according to various localization combinations, diverse measurements contain different credible information of the actual vehicle location due to the error of the algorithm and the complexity and uncertainty in the underwater environment. As a result of this difference, partial measurements acquired by the above method can belong to outliers. Therefore, a particle swarm optimization (PSO) algorithm based on the outliers elimination is proposed in this paper and the details will be given in the following paragraphs.

3. PSO Based on Outliers Elimination

Based on multiple-array positive sound source localization system, more accurate target position can be obtained by fusing multisource information; however, on the basis of the problem proposed in Section 2, it is necessary to eliminate the outliers that contain poor information of target position. The occurrence of outliers is mainly caused by the following factors, the measured values obtained by the sensor at a certain moment contain too much noise information thanks to the uncertain underwater environment, the packets Φ_i can also be broken or damaged when the transponders respond to the vehicle with Φ_i, and so on. In this paper, we adopt the anomaly detection algorithm called the Isolation Forest [23] to search for the outliers from multiple measurements of the undetermined vehicle location. Then the algorithm of particle swarm optimization based on simulated annealing is adopted to optimize measurements of the target location after the outliers have been eliminated.

3.1. Elimination of Outliers. The key to eliminate the outliers is to detect outliers, which is called the outlier detection and has wide application in the field of machine learning and research. An outlier is an observation which deviates so much from other observations as to arouse suspicions that it was generated by a different mechanism [24]. Isolation Forest (iForest) Algorithm [23] is an anomaly detector that measures without distance or density. It performs an operation to isolate each instance from the rest of instances in a given data set. Because outliers have characteristics of being 'few and different,' they are more susceptible to isolation in a tree structure than normal instances. Therefore, outliers have shorter average path lengths than those of normal instances over a collection of isolation trees (iTrees). Therefore, based on this consideration, we adopt the approach of iForest to eliminate the possible anomalies in the multiple-array underwater acoustic positioning.

The whole algorithm of the iForest mainly contains two stages. The first is training stage; iTrees are constructed by recursively partitioning the given training set until instances are isolated or a specific tree height is reached that will result in a partial model [24]. When building an iTree, first we randomly extract a batch of samples from the given training set and then randomly select an attribute and a value between the maximum value and the minimum value of the attribute. The data in the sample that are smaller than the value of the data are divided into the left tree, and the data that are greater than or equal to the data are divided into the right tree. Finally, in the left and right branches, repeat the above steps until the following conditions are met: the data cannot be divided, that is only one data, or all the same data; the binary tree reaches the limit of the maximum depth. Details of the training stage are described in [23].

The second stage is evaluating stage; an anomaly score Score(x) is derived from the expected path length $E(h(x))$ for each test instance. $E(h(x))$ are derived by passing instances through each iTree in an iForest; a single path length $h(x)$ is derived by counting the number of edges e from the root node to a terminating node as instance x traverses through an iTree. When x is terminated at an external node, where Size > 1, the return value is e plus an adjustment $C(T.\text{size})$. Given a data set of n instances, thus the path length $h(x)$ of the instance x in each iTree can be calculated as follows:

$$h(x) = e + C(T.\text{size}), \tag{16}$$

where e denotes the number of edges from the root node to a terminating node and $C(T.\text{size})$ is an adjustment and accounts for an unbuilt subtree beyond the tree height limit. $C(n)$ is calculated by

$$C(n) = 2H(n-1) - \frac{2(n-1)}{n}, \tag{17}$$

where $H(i)$ is the harmonic number and it can be estimated by $\ln(i) + 0.5772156649$ (Euler's constant). As $C(n)$ is the average of $h(x)$ given n, we use it to normalize $h(x)$. The anomaly score Score(x) of an instance x is defined as

$$\text{Score}(x, n) = 2^{-E(h(x))/C(n)}, \tag{18}$$

where $E(h(x))$ is the average of $h(x)$ from a collection of isolation trees. n is the sample size of training sample in single iTree. $C(n)$ is the average path length of the binary tree constructed by using n instances.

According to formula (18) of the anomaly score, when $E(h(x)) \rightarrow C(n)$, Score $\rightarrow 0.5$, then the entire sample does not really have any distinct anomaly; when $E(h(x)) \rightarrow 0$, Score $\rightarrow 1$, then they are definitely anomalies; when $E(h(x)) \rightarrow n-1$, Score $\rightarrow 0$, then they are quite safe to be regarded as normal instances.

The result of multiple-array localization will be taken as the evaluation sample and input to the trained network of the binary tree. Then it can be searched for outliers from the multiple-array localization and remove them to get relatively accurate measurements for the undetermined vehicle location. Though it is still unknown which one is closer to the real vehicle location, PSO is a population-based stochastic optimization technique which can effectively solve complex optimization problems [25]. Thus, an improved PSO algorithm will be adopted to optimize the relatively accurate measurements in multiple-array positive sound source localization.

3.2. Improved PSO Localization.

The basic idea of PSO approach based on outliers elimination is that the vehicle optimizes the result of multiple-array localization after eliminating outliers, and then it easily gets the suboptimal localization result for the real vehicle position by this optimization. In the interest of ensuring the convergence of PSO and eliminating the boundary, we use PSO to increase the contraction factor. The update equation of the speed and location is as follows:

$$\begin{aligned} v_{ij}(k+1) &= \chi\left[v_{ij}(k) + c_1 r_1\left(p_{ij}(k) - x_{ij}(k)\right)\right. \\ &\quad \left. + c_2 r_2\left(p_{gj}(k) - x_{ij}(k)\right)\right], \\ x_{ij}(k+1) &= x_{ij}(k) + v_{ij}(k+1), \end{aligned} \tag{19}$$

where c_1 and c_2 (nonnegative constants) are the learning factor. $\chi = 2/|2 - C - \sqrt{C^2 - 4C}|$ is the contraction factor, $C = c_1 + c_2 > 4$. r_1 and r_2 are the independent pseudo-random numbers obeyed uniform distribution of the $[0, 1]$. $i = [1, m]$, $j = [1, n]$, m and n represent the number and dimension of the particle, respectively. $p_{ij}(k)$ represents the best position that each particle has been experienced, and $p_{gj}(k)$ represents the best position that has been experienced globally.

Note that the particles are likely to move towards a local minimum by using the above formula to update. Then take the basic idea of the simulated annealing (SA) into account and determine a location $p'_{ij}(k)$ from numerous $p_{ij}(k)$ as the global optimum to replace the actual global optimum $p_{gj}(k)$ with the strategy of roulette. Thus, update the speed as follows:

$$\begin{aligned} v_{ij}(k+1) &= \chi\left[v_{ij}(k) + c_1 r_1\left(p_{ij}(k) - x_{ij}(k)\right)\right. \\ &\quad \left. + c_2 r_2\left(p'_{ij}(k) - x_{ij}(k)\right)\right]. \end{aligned} \tag{20}$$

The probability for each individual can be selected as the global optimum is calculated by

$$p_i = \frac{e^{-(f_{pi}-f_{pg})/t}}{\sum_{j=1}^{N} e^{-(f_{pj}-f_{pg})/t}}, \quad (21)$$

where N is the size of species. e is the constant. f_{pi} represents the fitness of the ith particle and f_{pg} represents the fitness of the global optimal particle undergone.

The single-way distance measurements are assumed to be disturbed by noise with Gaussian distribution due to the measurement error of the time and angle

$$d_i' = d(p) + w, \quad w \sim N(0, \mathrm{COV}), \quad (22)$$

where $d(p) = (d_1(p), d_2(p), \ldots, d_i(p), \ldots, d_N(p))^T$ is the vector of real distances between the vehicle and transponders S_i, $d_i(p) = \sqrt{(x_{R_t} - x_{s_i})^2 + (y_{R_t} - y_{s_i})^2 + (z_{R_t} - z_{s_i})^2}$, $p = (x_{R_t}, y_{R_t}, z_{R_t})^T$; $d' = (d_1', d_2', \ldots, d_N')^T$ is the measurement of $d(p)$; w represents the measurement noise vector which cause ranging errors; $\mathrm{COV} = \mathrm{diag}(\sigma_1^2, \sigma_2^2, \ldots, \sigma_i^2, \ldots, \sigma_N^2)$ is the covariance matrix of w, where σ_i^2 is the variance of the distance measurement noise based on transponder s_i.

In order to get the accurate vehicle position p, we take the sum square error function of distance measurements as the fitness function of PSO method

$$\mathrm{fitness}(p) = \left\| d(p) - d' \right\|^2, \quad (23)$$

where $\|\cdot\|$ represents the 2-norm.

Remark 1. Based on the proposed network model, the weight for distance measurement noise is comprised of time and angle measurement noise. In simulation experiments, their impact on localization accuracy will be further analyzed in the following paragraphs, respectively.

4. Simulation Experiments

In this paper, to test the performance of the proposed approach, some simulation experiments are conducted which were coded in MATLAB. In these experiments, a vehicle with some sensors moves in a predetermined trajectory, some single-way range localization algorithms are used to keep track of the robot position, and the vehicle travels freely with velocity with v_r and constrained by the current with v_f in a 3D underwater environment of 400 m × 400 m × 300 m. The space coordinate varies from [0, 0, 100] to [400, 400, 400]. Eight transponders are deployed in the space coordinate varying from [0, 0, 350] to [400, 400, 400] (see Figure 3). The noise model is unknown, for simplification without loss of generality, the noise is given out artificially in the simulation experiments. The sound speed is set to be 1500 m/s. The specific type of sensors and the noise function for a given sensor are ignored in this paper. To show the advantages of PSO localization approach based outliers elimination, it is compared with the RTR technology [17] and SL-STTS algorithm [18]. To remove the random effects on the single-way range localization and validate the performance of our algorithm, every experiment was conducted 20 times. The parameters in all of the experiments are the same, which are listed in Table 1.

TABLE 1: Simulation parameters of the used instruments for underwater vehicle localization.

Instrument	Parameters	Precision
Inclinometer	Pitch	0.01–1°
Z-axis gyro	Yaw	0.01–0.1°
Doppler velocity logger	Velocity	0.1–0.3 cm/s
Timer	Time	0.001–0.01 s

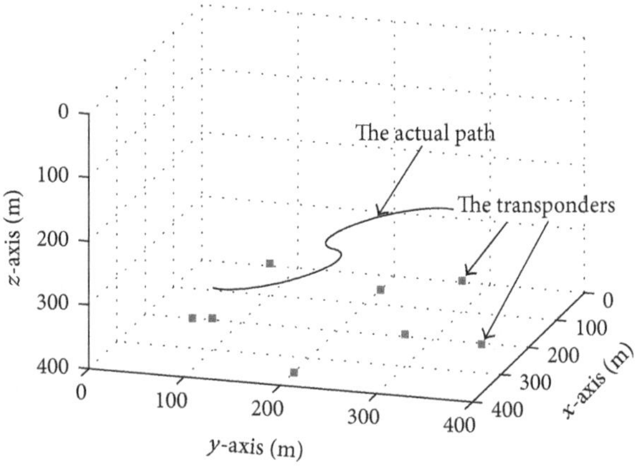

FIGURE 3: The simulation environment of the localization task.

ASDM noises w are mainly caused by the noises of time and angle of arrival measurement. Thus, the distance measurement performance of each algorithm under the same distance measurement noise condition will be firstly analyzed in Section 4.1; the impact of distance measurement noise w on the localization accuracy will be compared in Section 4.2; then this paper further investigates the impact of the time measurement noise w_t and the angle measurement noise w_a on the localization accuracy, respectively.

4.1. Performance Comparison of Range Measurement. In this experiment, the noise of the distance measurement is under Gaussian distribution condition; we set $w \sim N(0, 10^{-1})$, $v_f = 1.65$ m/s (in Figure 4), and $v_r = 2$ m/s (in Figure 5). The results of the distance measurement performance based on the proposed approach in this experiment are shown in Figures 4 and 5. The absolute difference between the real distance and its measurement is regarded as error standard of distance measurement. In Figure 4, for the simplicity, we compare the distance measurement performance under the different vehicle speed v_r due to the vehicle speed that can affect the time difference $|\varepsilon_i|$ and $|\varepsilon_i|$ increasing monotonically with v_r. In Figure 5, e_r and e_f represent the unit vector of v_r and v_f, respectively, where the distance measurement performance of the proposed approach under the different current speed v_f is compared to other methods,

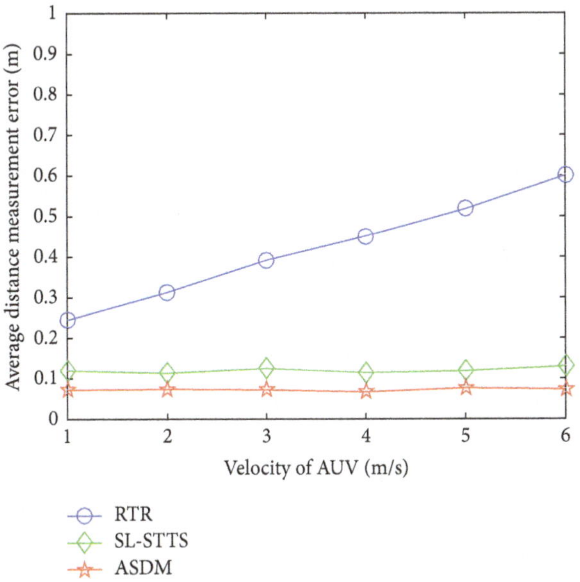

FIGURE 4: Average distance measurement error versus velocity of AUV.

and several typical current directions relative to the vehicle speed v_r are considered.

Remark 2. The average distance measurement error (ADME) is calculated by

$$\text{ADME} = \sqrt{\frac{1}{n}\sum_{i=1}^{n}\left\|d_i - d_i'\right\|^2}, \qquad (24)$$

where d_i and d_i' are the real value and the predicted value of single range from the vehicle to the transponder at the ith step, respectively; n is set as 20. $\|\cdot\|$ represents the 2-norm.

The results of ADME in Figure 4 show that the distance measurements calculated by the ASDM are closest to the actual distance from the vehicle to the transponders, the time difference $|\varepsilon_i|$ will simultaneously grow with v_r increasing, and the error of the proposed approach is less than both the RTR and SL-STTS methods, while the ASDM method is less affected, which shows that the ASDM approach can deal with the single-way ranging measurements stably. Because the size and direction of the current v_f are unknown but bounded, thus we adopt several typical directions for the current v_f relative to v_r, and the results of single-way distance measurement performance relative to the current v_f of the different size and direction in Figure 5 show that the error of the proposed approach is less than both RTR and SL-STTS approaches. It is easy to observe that the proposed method is less affected by the current v_f. The results in this experiment show that the proposed approach has a good performance to deal with the localization problem with $|\varepsilon_i|$ increasing.

4.2. Performance Comparison of Localization. To further validate the performance of the proposed approach, this experiment is conducted. We adopt the root mean square error (RMSE), the maximum absolute error (MAXE), and the minimum absolute error (MINE) as performance evaluation indexes. In this experiment, we firstly analyze how the range measurement noise affects localization accuracy. It should be noted that the accuracy of time Δt_i and angles dominates the precision of OWTT estimation and further affects localization performance. So how the measurement noise of time w_t and angles w_a impacts localization will be further analyzed, respectively. The results of the localization task based on the proposed approach are compared with the SL-RTR and SL-STTS algorithm in Figure 6, where the vehicle moves on the specified trajectory shown in Figure 3 and performs its self-localization, and the sampling time interval $T_s = 1(s)$. Figure 7 is the root mean square error comparison of the vehicle under different noise levels, and the performance indexes of the localization task under different noise levels are listed in Tables 2–4. For simplicity, the variance of distance measurement noise σ_i^2 based on transponder s_i is set equal and independent of each other; the time measurement Δt_i based on transponder s_i is assumed to be disturbed by Gaussian distributed noise with $w_t \sim N(0, \sigma_t^2)$; the input angles are disturbed by Gaussian distributed noise with $w_a \sim N(0, \sigma_a^2)$, and we set $v_r = 2\,(\text{m/s})$ and $v_f = 0.54\,(\text{m/s})$.

Remark 3. The root mean square error (RMSE) in this study is calculated by

$$\text{RMSE} = \sqrt{\frac{1}{n}\sum_{i=1}^{n}\left\|p - p_i'\right\|^2}, \qquad (25)$$

where p is the real position of the vehicle at time instant t and p' is the measurement of the vehicle position. $\|\cdot\|$ represents the 2-norm.

The results in Figure 6 show that the trajectory calculated by PSO-OE is closest to the actual trajectory; it means that the absolute error based on the proposed approach is less than both the SL-RTR and SL-STTS methods. And the fluctuation of errors in the proposed approach is very little, which shows that the proposed approach can deal with the localization problem stably. From the results shown in Figures 7(a) and 7(b), we can see that the values of RMSE for SL-RTR, SL-STTS, and PSO-OE have a growth, but the RMSE values of PSO-OE rise much slower (see Table 2). In Figures 7(c) and 7(d), the values of RMSE for all these three methods go up sharply. The reason is that the bias of time measurements in milliseconds may cause error of distance estimates in meters. The localization performance of PSO-OE is better compared with SL-RTR and SL-STTS under the same noise conditions (see Table 3). We can also observe that the RMSE of PSO-OE goes up slowly as σ_a^2 increases quickly (see Figures 7(e) and 7(f), and Table 4). In other words, the localization accuracy is sensitive to the time and angle measurement noise, which verifies the significance of precise ranging, and PSO-OE works better than the other two methods.

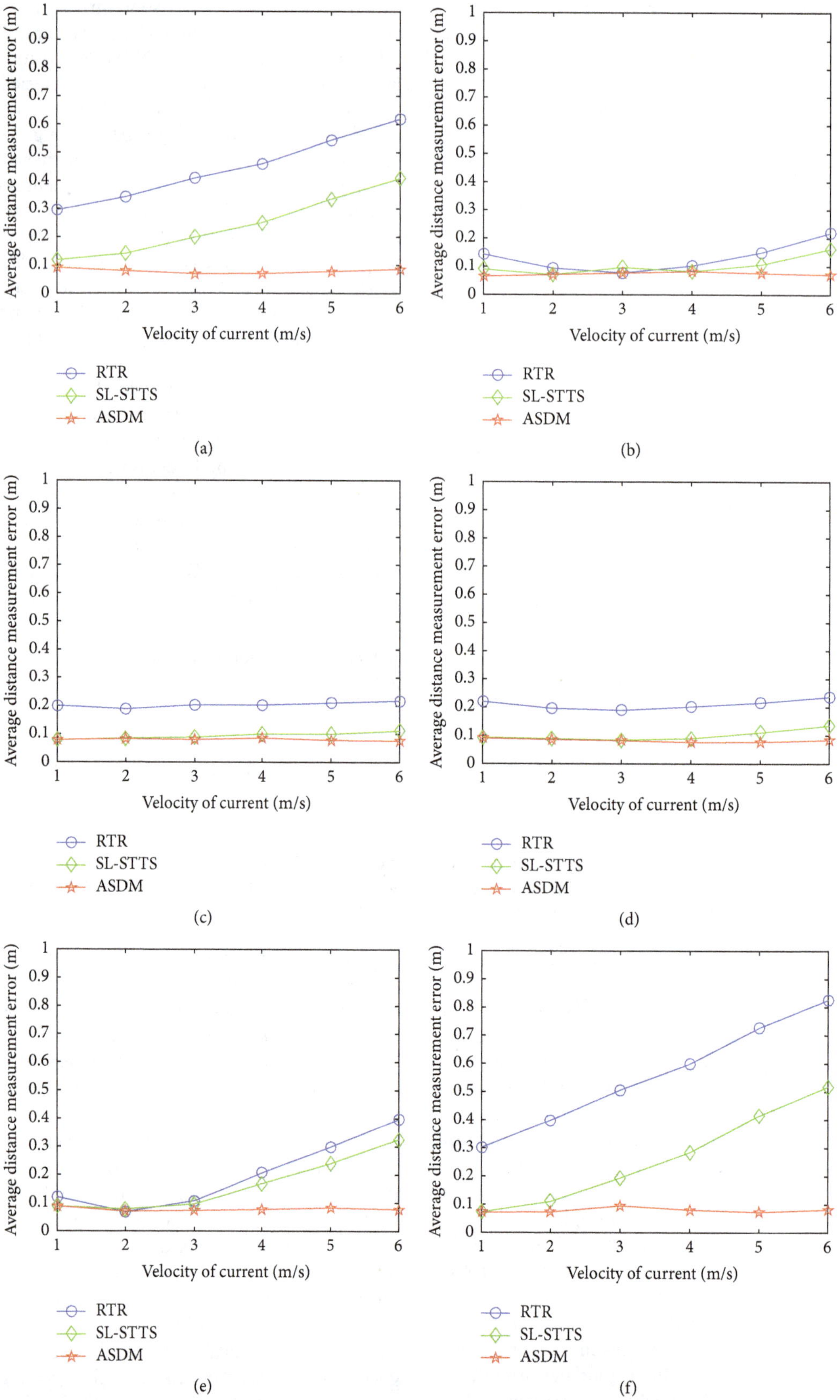

FIGURE 5: Average distance measurement error versus velocity of current: (a) the angle between e_r and e_f^a is 0; (b) the angle between e_r and e_f^b is π; and ((c), (d), (e), (f)) the angle between e_r and e_f^c is $\pi/2$; the differences are that $e_f^c = -e_f^d$, $e_f^e = -e_f^f$, and $e_f^c \perp e_f^e$.

TABLE 2: The localization error of the localization task by different distance measurement noise level.

The algorithms	With current	Indexes	Error (m)					
SL-RTR	No	RMSE	0.3300	0.3397	0.3773	0.5711	1.1476	3.6589
		MAXE	0.3615	0.3880	0.5499	0.8975	2.4403	5.7262
		MINE	0.2843	0.2920	0.2438	0.0850	0.3856	0.8939
	Yes	RMSE	0.3942	0.3904	0.4214	0.5272	1.1113	3.8004
		MAXE	0.4562	0.4414	0.5557	0.8022	2.5241	5.9087
		MINE	0.3592	0.3251	0.2770	0.0913	0.1631	1.1050
SL-STTS	No	RMSE	0.0203	0.0373	0.1140	0.3344	1.0994	3.6542
		MAXE	0.0650	0.0653	0.1974	0.7522	2.0986	5.6516
		MINE	0.0012	0.0093	0.0225	0.0855	0.2409	0.8670
	Yes	RMSE	0.0626	0.0665	0.1339	0.3565	1.0293	3.7895
		MAXE	0.1230	0.1096	0.2291	0.6043	2.2054	5.7935
		MINE	0.0276	0.0268	0.0251	0.1232	0.2359	1.1462
PSO-OE	No	RMSE	0.0107	0.0317	0.0943	0.3014	0.9502	3.1592
		MAXE	0.0194	0.0665	0.1527	0.4957	1.9450	4.6592
		MINE	0.0029	0.0081	0.0128	0.0768	0.1425	1.0667
	Yes	RMSE	0.0109	0.0358	0.0927	0.2937	0.8946	2.9731
		MAXE	0.0238	0.0617	0,1537	0.4837	1.8532	4.5431
		MINE	0.0011	0.0087	0.0118	0.0825	0.1537	0.9125

TABLE 3: The localization error of the localization task by different time measurement noise level.

The algorithms	With current	Indexes	Error (m)					
SL-RTR	No	RMSE	0.4281	0.4301	0.6281	1.0118	2.5554	5.4386
		MAXE	0.5277	0.5255	0.836	1.3430	3.1876	7.5779
		MINE	0.3196	0.1818	0.0407	0.0847	0.1889	0.9865
	Yes	RMSE	0.4352	0.4670	0.7001	1.1680	2.7793	6.1894
		MAXE	0.5436	0.5120	0.8558	1.4202	3.3948	8.9425
		MINE	0.341	0.1883	0.0551	0.0981	0.2189	1.1733
SL-STTS	No	RMSE	0.0638	0.1545	0.3888	0.9521	2.3790	5.4477
		MAXE	0.1245	0.2871	0.5028	1.1747	2.8768	8.1081
		MINE	0.0188	0.0256	0.0731	0.1543	0.2863	0.8151
	Yes	RMSE	0.0737	0.1697	0.4356	1.0346	2.5611	6.2015
		MAXE	0.1345	0.2927	0.5471	1.3098	3.1085	9.1164
		MINE	0.0213	0.0302	0.0977	0.1960	0.3283	0.9877
PSO-OE	No	RMSE	0.0414	0.1144	0.2981	0.8265	2.2672	5.2514
		MAXE	0.0982	0.1965	0.4385	1.1231	3.0361	8.8767
		MINE	0.0180	0.0259	0.0863	0.1805	0.3230	0.9246
	Yes	RMSE	0.0420	0.1131	0.2953	0.7958	2.1514	5.1646
		MAXE	0.1129	0.2037	0.4173	1.0574	2.8796	8.3295
		MINE	0.0135	0.0236	0.0675	0.1527	0.2769	0.7933

5. Discussions

The results of the simulation experiments in Section 4 show that the proposed approach can satisfy the localization task under various situations. In this section, the robustness and fault tolerance of the proposed approach are discussed, where the problem of noise or faults acting on transponders will be taken into account [26].

At first, the noise acting on transponders s_i of the proposed approach is discussed. Transponder time measurement noise has been taken into consideration in Section 4, where the experiments were based on the fact that transponders location is at the fixed inertial location. However, transponders location will inevitably drift with the accumulation of time caused by the current in the actual situation. Therefore, an experiment is conducted in the case of a drift in the transponder location, where the proposed approach is also compared with the SL-RTR and SL-STTS methods. For simplicity, σ_i^2 is set equal to each other and distance independent in this experiment, and we set $\sigma_i^2 = 10^{-2}$. Location noise w_l on

TABLE 4: The localization error of the localization task by different angle measurement noise level.

The algorithms	With Current	Indexes	Error (m)					
SL-RTR	No	RMSE	0.4670	0.4549	0.4752	0.4607	0.4729	0.4662
		MAXE	0.7283	0.7170	0.8583	0.8550	0.8233	0.8169
		MINE	0.1313	0.1153	0.0878	0.0757	0.1304	0.1669
	Yes	RMSE	0.5282	0.5488	0.5210	0.5159	0.5471	0.5493
		MAXE	0.8241	0.9676	0.8903	0.8941	0.8846	1.0504
		MINE	0.1337	0.1190	0.1049	0.0829	0.1805	0.1858
SL-STTS	No	RMSE	0.2531	0.2623	0.2728	0.3040	0.3888	0.4879
		MAXE	0.3885	0.4020	0.4354	0.5745	0.7503	1.0175
		MINE	0.0315	0.0344	0.0456	0.0424	0.0553	0.0528
	Yes	RMSE	0.2717	0.2823	0.2744	0.3097	0.4179	0.5738
		MAXE	0.5032	0.5841	0.5440	0.6563	0.8306	1.1287
		MINE	0.0574	0.0384	0.0454	0.0504	0.0588	0.0429
PSO-OE	No	RMSE	0.2463	0.2411	0.2575	0.2451	0.2648	0.2656
		MAXE	0.4412	0.4612	0.5195	0.5572	0.6023	0.8000
		MINE	0.0423	0.0299	0.0422	0.0460	0.0569	0.0660
	Yes	RMSE	0.2524	0.2549	0.2598	0.2574	0.2631	0.2647
		MAXE	0.4576	0.4781	0.5238	0.5624	0.5829	0.8251
		MINE	0.0541	0.0397	0.0410	0.0519	0.0585	0.0686

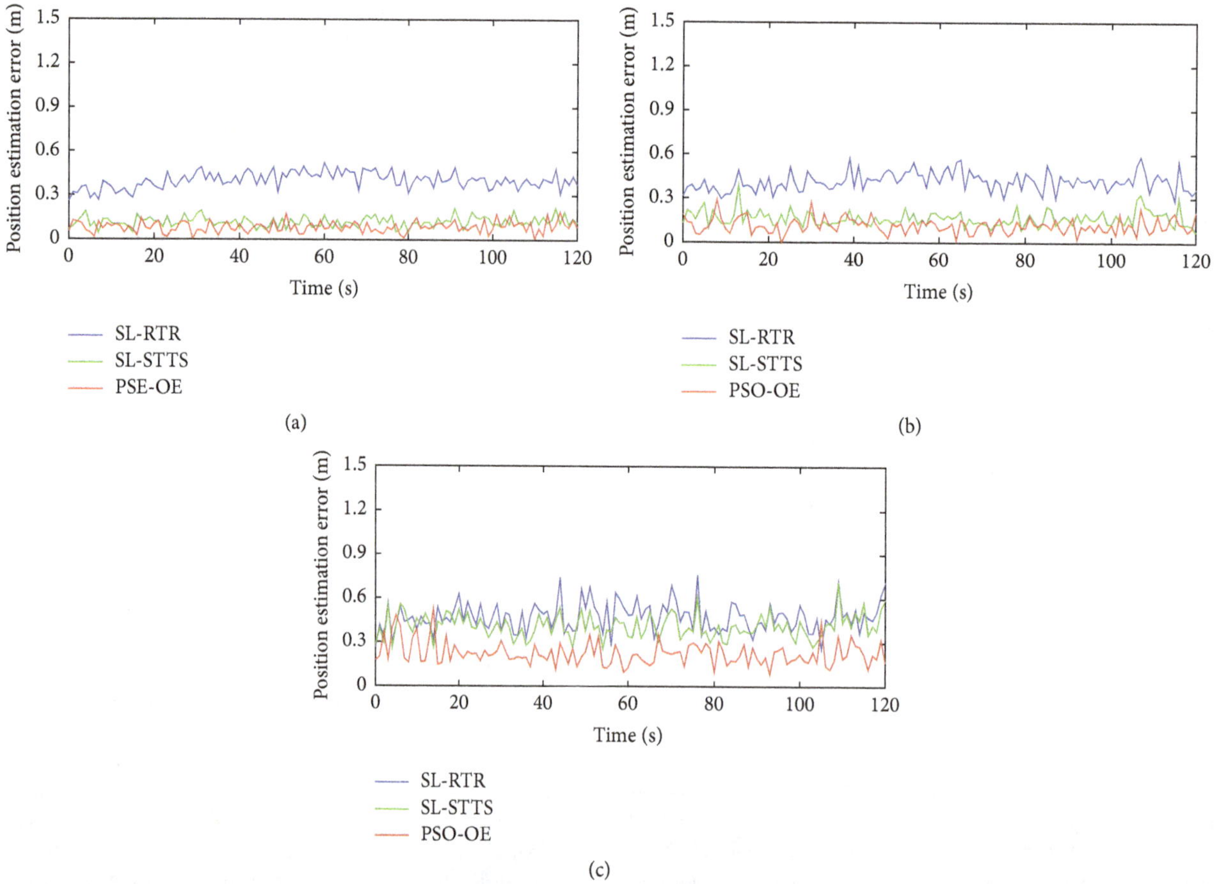

FIGURE 6: The localization error in the second experiment: (a) set $\sigma_i^2 = 10^{-2}$; (b) set $\sigma_t^2 = 10^{-8}$, $\sigma_a^2 = 0$; (c) set $\sigma_t^2 = 10^{-7}$, $\sigma_a^2 = 10^{-4}$.

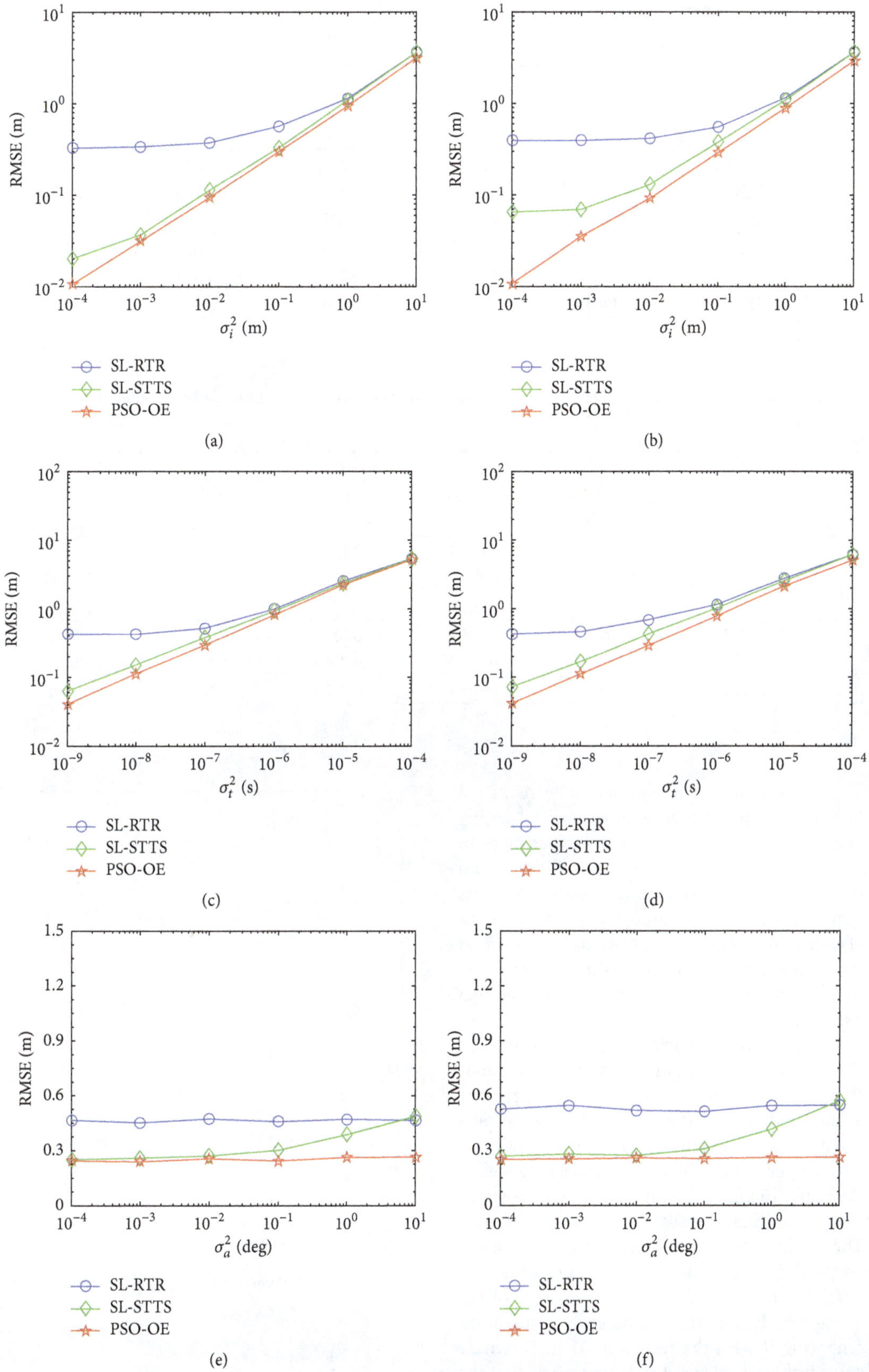

FIGURE 7: The root mean square error of different noise levels in the second experiment: ((a), (b)) RMSE versus variance of distance measurement noise; ((c), (d)) RMSE versus variance of time measurement noise, set $\sigma_a^2 = 0$; ((e), (f)) RMSE versus variance of angle measurement noise, set $\sigma_t^2 = 10^{-7}$. The only difference is that the current is not taken into account in (a), (c), and (e).

Table 5: The performance indexes of the localization task by different transponder location noise.

The algorithms	Indexes	Error (m)					
SL-RTR	RMSE	0.4559	0.4697	0.5932	0.7234	1.6093	4.0152
	MAXE	0.5919	0.6315	0.6980	0.8405	1.7699	4.1409
	MINE	0.2628	0.3264	0.3880	0.577	1.4271	3.8612
SL-STTS	RMSE	0.1175	0.1261	0.2302	0.3734	1.3499	3.7259
	MAXE	0.2053	0.2496	0.3706	0.4976	1.4972	3.857
	MINE	0.0222	0.0135	0.0231	0.2671	1.1917	3.5795
PSO-OE	RMSE	0.0578	0.0900	0.1561	0.3018	0.9805	3.2710
	MAXE	0.1163	0.2024	0.2613	0.4364	1.2310	3.4213
	MINE	0.0065	0.0169	0.0211	0.1953	0.9075	3.1085

Table 6: The performance indexes of the localization task by different distance measurement noise level, where two transponders failed.

The algorithms	Indexes	Error (m)					
SL-RTR	RMSE	0.4475	0.4408	0.4718	0.5743	1.3413	4.3516
	MAXE	0.4695	0.5046	0.6855	0.9845	2.9233	9.2810
	MINE	0.4263	0.3694	0.3098	0.1780	0.2130	0.4627
SL-STTS	RMSE	0.0650	0.0724	0.1502	0.3863	1.2818	4.3283
	MAXE	0.0866	0.1249	0.3401	0.6659	2.6442	8.9310
	MINE	0.0186	0.0314	0.0478	0.0816	0.3027	0.4491
PSO-OE	RMSE	0.0237	0.0449	0.1096	0.2997	0.9593	3.3533
	MAXE	0.0478	0.0923	0.2136	0.5223	2.0539	6.5690
	MINE	0.0098	0.0170	0.0259	0.0483	0.3870	0.4943

each dimension of transponder s_i is assumed to be disturbed by Gaussian distributed noise with $w_l \sim N(0, \sigma_l^2)$. The setting of other parameters is the same as the second experiment in Section 4. The experimental results are shown in Figure 8 and Table 5. The results show that the localization performance is sensitive to location noise of transponders, which verifies the significance of transponders location; however, the proposed approach contributes to accurate location estimates of vehicle and its sensitivity to transponders location noise is less than the other two methods.

To further discuss the robustness of the proposed approach under the condition of faults acting on transponders, a simulation experiment was conducted, where the ranging noise distribution and the parameters of the proposed approach are the same as the experiment in Section 4, except that the number of the transponders working well is different. In this experiment, it is assumed that two transponders will fail. The experimental results are shown in Figure 9 and Table 6. The results show that the accuracy of localization will decrease with the increase of ranging noises when two transponders fail. However, the localization accuracy of the proposed approach is higher than another two methods. Moreover, comparing it with the results in Table 2 (where all transponders work well), it is not difficult to find that the accuracy of the proposed approach decreased more slowly compared to the other two methods as two transponders malfunctioning: namely, the influence of transponder faults on the proposed approach is less than the other two methods.

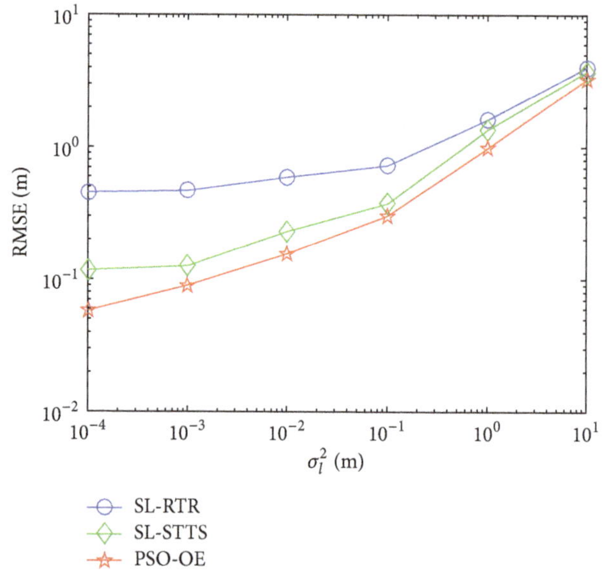

Figure 8: RMSE versus variance of transponders location noise.

6. Conclusions

Localization of autonomous underwater vehicles (AUVs) is investigated in this paper, and a single-way range localization approach based on PSO of outliers elimination is proposed.

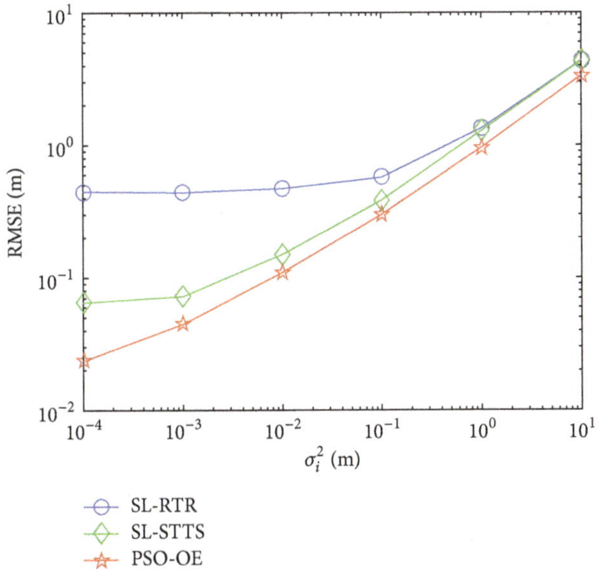

FIGURE 9: RMSE versus variance of distance measurement noise, where two transponders are failed.

In the proposed approach, various conditions are considered in the localization algorithm based on SL-STTS, including time-synchronization free localization and long propagation delay. Furthermore, an accurate single-way range localization algorithm based PSO of outliers elimination is proposed to improve the localization performance of the SL-STTS, which can reduce the error caused by the current and make full use of the vehicle position information from different measurements. The proposed approach can improve the accuracy of the single-way range localization, without any a priori knowledge of the noise model. Also, the proposed approach can deal with the searching and exploring problem in unknown environments, which has broad applications, such as the perceiving and detecting for the underwater dam crack.

Conflicts of Interest

The authors declare that there are no conflicts of interest regarding the publication of this paper.

Acknowledgments

The authors would like to thank the National Natural Science Foundation of China (61203365 and 61573128), the Jiangsu Province Natural Science Foundation (BK2012149), and the Fundamental Research Funds for the Central Universities (2015B20114) for their support of this paper.

References

[1] M. R. Azimi-Sadjadi, N. Klausner, and J. Kopacz, "Detection of underwater targets using a subspace-based method with learning," *IEEE Journal of Oceanic Engineering*, vol. 42, no. 4, pp. 869–879, 2017.

[2] B. J. Williamson, S. Fraser, P. Blondel, P. S. Bell, J. J. Waggitt, and B. E. Scott, "Multisensor Acoustic Tracking of Fish and Seabird Behavior Around Tidal Turbine Structures in Scotland," *IEEE Journal of Oceanic Engineering*, vol. 42, no. 4, pp. 948–965, 2017.

[3] Y. Lin, J. Hsiung, R. Piersall, C. White, C. G. Lowe, and C. M. Clark, "A Multi-Autonomous Underwater Vehicle System for Autonomous Tracking of Marine Life," *Journal of Field Robotics*, vol. 34, no. 4, pp. 757–774, 2017.

[4] M. A. Goodrich, B. S. Morse, D. Gerhardt et al., "Supporting wilderness search and rescue using a camera-equipped mini UAV," *Journal of Field Robotics*, vol. 25, no. 1-2, pp. 89–110, 2008.

[5] J. Ni, L. Yang, L. Wu, and X. Fan, "An Improved Spinal Neural System-Based Approach for Heterogeneous AUVs Cooperative Hunting," *International Journal of Fuzzy Systems*, pp. 1–15, 2017.

[6] L. Paull, S. Saeedi, M. Seto, and H. Li, "AUV navigation and localization: a review," *IEEE Journal of Oceanic Engineering*, vol. 39, no. 1, pp. 131–149, 2014.

[7] P. J. Zeno, S. Patel, and T. M. Sobh, "Review of neurobiologically based mobile robot navigation system research performed since 2000," *Journal of Robotics*, vol. 2016, Article ID 8637251, 17 pages, 2016.

[8] T. R. Wanasinghe, G. K. I. Mann, and R. G. Gosine, "Decentralized Cooperative Localization Approach for Autonomous Multirobot Systems," *Journal of Robotics*, vol. 2016, Article ID 2560573, 18 pages, 2016.

[9] D. De Palma, F. Arrichiello, G. Parlangeli, and G. Indiveri, "Underwater localization using single beacon measurements: Observability analysis for a double integrator system," *Ocean Engineering*, vol. 142, pp. 650–665, 2017.

[10] G. Indiveri, D. De Palma, and G. Parlangeli, "Single Range Localization in 3-D: Observability and Robustness Issues," *IEEE Transactions on Control Systems Technology*, vol. 24, no. 5, pp. 1853–1860, 2016.

[11] G. Vallicrosa and P. Ridao, "Sum of gaussian single beacon range-only localization for AUV homing," *Annual Reviews in Control*, vol. 42, pp. 177–187, 2016.

[12] G. Isbitiren and O. B. Akan, "Three-dimensional underwater target tracking with acoustic sensor networks," *IEEE Transactions on Vehicular Technology*, vol. 60, no. 8, pp. 3897–3906, 2011.

[13] W. Cheng, A. Thaeler, X. Cheng, F. Liu, X. Lu, and Z. Lu, "Time-Synchronization Free Localization in Large Scale Underwater Acoustic Sensor Networks," in *Proceedings of the 2009 29th IEEE International Conference on Distributed Computing Systems Workshops (ICDCS Workshops)*, pp. 80–87, Quebec, Canada, June 2009.

[14] D. Tollefsen, P. Gerstoft, and W. S. Hodgkiss, "Multiple-array passive acoustic source localization in shallow water," *The Journal of the Acoustical Society of America*, vol. 141, no. 3, pp. 1501–1513, 2017.

[15] X. Tan and J. Li, "Cooperative positioning in underwater sensor networks," *IEEE Transactions on Signal Processing*, vol. 58, no. 11, pp. 5860–5871, 2010.

[16] X. Cheng, H. Shu, Q. Liang, and D. H. Du, "Silent positioning in underwater acoustic sensor networks," *IEEE Transactions on Vehicular Technology*, vol. 57, no. 3, pp. 1756–1766, 2008.

[17] W. Gao, Y. Liu, B. Xu, and Y. Che, "An improved cooperative localization method for multiple autonomous underwater vehicles based on acoustic round-trip ranging," in *Proceedings of the 2014 IEEE/ION Position, Location and Navigation Symposium, PLANS 2014*, pp. 1420–1423, May 2014.

[18] J. Li, H. Gao, S. Zhang, S. Chang, J. Chen, and Z. Liu, "Self-localization of autonomous underwater vehicles with accurate sound travel time solution," *Computers and Electrical Engineering*, vol. 50, pp. 26–38, 2016.

[19] N. H. Kussat, C. D. Chadwell, and R. Zimmerman, "Absolute positioning of an autonomous underwater vehicle using GPS and acoustic measurements," *IEEE Journal of Oceanic Engineering*, vol. 30, no. 1, pp. 153–164, 2005.

[20] H. Ramezani, H. Jamali-Rad, and G. Leus, "Target localization and tracking for an isogradient sound speed profile," *IEEE Transactions on Signal Processing*, vol. 61, no. 6, pp. 1434–1446, 2013.

[21] B. Ferreira, A. Matos, and N. Cruz, "Optimal positioning of autonomous marine vehicles for underwater acoustic source localization using TOA measurements," in *Proceedings of the 2013 IEEE International Underwater Technology Symposium, UT 2013*, Japan, March 2013.

[22] S. Xu and K. Dogancay, "Optimal Sensor Placement for 3-D Angle-of-Arrival Target Localization," *IEEE Transactions on Aerospace and Electronic Systems*, vol. 53, no. 3, pp. 1196–1211, 2017.

[23] F. T. Liu, K. M. Ting, and Z.-H. Zhou, "Isolation forest," in *Proceedings of the 8th IEEE International Conference on Data Mining, ICDM 2008*, pp. 413–422, December 2008.

[24] S. Aryal, M. T. Kai, J. R. Wells, and T. Washio, "Improving iforest with relative mass," in *Proceedings of the Pacific-Asia Conference on Knowledge Discovery and Data Mining*, pp. 510–521, 2014.

[25] B. F. Gumaida and J. Luo, "An efficient algorithm for wireless sensor network localization based on hierarchical structure poly-particle swarm optimization," *Wireless Personal Communications*, vol. 97, 125, no. 1, p. 151, 2017.

[26] A. Buscarino, L. Fortuna, M. Frasca, M. Iachello, and V.-T. Pham, "Robustness to noise in synchronization of network motifs: Experimental results," *Chaos: An Interdisciplinary Journal of Nonlinear Science*, vol. 22, no. 4, 2012.

2

Kinematics and Transmission Performance Analyses of a 2T2R Type 4-DOF Spatial Parallel Manipulator

Haiyan An,[1] Bin Li,[1,2] Shoujun Wang,[1,2] and Weimin Ge[1,2]

[1] *Tianjin Key Laboratory for Advanced Mechatronic System Design and Intelligent Control, School of Mechanical Engineering, Tianjin University of Technology, Tianjin 300384, China*
[2] *National Demonstration Center for Experimental Mechanical and Electrical Engineering Education (Tianjin University of Technology), China*

Correspondence should be addressed to Bin Li; cnrobot@163.com

Academic Editor: Shahram Payandeh

A 2-RPU&2-SPS spatial parallel mechanism (SPM) is researched. Firstly, the number and property of degrees of freedom (DOF) of the SPM are analyzed by screw theory. There are two rotational and two translational movements (2R2T) of the mechanism that can be achieved. Secondly, the position analyses are researched. For the inverse position analysis, the explicit expression can be obtained from the independent motion parameters of the given mechanism, and the forward position problem is solved by calculating a set of nonlinear equation systems. Then we obtained the workspace of the mechanism based on the analytic formulae of the inverse displacement. Finally, by establishing the Jacobian matrix of the mechanism, the singularity of the mechanism is obtained, and the kinematics transmission performance of the mechanism is studied by using the index of the output efficiency of the limb output of the mechanism. This work will provide the theoretical basis for prototype development and application of the mechanism.

1. Introduction

In recent years, with the deepening of the relevant theoretical research of the parallel mechanism and the continuous expansion of the application field, the lower-mobility mechanisms have become the research hotspot in the field of parallel mechanisms. In the comparison between 6-DOF parallel mechanism and lower-mobility mechanism, the latter has the advantages of simpler mechanical design, low manufacturing cost, larger workspace, high accuracy, high velocity, and high stiffness and the accumulated error of positional is little. In the fields of mechanical processing, medical devices, space technology, and sensors and other fields, the lower-mobility mechanism has received increasing attention from industry and academia in various countries.

The 4-DOF parallel mechanism is the most important kind of mechanism in the lower-mobility mechanism, which can be divided into 3T1R, 2T2R, and 3R1T according to the property of DOF. At present, the research depth and breadth of the 4-DOF SPM are far from reaching the research level of 3-DOF SPM. One of the principal reasons is the case that the configuration of the 4-DOF SPM is much smaller than the 3-DOF SPM. In view of this situation, domestic and foreign scholars have carried on the related theoretical research and application to the 4-DOF SPM.

In 1983, Hunt [1] proposed a classical mixed 3-RPS PM, and many literatures could be found about the 3-RPS PM related research work. The kinematic analyses concerning instantaneous motions analysis, dynamic modeling, synthesis, workspace analysis, position analysis, and the like can be found in the literature [2–8]. The 2-DOF PMs of Diamond [9], the 3-DOF PMs of Trivariant [10], Tricept [11], and Delta [12] have been widely applied for industrial treatment. Zeng [13] proposed a series of 2-DOF rotational decoupled PMs based on the mechanism of four-bar linkage. Li [14] researched the type synthesis of the RPR equivalent PM and proposed a new architecture with encouraging potential in practice. Kong [15] firstly used the virtual chain method carried out the PMs type synthesis with a variety of operating modes. The type synthesis of 3-DOF cable driven PMs was also researched by Khakpour [16]. Compared with the 2-DOF or 3-DOF counterparts, the 4-DOF PMs are seldom used in practice. Rolland [17] used parallelogram to investigate two 4-DOF parallel mechanisms to eliminate the rotation. Zhao [18] investigated a 4-URU parallel mechanism with the symmetrical topology, which has one rotational DOF and three translational DOFs along the normal of the base. Araujo-Gómez [19] proposed a two translational and two rotational (2T2R) four-degrees-of-freedom (DOF) parallel kinematic mechanism (PKM) designed as a knee rehabilitation and diagnosis mechatronics system. Q. zhao [20] presents the evolution process of pressure angles from planar parallel mechanisms to spatial parallel mechanisms. L. M. Zhang [21] deals with dynamic dimensional synthesis of the Delta robot using the pressure/transmission angle constraints.

In this paper, a 2T2R 4-DOF 2-RPU & 2-SPS SPM is investigated and it has a lot of advantages. Firstly, the mechanism moving platform can realize the output of two rotational and two mobile decoupling motions. Secondly, because of the symmetrical arrangement of four moving limbs, the mechanism has a large translational motion space and rotational motion space. Thirdly, the inverse and forward analysis of the mechanism is simple, which simplifies the series of technical problems such as trajectory planning, control, and correction, which are conducive to the application of the mechanism in the field of parallel machine tool equipment. In this paper, firstly, the number and property of degrees of freedom (DOF) of the SPM are analyzed by screw theory. There are two rotational and two translational movements (2R2T) of the mechanism that can be achieved. Secondly, the inverse and forward position analyses of the mechanism are researched. Then we obtained the workspace of the mechanism based on the analytic formulae of the inverse displacement. Finally, by establishing the Jacobian matrix of the mechanism, the singularity of the mechanism is obtained, and the kinematics transmission performance of the mechanism is studied by using the index of the output efficiency of the limb output of the mechanism. This work will provide the theoretical basis for prototype development and application of the mechanism.

2. Structure Characteristics of the 2-RPU&2-SPS SPM

As shown in Figure 1, the investigated 4-DOF 2-RPU&2-SPS SPM composed of a moving platform and a base platform and both platforms are attached by two same SPS limbs and two same RPU limbs. The spherical, revolute, prismatic, and universal joints can be abbreviated as S, R, P, and U, respectively, and the underlined p represents actuated joints. Right for the researched SPM, the two same RPU limbs connect the base platform to the moving platform with R, P, and U joints in sequence. Similarly, two identical SPS limbs connect the base platform to the moving platform by S, P, and S joints in sequence.

In order to establish the position analysis model of the mechanism as shown in Figure 2, place the reference frame $O-xyz$ attached to the base platform with O located in the center of this quadrilateral, and the x axis is collinear with B_2B_4 and points from point O to point B_2. The y axis is collinear with B_1B_3 and points from point O to point B_3, and

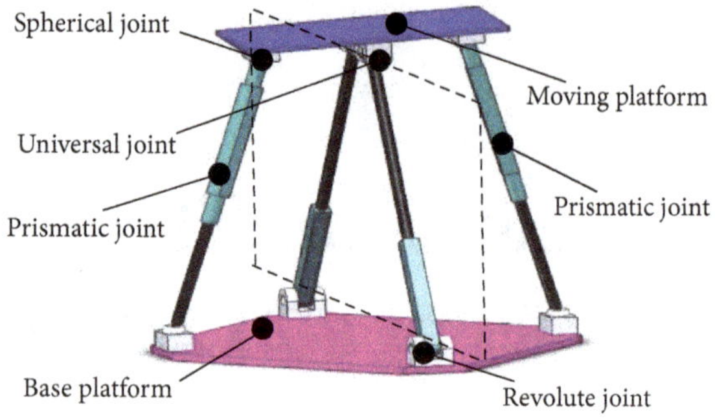

FIGURE 1: CAD model of the proposed 4-DOF SPM 2-RPU&2-SPS.

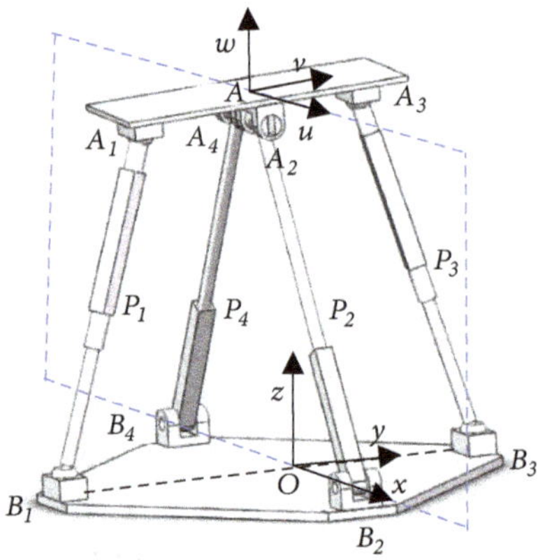

FIGURE 2: Schematic model of the 2-RPU&2-SPS SPM.

the z axis is perpendicular to the base platform. The moving frame $A-uvw$ is established on the moving platform, where point A is the center of the moving platform, the u axis is collinear with A_2A_4 and points from point A to point A_2, the v axis is collinear with A_1A_3 and points from point A to point A_3, and the w axis is perpendicular to the moving platform. Here, B_2 and B_4 are the rotational axes of the revolute joints, A_1, A_3, B_1, and B_3 are the centers of the spherical joint, and A_2 and A_4, respectively, represent the universal joints center.

Note that, for the RPU limb, the axis of the revolute joint connected to the base platform is parallel to the y axis and the axis of the universal joint connected to the moving platform is also parallel to the y axis. The other axis of the universal joint is parallel to the u axis; the two axes of the universal joints are perpendicular to each other. For the SPS limb, the line connecting the two spherical joints connected to the base platform coincides with the y axis and the connection is symmetrical about the x axis on the base platform. The line connecting the two spherical joints connected to the moving platform coincides with the v axis and the connection is symmetrical about the u axis on the moving platform.

Through the spatial arrangement of the abovementioned kinematic pair, the two RPU limbs can only move in the plane of xoz and the movement of the two SPS limbs can make the moving platform produce a rotation around the u axis from the perspective of geometric constraints. This structure determines that the motion trajectory of the moving platform is the motion of a plane and the two rotations of the v axis and the u axis, respectively, thus forming a 2T2R 4-DOF SPM.

In order to research the 2-RPU&2-SPS SPM conveniently, the moving coordinate system $A - uvw$ is rotated relative to the reference coordinate system $O - xyz$ by $x - z - y$, the corresponding Euler angles of the rotation transformation of each coordinate system are ψ, ϕ, and θ, and the composition of the rotation matrix is

$$\mathbf{R} = \begin{bmatrix} c\theta c\phi & s\psi s\theta - c\psi c\theta s\phi & c\psi s\theta + s\psi c\theta s\phi \\ s\phi & c\psi c\phi & -s\psi c\phi \\ -s\theta c\phi & s\psi c\theta + c\psi s\theta s\phi & c\psi c\theta - s\psi s\theta s\phi \end{bmatrix} = \begin{bmatrix} u_x & v_x & w_x \\ u_y & v_y & w_y \\ u_z & v_z & w_z \end{bmatrix} = [\mathbf{u} \ \mathbf{v} \ \mathbf{w}] \tag{1}$$

where sin and cos can be abbreviated to "s" and "c", respectively, and $\mathbf{u}$, $\mathbf{v}$, and $\mathbf{w}$ represents the measure of the unit vector of the three coordinate axes of the moving coordinate system in the reference coordinate system.

3. The DOF Analysis of the 2-RPU&2-SPS SPM

The number of DOF of the 4-DOF 2-RPU&2-SPS SPM can be obtained by the calculation of the criterion *Kutzbach – Grübler*

$$F = \lambda (n - g - 1) + \sum_{i=1}^{g} f_i + v - \zeta = 6 \times (10 - 12 - 1) + 22 + 2 - 2 = 4 \tag{2}$$

where λ denotes the task space order, n stands for the links number, g is the kinematic pairs number, f_i represents the degrees of freedom of joint i, v denotes the number of redundant constraints, and ζ represents the local degree of freedom.

We can quickly find the number of DOF of the 2-RPU&2-SPS SPM by using (2). However, it should be pointed out that one shortcoming of the criterion *Kutzbach – Grübler* is that it can only obtain the number of DOFs of the mechanism rather than indicate the attributes of the DOF, whether they are translational or rotational DOFs. Otherwise, we can analyze the motion of a 2-RPU&2-SPS SPM via the analysis of the screw theory effectively; then we can easily obtain motions of translation and rotation in three-dimensional space. In Figure 3 through the analysis of the screw theory, the following can be drawn:

$$\begin{aligned} \$_1 &= (1 \ 0 \ 0 \ 0 \ 0 \ 0) \\ \$_2 &= (0 \ 1 \ 0 \ 0 \ 0 \ 0) \\ \$_3 &= (0 \ 0 \ 0 \ 0 \ -\sin(\alpha) \ -\cos(\alpha)) \\ \$_4 &= (1 \ 0 \ 0 \ 0 \ -q_{2(4)}\cos(\alpha) \ q_{2(4)}\sin(\alpha)) \end{aligned} \tag{3}$$

where $q_i = \|A_iB_i\|$, $i = 2, 4$.

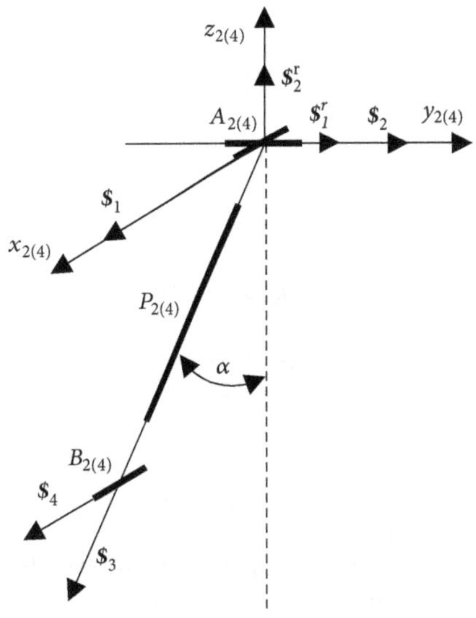

FIGURE 3: Twist system of the RPU limb.

Considering the reciprocity between twist and wrench, the constraint system for the RPU limb can be described as

$$\begin{aligned} \$_1^r &= (0 \ 1 \ 0 \ 0 \ 0 \ 0) \\ \$_2^r &= (0 \ 0 \ 0 \ 0 \ 0 \ 1) \end{aligned} \tag{4}$$

Similarly, kinematics of screw and reverse screw of the SPS limbs can also be obtained; since the kinematics screw of SPS limb is composed of six linear independence screws without constraint reverse screw, the two SPS limbs do not have constraints on the mechanism.

It can be seen from (4) that a plane along the y axis and a rotation around the z axis are constrained so that the 2-RPU & 2-SPS SPM has two translational degrees of freedom (along the x and z axes direction) and two rotational degrees of freedom (rotation around the x and y axes).

4. Mechanism Position Analysis

4.1. Mechanism Analysis of the Inverse Position. As shown in Figure 2, the closed-loop vector equation is constructed, and the position vector $\mathbf{r} = (x \ y \ z)^T$ of point A can be represented in the reference coordinate system $O - xyz$ which gives

$$\mathbf{r} = \mathbf{b}_i + q_i\mathbf{w}_i - \mathbf{a}_i \quad i = 1, 2, 3, 4 \tag{5}$$

where $\mathbf{a}_i$, $\mathbf{b}_i$ are the position vectors of A_i and B_i, respectively. $\mathbf{w}_i$ is the unit vector of limb i. $\mathbf{a}_i = \mathbf{R}\mathbf{a}_{i0}$, $\mathbf{R}$ is the rotation matrix of the mechanism, and $\mathbf{a}_{i0}$ is the measure of $\mathbf{a}_i$ in the coordinate system. From the structure of the mechanism, we can see

$$\begin{aligned}
\mathbf{a}_{10} &= a_1 (0 \ \ -1 \ \ 0)^T, \\
\mathbf{a}_{20} &= a_2 (1 \ \ 0 \ \ 0)^T \\
\mathbf{a}_{30} &= a_3 (0 \ \ 1 \ \ 0)^T, \\
\mathbf{a}_{40} &= a_4 (-1 \ \ 0 \ \ 0)^T \\
\mathbf{b}_{10} &= b_1 (0 \ \ -1 \ \ 0)^T, \\
\mathbf{b}_{20} &= b_2 (1 \ \ 0 \ \ 0)^T \\
\mathbf{b}_{30} &= b_3 (0 \ \ 1 \ \ 0)^T, \\
\mathbf{b}_{40} &= b_4 (-1 \ \ 0 \ \ 0)^T
\end{aligned} \tag{6}$$

where a_1, a_2, a_3, and a_4, respectively, indicate the distance from point A to points A_1, A_2, A_3, and A_4, and b_1, b_2, b_3, and b_4, respectively, denote the distance from point B to points B_1, B_2, B_3, and B_4.

Right for the R$\underline{\text{P}}$U limb, making the dot product with $\mathbf{c}_i$ on both sides of (5) leads to

$$(\mathbf{r} + \mathbf{a}_i)^T \mathbf{c}_i = 0, \quad i = 2, 4 \tag{7}$$

where $\mathbf{c}_2 = \mathbf{c}_4 = (0 \ \ 1 \ \ 0)^T$.

In the same way, adapting to the S$\underline{\text{P}}$S limb, making the dot product with $\mathbf{c}_i$ on both sides of (5), yields

$$(\mathbf{r} - \mathbf{b}_i)^T \mathbf{c}_i = 0, \quad i = 1, 3 \tag{8}$$

where $\mathbf{c}_1 = \mathbf{R}\mathbf{c}_{10}$, $\mathbf{c}_3 = \mathbf{R}\mathbf{c}_{30}$ and $\mathbf{c}_{10} = \mathbf{c}_{30} = (1 \ \ 0 \ \ 0)$.

As shown in Figure 2, based on the mechanism geometric constraint, the two R$\underline{\text{P}}$U limbs can just only move in the plane xoz, so we can get

$$u_y = 0 \tag{9}$$

Through (7), (8), and (9), we can obtain

$$\begin{aligned}
\phi &= 0 \\
x &= z \tan\theta \\
y &= 0
\end{aligned} \tag{10}$$

The position vector $\mathbf{r} = (x \ \ y \ \ z)^T$ for a given reference point A can be determined by (10) for the corresponding attitude angle and the corresponding moving platform relative to the rotation table R of the base platform, and the length of each limb can be obtained by

$$q_i = |\mathbf{r} - \mathbf{b}_i + \mathbf{a}_i|, \quad i = 1, 2, 3, 4 \tag{11}$$

4.2. Mechanism Analysis of the Forward Position. The relative position of the moving coordinate system in the reference coordinate system is obtained from the given length of the mechanism and a certain mathematical operation. Then according to (11) the forward position analysis of the 2-R$\underline{\text{P}}$U&2-S$\underline{\text{P}}$S SPM can be obtained

TABLE 1: Mechanism analysis of the inverse position.

Case (a)	Inputs			
	ψ (deg.)	θ (deg.)	z (mm)	x (mm)
	25	15	650	174.1670
	Outputs			
	q_1 (mm)	q_2 (mm)	q_3 (mm)	q_4 (mm)
	736.8688	639.0079	844.7550	807.5673
Case (b)	Inputs			
	ψ (deg.)	θ (deg.)	z (mm)	x (mm)
	25	-15	650	-74.1670
	Outputs			
	q_1 (mm)	q_2 (mm)	q_3 (mm)	q_4 (mm)
	736.8688	807.5673	844.7550	639.0079

Then according to (11) the mechanism analysis of the forward position can be obtained as

$$\begin{aligned}
q_1^2 &= (z * \tan(\theta) - a_1 * \sin(\psi) * \sin(\theta))^2 \\
&\quad + (b_1 - a_1 * \cos(\psi))^2 \\
&\quad + (z - a_1 * \sin(\psi) * \cos(\theta))^2
\end{aligned} \tag{12}$$

$$\begin{aligned}
q_2^2 &= (z * \tan(\theta) - b_2 + a_2 * \cos(\theta))^2 \\
&\quad + (z - a_2 * \sin(\theta))^2
\end{aligned} \tag{13}$$

$$\begin{aligned}
q_3^2 &= (z * \tan(\theta) + a_3 * \sin(\psi) * \sin(\theta))^2 \\
&\quad + (a_3 * \cos(\psi) - b_3)^2 \\
&\quad + (z + a_3 * \sin(\psi) * \cos(\theta))^2
\end{aligned} \tag{14}$$

$$\begin{aligned}
q_4^2 &= (z * \tan(\theta) + b_4 - a_4 * \cos(\theta))^2 \\
&\quad + (z + a_4 * \sin(\theta))^2
\end{aligned} \tag{15}$$

Since the positive solution constrained equations of the 4-DOF SPM are complex, this paper will directly call the mathematical calculation software to solve the numerical solution of the nonlinear Equations (12)-(15).

4.3. Numerical Examples. According to the forward and inverse position analysis of Sections 4.1 and 4.2, the selections of the mechanism parameters are $a_1 = a_3 = 150$ mm, $a_2 = a_4 = 75$ mm, $b_1 = b_3 = 550$ mm, $b_2 = b_4 = 350$ mm. Right for the inverse position analysis, the calculated output results are shown in Table 1 and Figure 4 by calculating two given different sets of inputs ($\psi \ \ \theta \ \ z$). Adapting to the mechanism analysis of the forward position, for the purpose of testifying the correlation of the analyses of the inverse and forward position, the outputs value of the inverse position analysis are served as the inputs ($q_1 \ \ q_2 \ \ q_3 \ \ q_4$) of the forward position analysis, and the output results can be obtained by solving the nonlinear equation system as shown in Table 2 and Figure 5.

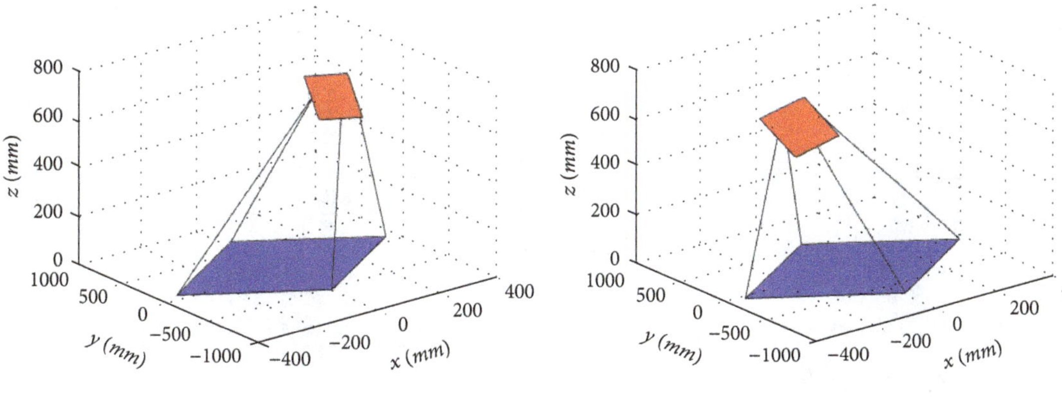

FIGURE 4: Different configurations for inverse position.

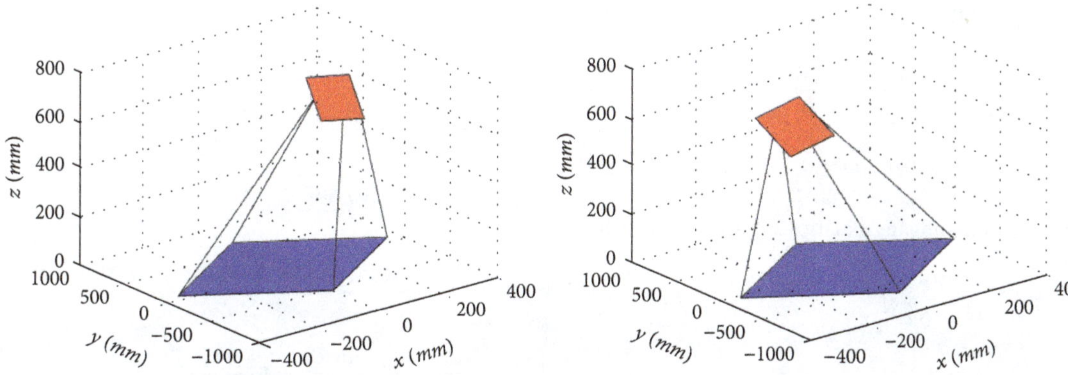

FIGURE 5: Different configurations for forward position.

TABLE 2: Mechanism analysis of the forward position.

Case (a)	Inputs			
	q_1 (mm)	q_2 (mm)	q_3 (mm)	q_4 (mm)
	736.8688	639.0079	844.7550	807.5673
	Outputs			
	ψ (deg.)	θ (deg.)	z (mm)	x (mm)
	25	15	650	174.1670
Case (b)	Inputs			
	q_2 (mm)	q_2 (mm)	q_3 (mm)	q_4 (mm)
	736.8688	807.5673	844.7550	639.0079
	Outputs			
	ψ (deg.)	θ (deg.)	z (mm)	x (mm)
	25	-15	650	-74.1670

As shown in Figures 4 and 5, we can see that the output of the positive and inverse solution of the mechanism is consistently, which proves the accuracy of the calculation.

5. The Overall Jacobian Matrix

5.1. The Overall Jacobian Matrix of the Mechanism. Equation (5) at time differentiating can be written as

$$\dot{\mathbf{r}} = \dot{q}_i\mathbf{w}_i + \boldsymbol{\omega}_i \times q_i\mathbf{w}_i - \boldsymbol{\omega} \times \mathbf{a}_i \quad i = 1, 2, 3, 4 \tag{16}$$

where "×" represents the cross product of the vectors, $\dot{q}_i$ denotes the velocity of the linear actuator i, and $\mathbf{w}_i$ stands for the angular velocity of link, and A_iB_i, $\dot{\mathbf{r}} = (\dot{x}\ \ \dot{y}\ \ \dot{z})^T$, and $\boldsymbol{\omega} = (\omega_x\ \ \omega_y\ \ \omega_z)$, respectively, stand for the spatial linear velocity and the spatial angular velocity of the moving platform.

Through dot multiplying on both sides of (16) with $\mathbf{w}_i$, the passive variables $\mathbf{w}_i$ are eliminated; this leads to

$$\dot{q}_i = \dot{\mathbf{w}}_i\dot{\mathbf{r}} + \boldsymbol{\omega}\left(\mathbf{a}_i \times \mathbf{w}_i\right) \quad i = 1, 2, 3, 4 \tag{17}$$

When the SPM is not in a singular configuration, (17) can be expressed as

$$\dot{\mathbf{q}}_a = \mathbf{J}_a\dot{\mathbf{X}} \tag{18}$$

where

$$\dot{\mathbf{q}}_a = \begin{bmatrix} \dot{q}_1 \\ \dot{q}_2 \\ \dot{q}_3 \\ \dot{q}_4 \end{bmatrix},$$

$$\dot{\mathbf{X}} = \begin{bmatrix} \dot{\mathbf{r}} \\ \boldsymbol{\omega} \end{bmatrix}, \tag{19}$$

$$\mathbf{J}_a = \begin{bmatrix} \mathbf{w}_1^T & (\mathbf{a}_1 \times \mathbf{w}_1)^T \\ \mathbf{w}_2^T & (\mathbf{a}_2 \times \mathbf{w}_2)^T \\ \mathbf{w}_3^T & (\mathbf{a}_3 \times \mathbf{w}_3)^T \\ \mathbf{w}_4^T & (\mathbf{a}_4 \times w_4)^T \end{bmatrix}$$

Equation (18) denotes the solution of the inverse velocity for the 2-R$\underline{\text{P}}$U&2-S$\underline{\text{P}}$S SPM.

Equations (7) and (8) at time differentiating, respectively, lead to

$$\mathbf{c}_i \cdot \dot{\mathbf{r}} + (\mathbf{c}_i \times \mathbf{a}_i) \cdot \boldsymbol{\omega} = 0 \quad i = 2, 4 \tag{20}$$

$$\mathbf{c}_i \cdot \dot{\mathbf{r}} + [\mathbf{c}_i \times (\mathbf{r} - \mathbf{b}_i)] \cdot \boldsymbol{\omega} = 0 \quad i = 1, 3 \tag{21}$$

When the mechanism is no singularity, (20) and (21) can be expressed as

$$\mathbf{0} = \mathbf{J}_c \dot{\mathbf{X}} \tag{22}$$

where $\mathbf{J}_c = \begin{bmatrix} \mathbf{c}_1^T & [\mathbf{c}_1 \times (\mathbf{r}-\mathbf{b}_1)]^T \\ \mathbf{c}_2^T & (\mathbf{c}_2 \times \mathbf{a}_2)^T \\ \mathbf{c}_3^T & [\mathbf{c}_3 \times (\mathbf{r}-\mathbf{b}_3)]^T \\ \mathbf{c}_4^T & (\mathbf{c}_4 \times \mathbf{a}_4)^T \end{bmatrix}$ and 0 represents a zero matrix of 4×1.

Simultaneously to (18) and (22) we can get

$$\dot{\mathbf{q}} = \mathbf{J}_0 \dot{\mathbf{X}} \tag{23}$$

where $\dot{\mathbf{q}} = \begin{bmatrix} \dot{\mathbf{q}}_a \\ \mathbf{0} \end{bmatrix}$, $\mathbf{J}_0 = \begin{bmatrix} \mathbf{J}_a \\ \mathbf{J}_c \end{bmatrix}$ is the overall Jacobian matrix of the 2-R$\underline{\text{P}}$U&2-S$\underline{\text{P}}$S SPM.

Equation (10) at time differentiating can be written as

$$\begin{aligned} \dot{\phi} &= 0 \\ \dot{x} &= \dot{z} \tan(\theta) + z \sec^2(\theta) \dot{\theta} \\ \dot{y} &= 0 \end{aligned} \tag{24}$$

The constrained Jacobian matrix of the2-R$\underline{\text{P}}$U&2-S$\underline{\text{P}}$S SPM is

$$\mathbf{J}_0 = \begin{bmatrix} J_{11} & J_{12} & J_{13} & J_{14} \\ J_{21} & J_{22} & J_{23} & J_{24} \\ J_{31} & J_{32} & J_{33} & J_{34} \\ J_{41} & J_{42} & J_{43} & J_{44} \end{bmatrix} \tag{25}$$

where

$$\begin{aligned}
J_{11} &= \frac{x - a_1 \sin(\psi) \sin(\theta)}{L_1} \\
J_{12} &= \frac{z - a_1 \sin(\psi) \cos(\theta)}{L_1} \\
J_{13} &= \frac{z a_1 \sin(\psi) \sin(\theta) - a_1 x \sin(\psi) \cos(\theta)}{L_1} \\
J_{14} & \\
&= \frac{-a_1 x \cos(\psi) \sin(\theta) + a_1 b_1 \sin(\psi) - a_1 z \cos(\psi) \cos(\theta)}{L_1} \\
J_{21} &= \frac{x - b_2 + a_2 \cos(\theta)}{L_2}, \\
J_{22} &= \frac{z - a_2 \sin(\theta)}{L_2} \\
J_{23} &= \frac{-a_2 x \sin(\theta) + a_2 b_2 \sin(\theta) - a_2 z \cos(\theta)}{L_2} \\
J_{24} &= 0 \\
J_{31} &= \frac{x + a_3 \sin(\psi) \sin(\theta)}{L_3} \\
J_{32} &= \frac{z + a_3 \sin(\psi) \cos(\theta)}{L_3} \\
J_{33} &= \frac{-z a_3 \sin(\psi) \sin(\theta) + a_3 x \sin(\psi) \cos(\theta)}{L_3} \\
J_{34} &= \frac{a_3 x \cos(\psi) \sin(\theta) + a_3 b_3 \sin(\psi) + a_3 z \cos(\psi) \cos(\theta)}{L_3} \\
J_{41} &= \frac{x + b_4 - a_4 \cos(\theta)}{L_4}, \\
J_{42} &= \frac{z + a_4 \sin(\theta)}{L_4} \\
J_{43} &= \frac{a_4 x \sin(\theta) + a_4 b_4 \sin(\theta) + a_4 z \cos(\theta)}{L_4} \\
J_{44} &= 0 \\
L_1 &= \left(z^2 \sec^2(\theta) + a_1^2 + b_1^2 - 2 a_1 z \sin(\psi) \sec(\theta) \right. \\
&\quad \left. - 2 a_1 b_1 \cos(\psi)\right)^{1/2} \\
L_2 &= \left(z^2 \sec^2(\theta) + a_2^2 + b_2^2 - 2 b_2 z \tan(\theta) - 2 a_2 b_2 \cos(\theta)\right)^{1/2} \\
L_3 &= \left(z^2 \sec^2(\theta) + a_3^2 + b_3^2 + 2 a_3 z \sin(\psi) \sec(\theta) \right. \\
&\quad \left. - 2 a_3 b_3 \cos(\psi)\right)^{1/2} \\
L_4 &= \left(z^2 \sec^2(\theta) + a_4^2 + b_4^2 + 2 b_4 z \tan(\theta) - 2 a_4 b_4 \cos(\theta)\right)^{1/2}
\end{aligned} \tag{26}$$

In combination with the geometric constraints of the 2-R$\underline{\text{P}}$U&2-S$\underline{\text{P}}$S SPM, the input parameters of the mechanism are set as $-40^\circ \leqslant \psi \leqslant 40^\circ$, $-40 \leqslant \theta \leqslant 40^\circ$ and the overall Jacobian of the mechanism is shown in Figure 6.

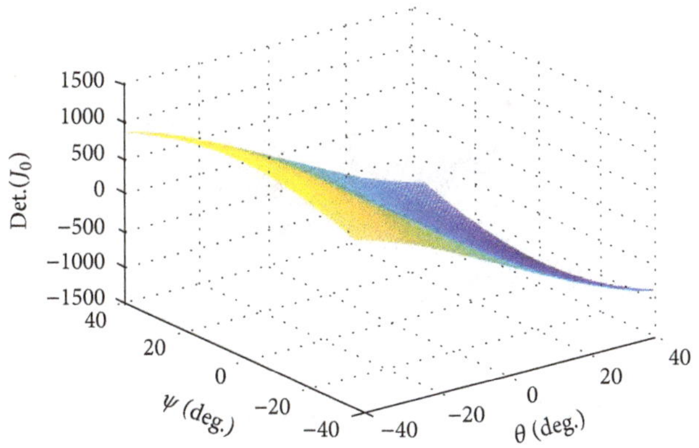

FIGURE 6: The overall Jacobian diagram of the mechanism.

5.2. Singularity Analysis. Singular configuration is in the process of the parallel mechanism operation to achieve a state of special configuration; under this kind of posture or nearby, mechanism can get out of control, mainly for mechanism freedom change; stiffness and transmission performance are reduced, which more serious when it causes the damage of organization structure, so the singularity is necessary.

(1) When |A|≠0 and |D|=0, which means that the determinant value of the input coefficient matrix is 0, then, the mechanism kinematic limb is in a boundary singular positon. That is, the output vector corresponding to a nonzero input vector is 0, so no velocity vector is generated at the output. In this configuration, the output element will lose one or more degrees of freedom while resisting one or more forces or moments when no torque is applied to the input.

(2) When |A|=0 and |D|≠0, that is, when the determinant value of the output coefficient matrix is 0, the corresponding mechanism reaches a singular position in the workspace, which is called actuated singularity of the mechanism. In other words, when all the actuated elements are locked, the output elements of the mechanism can still be partially moved. In this configuration, the output element will gain one or more degrees of freedom while being unable to resist one or more forces or moments when the input is locked.

(3) When |A|=0 and |D|=0, the input-output coefficient matrix determinant value is also 0. If the mechanism result parameter satisfies certain specific conditions and the positional relationship reached by the limb satisfies the kinematics equation configuration, the configuration of the mechanism is called hybrid singularity. This singularity is characterized by the ability of the limb to withstand limited movement when the actuated element is locked or the limited input does not produce any output.

Equation (23) shows the relation of the singularity configuration of the mechanism:

$$A\dot{\mathbf{X}} + D\dot{\mathbf{q}}_i = 0 \tag{27}$$

where

$$A = \begin{bmatrix} J_{11}\left|L_1\right| & J_{12}\left|L_1\right| & J_{13}\left|L_1\right| & J_{14}\left|L_1\right| \\ J_{21}\left|L_2\right| & J_{22}\left|L_2\right| & J_{23}\left|L_2\right| & J_{24}\left|L_2\right| \\ J_{31}\left|L_3\right| & J_{32}\left|L_3\right| & J_{33}\left|L_3\right| & J_{34}\left|L_3\right| \\ J_{41}\left|L_4\right| & J_{42}\left|L_4\right| & J_{43}\left|L_4\right| & J_{44}\left|L_4\right| \end{bmatrix}$$
$$D = -\begin{bmatrix} \left|L_1\right| & 0 & 0 & 0 \\ 0 & \left|L_2\right| & 0 & 0 \\ 0 & 0 & \left|L_3\right| & 0 \\ 0 & 0 & 0 & \left|L_4\right| \end{bmatrix} \tag{28}$$

A is the output coefficient matrix. D is the input coefficient matrix.

According to (27), the singularity configuration of the mechanism may be obtained:

(1) Because the length of the limb is not zero, there is no configuration singularity and hybrid singularity.

(2) When |A|=0, |D|≠0, there is an actuated singularity.
When

$$|A| = \begin{vmatrix} J_{11}\left|L_1\right| & J_{12}\left|L_1\right| & J_{13}\left|L_1\right| & J_{14}\left|L_1\right| \\ J_{21}\left|L_2\right| & J_{22}\left|L_2\right| & J_{23}\left|L_2\right| & J_{24}\left|L_2\right| \\ J_{31}\left|L_3\right| & J_{32}\left|L_3\right| & J_{33}\left|L_3\right| & J_{34}\left|L_3\right| \\ J_{41}\left|L_4\right| & J_{42}\left|L_4\right| & J_{43}\left|L_4\right| & J_{44}\left|L_4\right| \end{vmatrix} = 0 \tag{29}$$

there is an actuated singularity, then we obtain $\theta = 77.63^\circ$ or 0°.

5.3. Velocity and Acceleration. Simultaneous (23) and (24) can obtain the spatial linear velocity and the spatial angular velocity of the moving platform.

$$\boldsymbol{\omega} = \left[\omega_x \;\; \omega_y \;\; \omega_z\right]^T = \left[\dot{\psi}\, c\theta \;\; \dot{\theta} \;\; -\dot{\psi}\, s\theta\right]^T \tag{30}$$

$$\begin{aligned} \dot{\mathbf{r}} &= \left[\dot{x} \;\; \dot{y} \;\; \dot{z}\right]^T \\ &= \left[\dot{z} * \tan(\theta) + z * \left(\sec^2(\theta)\right) * \dot{\theta} \;\; 0 \;\; 1\right]^T \end{aligned} \tag{31}$$

Differentiating (30) and (31) with respect to time, respectively, can obtain the spatial linear acceleration and the spatial angular acceleration of the moving platform.

$$\begin{aligned} \ddot{\mathbf{r}} &= \left[\ddot{z} * \tan(\theta) + 2 * \dot{z} * \sec^2(\theta) * \dot{\theta} + z * \sec^2(\theta) * \ddot{\theta} + 2 * z * \left(\dot{\theta}\right)^2 * \sec^2(\theta) * \tan(\theta) \;\; 0 \;\; 0\right]^T \\ \dot{\boldsymbol{\omega}} &= \left[\ddot{\psi} * \cos(\theta) - \dot{\psi} * \sin(\theta) * \dot{\theta} \;\; \ddot{\theta} \;\; -\ddot{\psi} * \sin(\theta) - \dot{\psi} * \cos(\theta) * \dot{\theta}\right]^T \end{aligned} \tag{32}$$

6. Workspace Analysis

The working space of the parallel mechanism is one of the indexes of the comprehensive performance of the mechanism. It directly reflects the working ability of the mechanism, and its analysis results provide a theoretical basis for the design and application of the 2-RPU&2-SPS SPM. In Section 3, based on the position analysis, the working space of the SPM is obtained by utilizing the search algorithm combined with the structural characteristics of 2-RPU&2-SPS SPM, the inverse kinematic solution, and the limitation of the limb length.

6.1. Constraints on Workspace. The parallel mechanism moving platform and base platform are relate to the limb, which needs to use the spherical joint and revolute joint; then the conditions of the spherical joint and revolute joint angle range need to be set.

(1) Limit of Limb

$$q_{i\min} \leq q_i \leq q_{i\max} \quad i = 1, 2, 3, 4 \tag{33}$$

where q_i represents the length of limb i, $q_{i\min}$ denotes the minimum length of limb i, and $q_{i\max}$ stands for the maximum length of limb i.

(2) Limit of Rotation Angle

$$\theta_{i\min} \leq \theta_i \leq \theta_{i\max} \tag{34}$$

where θ_i represents the angle of revolute joint i, $\theta_{i\min}$ denotes the minimum angle of revolute joint i, and $\theta_{i\max}$ stands for the maximum angle of revolute joint i.

$$\psi_{i\min} \leq \psi_i \leq \psi_{i\max} \tag{35}$$

where ψ_i represents the angle of revolute joint i, $\psi_{i\min}$ denotes the minimum angle of revolute joint i, and $\psi_{i\max}$ stands for the maximum angle of revolute joint i.

6.2. Workspace Analysis. According to the structural characteristics of the 2-RPU & 2-SPS SPM, the scale parameters of the mechanism are selected as $a_1 = a_3 = 150$ mm, $a_2 = a_4 = 75$ mm, $b_1 = b_3 = 550$ mm, and $b_2 = b_4 = 350$ mm, and the constraints of the mechanism are $-30^\circ \leq \psi_i \leq 30^\circ$, $-45^\circ \leq \theta_i \leq 45^\circ$, and $700 \leq q_i \leq 1000$, and the working space is shown in Figure 7.

Because the 2-RPU&2-SPS SPM can only move along the x and z axes in the xoz plane and the rotation of the x and y axes, the working space of the mechanism in the xoz plane is about the symmetric distribution of the z axis in Figure 8, and the working space is roughly in the range of $-410 \leq x \leq 410$ (mm) and $655 \leq z \leq 860$ (mm); compared to the scale parameters of the mechanism, the work space of the mechanism is larger. The workspaces of the mechanism in xoy plane and yoz plane are shown in Figure 9 and Figure 10.

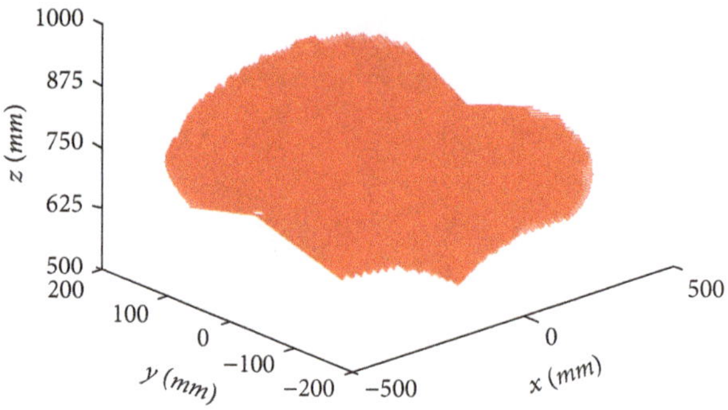

FIGURE 7: The entire workspace of the mechanism.

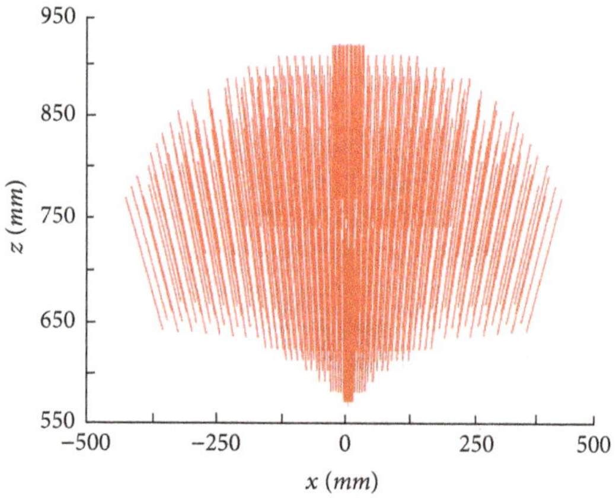

FIGURE 8: The workspace of the mechanism in $x - z$.

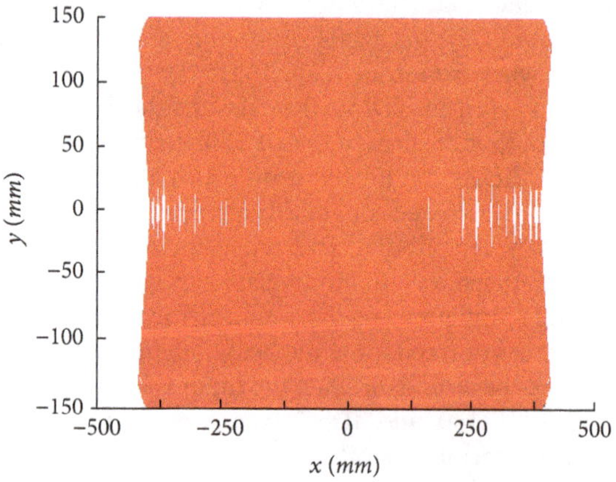

FIGURE 9: The workspace of the mechanism in $x - y$.

7. The Analysis of the Transmission Performance

One of the most important factors in designing and analyzing of the 2-RPU&2-SPS SPM is the evaluation of the performance, and there are many indicators of performance evaluation, such as dexterity, transmission angle, and torque/force transmission performance. Traditional transmission pressure angle is usually used to assess the limb of a single loop and

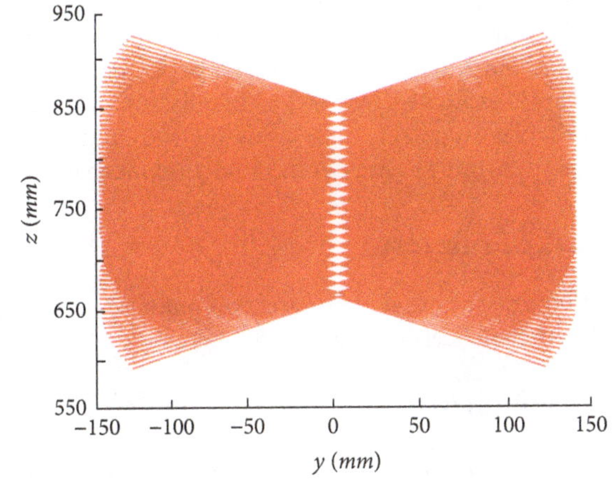

FIGURE 10: The workspace of the mechanism in y - z.

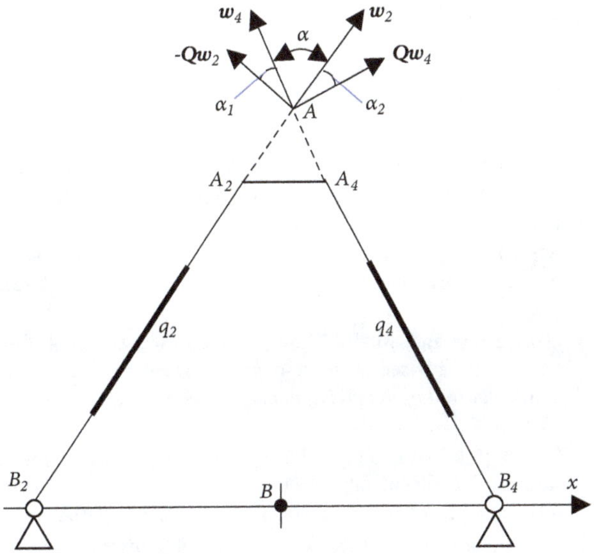

FIGURE 11: Pressure angles between RPU limbs.

should not be utilized for the evaluation of spatial parallel mechanism. However, in this paper, the 2-RPU&2-SPS SPM with two identical RPU limbs and two identical SPS libms and the two RPU limbs move in the same plane, due to the these two rotating joint axes being parallel to each other. Based on the special structure of the research mechanism, the pressure angle can be served to evaluate the transmission performance of the SPM torque/force.

The analysis of the 2-RPU&2-SPS SPM transmission performance in this paper is based on the pressure angle of the limbs, and the definition of the pressure angle is shown in Figures 11 and 12.

Right for the RPU limb, the acute angle α_i shown in Figure 11 is angle between the force $(\mathbf{w}_2(\mathbf{w}_4))$ generated at the same point of the two RPU limbs and the velocity $(\mathbf{Qw}_4(-\mathbf{Qw}_2))$ of point A. It indicates that when the actuated joint is locked, there is transmission capability of the force/motion from limb 2(4) to limb 4(2); α is the angle between the $\mathbf{w}_2$ and $\mathbf{w}_4$.

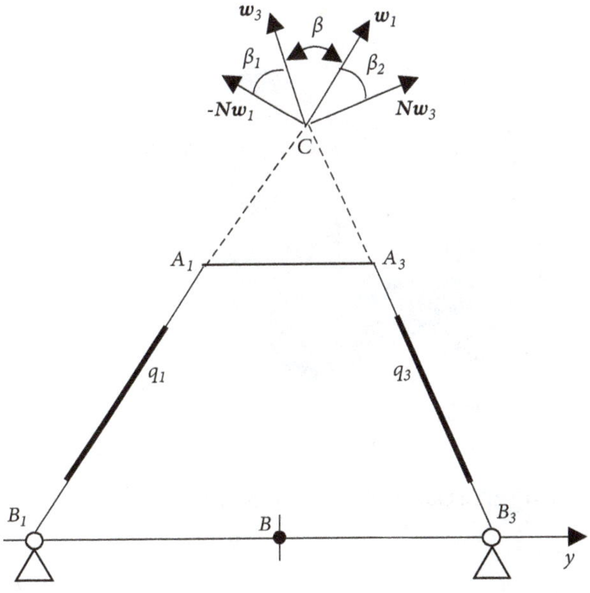

FIGURE 12: Pressure angles between SPS limbs.

Right for the SPS limb, the acute angle β_i shown in Figure 12 is angle between the force $(\mathbf{w}_1(\mathbf{w}_3))$ generated at the same point of the two SPS limbs and the velocity $(\mathbf{Nw}_1(-\mathbf{Nw}_3))$ of point C. It indicates that when the actuated joint is locked, there is transmission capability of the force/motion from limb 1(3) to limb 3(1). β is the angle between the $\mathbf{w}_1$ and $\mathbf{w}_3$.

For evaluating the transmission performance of the mechanism, the definitions of the transmission performance indicators are

$$\lambda_1 = \cos(\alpha), \qquad \lambda_2 = \cos(\beta) \tag{36}$$

When the transmission efficiency index λ_i of the limb is closer to 1, it indicates that the transmission efficiency of the mechanism is higher. On the contrary it is relatively low, and the range of transmission efficiency index is $0 \sim 1$.

Through the research of the 2-RPU&2-SPS SPM and the definition and analysis of the transmission performance, the *Matlab* software is used to simulate the transmission performance of the mechanism. According to the structural characteristics of the 2-RPU & 2-SPS SPM, the scale parameters of the mechanism are selected as $a_1 = a_3 = 150$ mm, $a_2 = a_4 = 75$ mm, $b_1 = b_3 = 550$ mm, and $b_2 = b_4 = 350$ mm, and the basic working parameters are selected as $-40^\circ \le \psi(\text{deg.}) \le 40^\circ$, $10^\circ \le \theta(\text{deg.}) \le 50^\circ$, and $700 \le z(\text{mm}) \le 900$. The simulation results shown in Figures 13 and 14.

As shown in Figures 13 and 14, the output transmission performance of each limb increases with the increase of z under the condition of the scale parameter and working parameter of a given mechanism, which shows that the output transmission performance of the mechanism increases with the increase of z value which is big and getting better. When the z value is a certain value, the output transmission performance of the RPU limb increases with the increase

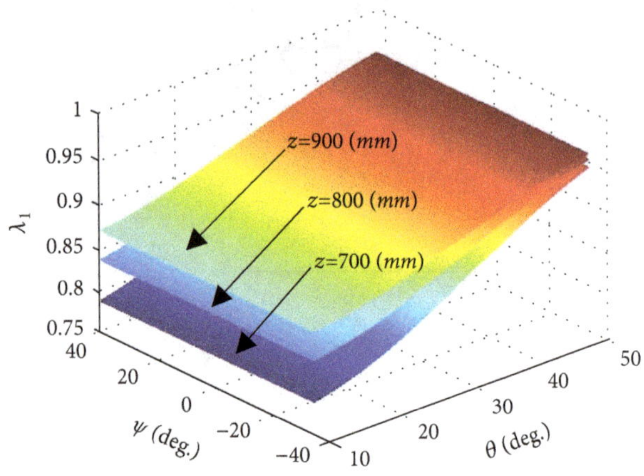

FIGURE 13: Output efficiency coefficient of the limb R̲PU.

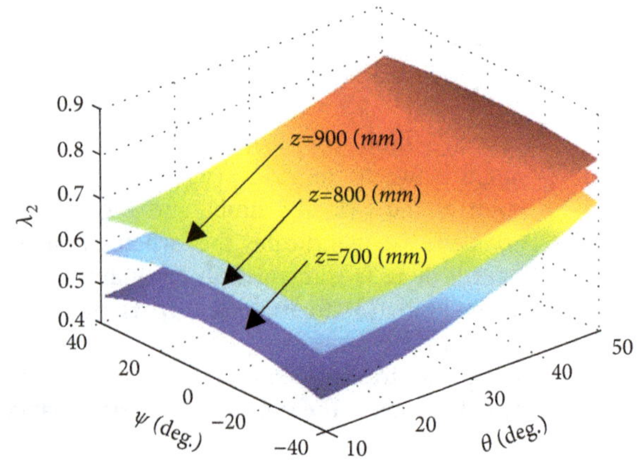

FIGURE 14: Output efficiency coefficient of the limb SP̲S.

of the angle θ. When $\theta = 10^\circ$, the limb output transmission performance is the worst and the output transmission performance of the limb is independent of the size of the angle ψ. Similarly, for the SP̲S limb, when the value of z is certainly, the output transmission performance of the limb decreases with the increase of the angle ψ. when $\psi = 0^\circ$, the output transmission performance of the limb is the best, at the same time, the output transmission performance of the limb increases with the increase of θ. when $\theta = 10^\circ$, the output transmission performance of the mechanism is the worst.

8. Conclusions

In this paper, the DOF of the 2-RP̲U&2-SP̲S SPM is analyzed by using the screw theory. At the same time, the position analysis of the 2-RP̲U&2-SP̲S SPM is obtained by utilizing the vector method, and the inverse kinematic solution of the 2-RP̲U&2-SP̲S SPM is acquired. The forward kinematic solution of the 2-RP̲U&2-SP̲S SPM is obtained by using the mathematical calculation software. Based on the screw theory, the entire Jacobian matrix of the 2-RP̲U&2-SP̲S SPM is established and the mechanism working space is studied. Finally, through the definition of output efficiency index of limb output, the kinematic transmission performance of the mechanism is evaluated by MATLAB software simulation, which provides some reference value for the optimization design and dynamic research of the follow-up mechanism.

Conflicts of Interest

The authors declare that there are no conflicts of interest regarding the publication of this paper.

Acknowledgments

This work was supported by the Intelligent Manufacturing Technology Major Project of Tianjin (nos. 15ZXZNGX00270 and 16ZXZNGX00070) and the National Key Research and Development Program of China (nos. 2017YFB1302103 and 2017YFB1303502).

References

[1] K. H. Hunt, "Structural kinematics of in-parallel-actuated robot-arms," *Journal of Mechanisms, Transmissions, and Automation in Design*, vol. 105, no. 4, pp. 705–712, 1983.

[2] K. J. Waldron, M. Raghavan, and B. Roth, "Kinematics of a hybrid series-parallel manipulation system," *Journal of Dynamic Systems, Measurement, and Control*, vol. 111, no. 2, pp. 211–221, 1989.

[3] S.-M. Song and M.-D. Zhang, "Study of reactional force compensation based on three-degree-of freedom parallel platforms," *Journal of Robotic Systems*, vol. 12, no. 12, pp. 783–794, 1995.

[4] Z. Huang, J. Wang, and Y. F. Fang, "Analysis of instantaneous motions of deficient-rank 3-RPS parallel manipulators," *Mechanism and Machine Theory*, vol. 37, no. 2, pp. 229–240, 2002.

[5] Y. Fang and Z. Huang, "Kinematics of a three-degree-of-freedom in-parallel actuated manipulator mechanism," *Mechanism and Machine Theory*, vol. 32, no. 7, pp. 789–796, 1997.

[6] A. Sokolov and P. Xirouchakis, *Kinematics of a 3-DOF Parallel Manipulator with an R-P-S Joint Structure*, Cambridge University Press, 2005.

[7] J. Schadlbauer, D. R. Walter, and M. L. Husty, "The 3-RPS parallel manipulator from an algebraic viewpoint," *Mechanism and Machine Theory*, vol. 75, pp. 161–176, 2014.

[8] N. Farhat, V. Mata, Á. Page, and F. Valero, "Identification of dynamic parameters of a 3-DOF RPS parallel manipulator," *Mechanism and Machine Theory*, vol. 43, no. 1, pp. 1–17, 2008.

[9] H. Li, Z. Yang, T. Huang et al., "Dynamics and optimization of a 2-Dof parallel robot with flexible links," in *Proceedings of the 7th World Congress on Intelligent Control and Automation*, vol. 27, pp. 1000–1003, 2008.

[10] R. Clavel, "A fast robot with parallel geometry," in *Proceedings of the International Symposium on Industrial Robots*, pp. 91–100, 1988.

[11] M. A. Hosseini, H.-R. M. Daniali, and H. D. Taghirad, "Dexterous workspace optimization of a Tricept parallel manipulator," *Advanced Robotics*, vol. 25, no. 13-14, pp. 1697–1712, 2011.

[12] H. T. Liu, T. Hung, J. P. Mei et al., "Kinematic design of a 5-DOF hybrid robot with large workspace/limb–stroke ratio," *Journal of Mechanical Design*, vol. 129, no. 5, pp. 530–537, 2006.

[13] D. Zeng, Y. Hou, Z. Huang, and W. Lu, "Type synthesis and characteristic analysis of a family of 2-DOF rotational decoupled parallel mechanisms," *Chinese Journal of Mechanical Engineering*, vol. 22, no. 6, pp. 833–840, 2009.

[14] Q. Li and J. M. Hervé, "Type synthesis of 3-DOF RPR-equivalent parallel mechanisms," *IEEE Transactions on Robotics*, vol. 30, no. 6, pp. 1333–1343, 2017.

[15] X. Kong, "Type synthesis of 3-DOF parallel manipulators with both a planar operation mode and a spatial translational operation mode," *Journal of Mechanisms and Robotics*, vol. 5, no. 4, Article ID 041015, 2013.

[16] H. Khakpour, L. Birglen, and S.-A. Tahan, "Synthesis of differentially driven planar cable parallel manipulators," *IEEE Transactions on Robotics*, vol. 30, no. 3, pp. 619–630, 2014.

[17] L. H. Rolland, "The manta and the kanuk: novel 4-DOF parallel mechanisms for industrial handling," in *Proceedings of the ASME Dynamic Systems and Control Division*, vol. 67, pp. 831–844, 1999.

[18] T. S. Zhao and Z. Huang, "A novel spatial 4-DOF parallel mechanism and its position analysis," *Mechanical Science and Technology*, vol. 19, no. 6, pp. 927–929, 2000.

[19] P. Araujo-Gómez, V. Mata, M. Díaz-Rodríguez, A. Valera, and A. Page, "Design and kinematic analysis of a novel 3UPS/RPU parallel kinematic mechanism with 2T2R motion for knee diagnosis and rehabilitation tasks," *Journal of Mechanisms and Robotics*, vol. 9, no. 6, Article ID 061004, 2017.

[20] Q. Zhao, J. Mei, T. Song, and S. Liu, "Pressure angle in parallel mechanisms: from planar to spatial," *Transactions of Tianjin University*, vol. 22, no. 5, pp. 411–418, 2016.

[21] L. M. Zhang et al., "Dimensional synthesis of the Delta robot using transmission angle constraints Dimensional synthesis of the Delta robot using transmission angle constraints," *Robotica*, vol. 30, no. 3, pp. 343–349, 2012.

3

Kinematics and Dynamics of a Tensegrity-Based Water Wave Energy Harvester

Min Lin,[1] Tuanjie Li,[1] and Zhifei Ji[2]

[1] *School of Electro-Mechanical Engineering, Xidian University, Xi'an 710071, China*
[2] *College of Mechanical and Energy Engineering, Jimei University, Xiamen 361021, China*

Correspondence should be addressed to Zhifei Ji; zfji18@163.com

Academic Editor: Shahram Payandeh

A tensegrity-based water wave energy harvester is proposed. The direct and inverse kinematic problems are investigated by using a geometric method. Afterwards, the singularities and workspaces are discussed. Then, the Lagrangian method was used to develop the dynamic model considering the interaction between the harvester and water waves. The results indicate that the proposed harvester allows harvesting 13.59% more energy than a conventional heaving system. Therefore, tensegrity systems can be viewed as one alternative solution to conventional water wave energy harvesting systems.

1. Introduction

Tensegrity systems are formed by a combination of rigid elements (struts) under compression and elastic elements (cables or springs) under tension. The use of cables or springs as tensile components leads to an important reduction in the weight of the systems. Due to this attractive nature, tensegrity systems have been proposed to be used in many disciplines. Moreover, a detailed description of the history of tensegrity systems is provided in [1, 2].

The first research work that deals with tensegrity systems was completed by Calladine [3]. Since then, tensegrity systems have been rapidly applied as structures in the architectural context. A tensegrity dome was proposed by Pellegrino [4]. Some design methods for tensegrity domes are proposed by Fu [5]. Afterwards, tensegrity structures have been also proposed to be served as bridges [6–9]. Moreover, the use of cables or springs in tensegrities allows them to be deployable [10, 11]. Due to this nature, some research works are found towards their use as antennas [12, 13]. For static applications, the subject of form-finding of tensegrities has attracted the attention of several researchers [14, 15]. Moreover, a review of form-finding methods was provided by Tibert and Pellegrino [16]. The basic issues about the statics of tensegrity structures were reviewed by Juan and Tur [17].

From an engineering point of view, tensegrities are a special class of structures whose components may simultaneously perform the purposes of structural force, actuation, sense, and feedback control. For such kind of structure, pulleys or other kinds of actuators may stretch/shorten some of the constituting components in order to substantially change their forms with a little variation of the structure's energy. Ingber [18] has demonstrated that tensegrity structures are very similar to cytoskeleton structures of unicellular organisms. Afterwards, the cellular tensegrity model is used to understand the cell structure, biological networks, and mechanoregulation [19, 20]. Tensegrity structures are also very similar to muscle-skeleton structures of high efficiency land animals whose speeds can reach up to 60 mph. The muscle-skeleton systems of these beings are composed of only tensional and compressional components. They thus have the ability to run with high speed [21].

Another interesting application of tensegrities is their development for use as mechanisms. Oppenheim and Williams [22] were the first to consider the actuation of tensegrity systems by modifying the lengths of their components in order to obtain tensegrity mechanisms. Afterwards, several mechanisms based on tensegrity systems were proposed, such as a flight simulator [23], a space telescope

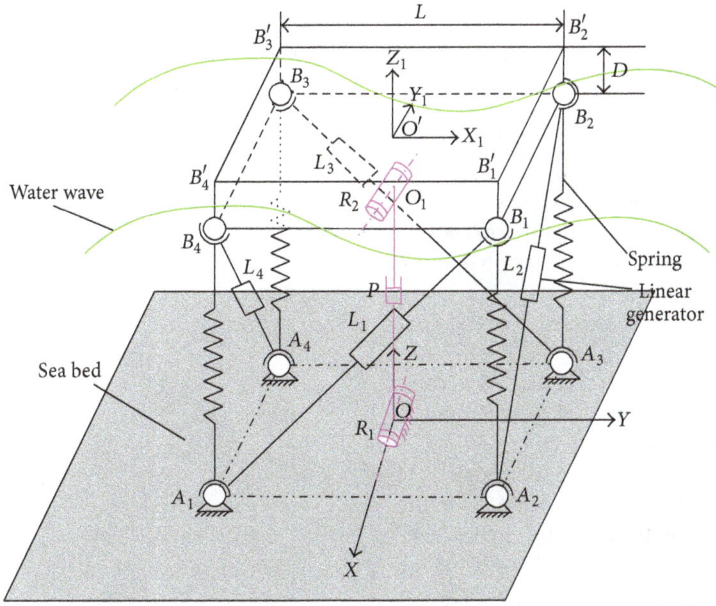

FIGURE 1: A tensegrity-based water wave energy harvester.

[24], and a tensegrity walking robot [25–27]. For tensegrity mechanisms, an interesting topic named tensegrity parallel mechanism has been proposed recently. The concept of tensegrity parallel mechanism was introduced by Marshall [28]. Then, Shekarforoush et al. [29] presented the statics of a 3-3 tensegrity parallel mechanism. Afterwards, Crane III et al. [30] proposed a planar tensegrity parallel mechanism and completed its equilibrium analysis. Tensegrity systems have been identified as one of three main research trends in mechanisms and robotics for the second decade of the 21st century [31]. However, just a few references have stated the possibility of using tensegrity systems as water wave energy harvesters. Scruggs and Skelton [32] made a preliminary investigation on the potential use of controlled tensegrity structures to harvest energy. Sunny et al. [33] studied the feasibility of harvesting energy using polyvinylidene fluoride patches mounted on vibrating prestressed membrane. Vasquez et al. [34] stated the possibility of using a planar tensegrity mechanism in ocean applications. This application is attractive since it can play an important role in the expansion of clean energy technologies that help the world's sustainable development.

This work presents the analysis of a tensegrity-based water wave energy harvester. Since this is the first stage for the development of a new application for tensegrity systems, a simplified linear model of sea waves was used to analyze the proposed harvester. The analytical solutions to the direct and inverse kinematic problems are found using a geometric method. Based on the obtained relationships between the input and output variables, the singular configurations have been discussed. The workspaces of the proposed mechanism have subsequently been computed. Afterwards, the dynamics were investigated. Finally, the energy harvesting capabilities of the tensegrity-based harvester are compared with a conventional heaving system.

2. Geometry of the Water Wave Energy Harvester

A diagram of the tensegrity-based water wave energy harvester is shown in Figure 1. It consists of a float, four springs, four linear generators, and one kinematic chain. The linear generators are joining node pairs A_iB_i ($i = 1, 2, 3, 4$) while the springs are joining node pairs A_1B_4, A_2B_1, A_3B_2, and A_4B_3. The float of height D is denoted by $B_1B_2B_3B_4$. This harvester is obtained from a square tensegrity parallel prism [10] by connecting the top of the latter to a float.

From Figure 1, it can be seen that the sides of the squares formed by nodes $A_1A_2A_3A_4$, $B_1B_2B_3B_4$, and $B_1'B_2'B_3'B_4'$ have the same length L. Moreover, the length of the linear generator joining node pairs A_iB_i is denoted by L_i. As illustrated in Figure 1, the springs and the linear generators are connected to the float and the sea bed at nodes A_i and B_i by spherical joints without friction. The sea bed is considered to be parallel to the horizontal plane. A fixed reference frame A (X, Y, Z) is located at the center of the square $A_1A_2A_3A_4$ with its X axis parallel to the line joining nodes A_4 and A_1 and its Z axis perpendicular to the sea bed, while a moving reference frame B (X_1, Y_1, Z_1) is located at the mass center of the float with its X_1 axis parallel to the line joining nodes B_4 and B_1 and its Z_1 axis perpendicular to the plane formed by nodes B_1, B_2, B_3, and B_4. Moreover, the vectors specifying the positions of nodes A_i and B_i in the fixed reference frame are defined as ${}^{A}\mathbf{a}_i$ and ${}^{A}\mathbf{b}_i$, respectively. Also, the vectors specifying the positions of nodes B_i in the moving reference frame are defined as ${}^{B}\mathbf{b}_i$.

In order to obtain an appropriate kinematic model of the harvester, the following hypotheses are made:

(i) The springs are linear with stiffness K and lengths l_j ($j = 1, 2, 3, 4$) and all the springs have the same free length L_0.

(ii) The water waves are traveling along the Y axis.

In Figure 1, a passive kinematic chain denoted by R_1PR_2 is used to connect nodes O and O_1. Nodes O and O_1 represent the centers of the squares $A_1A_2A_3A_4$ and $B_1B_2B_3B_4$, respectively. Considering the constraints introduced by this kinematic chain, the possible movements of the float driven by water waves are rotations about the X axis and translations along the Y and Z axes. Therefore, the harvester has three degrees of freedom.

The Cartesian coordinates of the mass center of the float in the fixed reference frame are defined as (x, y, z). From Figure 1, it can be seen that $x = 0$ is always satisfied. Moreover, the angle θ is used to specify the rotation of the float about X axis. Meanwhile, the range of θ is assumed to be $[-\pi/2 \ \ \pi/2]$. The variables y, z, and θ are driven by the water waves. As a consequence, they are thus chosen as the inputs of the system. Furthermore, only three of the four linear generators' lengths are independent. For this reason, the lengths of the generators joining nodes A_1B_1, A_2B_2, and A_3B_3 are chosen as the outputs of the system. It follows that the harvester's output vector is $\mathbf{O} = [L_1, L_2, L_3]^T$ while its input vector is $\mathbf{I} = [y, z, \theta]^T$.

3. Kinematic Analysis

For the harvester, the linear generators are used to convert wave motion cleanly into electricity. Generally, the efficiency of electricity generation of the system is highly dependent on the motions of linear generators. To provide great insight into the kinematics of the harvester, the relationship between the input and output vectors is developed in this section.

3.1. Direct Kinematic Analysis. The direct kinematic analysis consists in computing the output vector $\mathbf{O}$ for the given input vector $\mathbf{I}$. According to [35], the most convenient approach to set an algebraic equation system for kinematic problem of a parallel mechanism is to use the rotation matrix parameters and the position vector of the moving platform. This approach is used in this work to deal with the kinematic problems of the harvester. The position and orientation of the float are described by the position vector $\mathbf{P} = \mathbf{OO}' = [0, y, z]^T$ and the rotation matrix ${}^A\mathbf{R}_B$ with respect to the fixed reference frame. From Figure 1, it can be seen that the rotation matrix ${}^A\mathbf{R}_B$ can be defined by rotating the moving reference frame $\pi/2$ about Z_1 axis followed by θ about Y_1 axis. ${}^A\mathbf{R}_B$ thus takes the following form:

$$ {}^A\mathbf{R}_B = \begin{bmatrix} 0 & -1 & 0 \\ \cos\theta & 0 & \sin\theta \\ -\sin\theta & 0 & \cos\theta \end{bmatrix}. \tag{1} $$

Then, the position vectors of points B_i $(i = 1, 2, 3, 4)$ with respect to the fixed reference frame can be obtained:

$$ {}^A\mathbf{b}_i = \mathbf{P} + {}^A\mathbf{R}_B {}^B\mathbf{b}_i, \quad i = 1, 2, 3, 4. \tag{2} $$

The vectors specifying the positions of nodes B_i in the moving reference frame can be easily derived:

$$ \begin{gathered} {}^B\mathbf{b}_1 = \begin{bmatrix} \dfrac{L}{2} \\ -\dfrac{L}{2} \\ -\dfrac{D}{2} \end{bmatrix}, \\ {}^B\mathbf{b}_2 = \begin{bmatrix} \dfrac{L}{2} \\ \dfrac{L}{2} \\ -\dfrac{D}{2} \end{bmatrix}, \\ {}^B\mathbf{b}_3 = \begin{bmatrix} -\dfrac{L}{2} \\ \dfrac{L}{2} \\ -\dfrac{D}{2} \end{bmatrix}, \\ {}^B\mathbf{b}_4 = \begin{bmatrix} -\dfrac{L}{2} \\ -\dfrac{L}{2} \\ -\dfrac{D}{2} \end{bmatrix}. \end{gathered} \tag{3} $$

Substituting (3) into (2), we have

$$ \begin{gathered} {}^A\mathbf{b}_1 = \begin{bmatrix} \dfrac{L}{2} \\ y + \dfrac{L}{2}\cos\theta - \dfrac{D}{2}\sin\theta \\ z - \dfrac{D}{2}\cos\theta - \dfrac{L}{2}\sin\theta \end{bmatrix}, \\ {}^A\mathbf{b}_2 = \begin{bmatrix} -\dfrac{L}{2} \\ y + \dfrac{L}{2}\cos\theta - \dfrac{D}{2}\sin\theta \\ z - \dfrac{D}{2}\cos\theta - \dfrac{L}{2}\sin\theta \end{bmatrix}, \\ {}^A\mathbf{b}_3 = \begin{bmatrix} -\dfrac{L}{2} \\ y - \dfrac{L}{2}\cos\theta - \dfrac{D}{2}\sin\theta \\ z - \dfrac{D}{2}\cos\theta + \dfrac{L}{2}\sin\theta \end{bmatrix}, \\ {}^A\mathbf{b}_4 = \begin{bmatrix} \dfrac{L}{2} \\ y - \dfrac{L}{2}\cos\theta - \dfrac{D}{2}\sin\theta \\ z - \dfrac{D}{2}\cos\theta + \dfrac{L}{2}\sin\theta \end{bmatrix}. \end{gathered} \tag{4} $$

From Figure 1, it can also be seen that ${}^{A}\mathbf{a}_1 = [L/2, -L/2, 0]^T$, ${}^{A}\mathbf{a}_2 = [L/2, L/2, 0]^T$, ${}^{A}\mathbf{a}_3 = [-L/2, L/2, 0]^T$, and ${}^{A}\mathbf{a}_4 = [-L/2, -L/2, 0]^T$. With the position vectors of points A_i and B_i now known, the vector equation of the ith linear generator can be written as

$$\mathbf{L}_i = {}^{A}\mathbf{b}_i - {}^{A}\mathbf{a}_i, \quad i = 1, 2, 3, 4. \tag{5}$$

By using (5), the solution to the direct kinematic problem is found as follows:

$$L_1 = \left[\left(z - \frac{D}{2}\cos\theta - \frac{L}{2}\sin\theta\right)^2 + \left(y + \frac{L}{2}\cos\theta - \frac{D}{2}\sin\theta + \frac{L}{2}\right)^2\right]^{1/2}, \tag{6}$$

$$L_2 = \left[\left(z - \frac{D}{2}\cos\theta - \frac{L}{2}\sin\theta\right)^2 + \left(y + \frac{L}{2}\cos\theta - \frac{D}{2}\sin\theta - \frac{L}{2}\right)^2 + L^2\right]^{1/2}, \tag{7}$$

$$L_3 = \left[\left(z - \frac{D}{2}\cos\theta + \frac{L}{2}\sin\theta\right)^2 + \left(y - \frac{L}{2}\cos\theta - \frac{D}{2}\sin\theta - \frac{L}{2}\right)^2\right]^{1/2}. \tag{8}$$

Here, for the latter use, the length of the linear generator joining nodes A_4B_4 is also presented:

$$L_4 = \left[\left(z - \frac{D}{2}\cos\theta + \frac{L}{2}\sin\theta\right)^2 + \left(y - \frac{L}{2}\cos\theta - \frac{D}{2}\sin\theta + \frac{L}{2}\right)^2 + L^2\right]^{1/2}. \tag{9}$$

3.2. Inverse Kinematic Analysis. The inverse kinematic problem corresponds to the computation of the input vector **I** for the given output vector **O**. The solution to this problem can be found by solving (6)–(8) for the input variables y, z, and θ. Subtracting the square of (7) from that of (6) yields

$$2Ly + L^2\cos\theta - L^2 - L_1^2 + L_2^2 - DL\sin\theta = 0. \tag{10}$$

Subtracting the square of (8) from that of (6), we obtain

$$2Ly(\cos\theta + 1) - DL\sin\theta - L_1^2 + L_3^2 - 2Lz\sin\theta = 0. \tag{11}$$

From (10) and (11), the following expressions can be derived:

$$y = \frac{1}{2L}\left(L^2 + L_1^2 - L_2^2 - L^2\cos\theta + DL\sin\theta\right),$$
$$z = \frac{1}{2L\sin\theta}\left[L^2\sin^2\theta + DL\sin\theta\cos\theta + \left(L_1^2 - L_2^2\right)\cos\theta - L_2^2 + L_3^2\right]. \tag{12}$$

By substituting (12) into (6), the following equation is obtained:

$$4\left(L^2 - L_2^2\right)L^2\cos^2\theta - \left[\left(L_1^2 - L_2^2\right)^2 + \left(L_2^2 - L_3^2\right)^2 - \left(L_1^2 - L_3^2\right)^2\right]\cos\theta - 4L^2\left(L^2 - L_2^2\right) - \left(L_1^2 - L_2^2\right)^2 - \left(L_2^2 - L_3^2\right)^2 = 0. \tag{13}$$

Because of the range imposed on θ, four solutions for θ can be arrived at by solving (13). Furthermore, by substituting these results into (12), the solutions to the inverse kinematic problem are found.

4. Singularity Analysis

4.1. Jacobian Matrix. The Jacobian matrix of the harvester is defined as the relationships between a set of infinitesimal changes of its input vectors and the corresponding infinitesimal changes of its output vectors. The Jacobian matrix, **J**, relates $\delta\mathbf{I}$ to $\delta\mathbf{O}$ such that $\delta\mathbf{O} = \mathbf{J}\delta\mathbf{I}$. **J** can be rewritten in terms of matrices **C** and **D** such that $\mathbf{C}\delta\mathbf{O} = \mathbf{D}\delta\mathbf{I}$. From (6)–(8), the elements of **C** and **D** can be computed and written in terms of the input variables as follows:

$$C_{11} = \left[\left(z - \frac{D}{2}\cos\theta - \frac{L}{2}\sin\theta\right)^2 + \left(y + \frac{L}{2}\cos\theta - \frac{D}{2}\sin\theta + \frac{L}{2}\right)^2\right]^{1/2},$$

$$C_{22} = \left[\left(z - \frac{D}{2}\cos\theta - \frac{L}{2}\sin\theta\right)^2 + \left(y + \frac{L}{2}\cos\theta - \frac{D}{2}\sin\theta - \frac{L}{2}\right)^2 + L^2\right]^{1/2},$$

$$C_{33} = \left[\left(z - \frac{D}{2}\cos\theta + \frac{L}{2}\sin\theta\right)^2 + \left(y - \frac{L}{2}\cos\theta - \frac{D}{2}\sin\theta - \frac{L}{2}\right)^2\right]^{1/2},$$

$$C_{12} = C_{13} = C_{21} = C_{23} = C_{31} = C_{32} = 0,$$

$$D_{11} = y + \frac{L}{2}\cos\theta - \frac{D}{2}\sin\theta + \frac{L}{2},$$

$$D_{12} = D_{22} = z - \frac{D}{2}\cos\theta - \frac{L}{2}\sin\theta,$$

$$D_{13} = \left(z - \frac{D}{2}\cos\theta - \frac{L}{2}\sin\theta\right)\left(\frac{D}{2}\sin\theta - \frac{L}{2}\cos\theta\right) - \left(y + \frac{L}{2}\cos\theta - \frac{D}{2}\sin\theta + \frac{L}{2}\right)\left(\frac{L}{2}\sin\theta - \frac{D}{2}\cos\theta\right),$$

$$D_{21} = y + \frac{L}{2}\cos\theta - \frac{D}{2}\sin\theta - \frac{L}{2},$$

$$D_{31} = y - \frac{L}{2}\cos\theta - \frac{D}{2}\sin\theta - \frac{L}{2},$$

$$D_{32} = z - \frac{D}{2}\cos\theta + \frac{L}{2}\sin\theta,$$

$$D_{23} = \left(z - \frac{D}{2}\cos\theta - \frac{L}{2}\sin\theta\right)\left(\frac{D}{2}\sin\theta - \frac{L}{2}\cos\theta\right) - \left(y + \frac{L}{2}\cos\theta - \frac{D}{2}\sin\theta - \frac{L}{2}\right)\left(\frac{L}{2}\sin\theta + \frac{D}{2}\cos\theta\right),$$

$$D_{33} = \left(z - \frac{D}{2}\cos\theta + \frac{L}{2}\sin\theta\right)\left(\frac{D}{2}\sin\theta + \frac{L}{2}\cos\theta\right) + \left(y - \frac{L}{2}\cos\theta - \frac{D}{2}\sin\theta - \frac{L}{2}\right)\left(\frac{L}{2}\sin\theta - \frac{D}{2}\cos\theta\right). \tag{14}$$

For (14), it is noted that C_{ij} and D_{ij} are the elements located on the ith line and jth column of $\mathbf{C}$ and $\mathbf{D}$, respectively.

4.2. Singular Configurations. The singular configurations of the harvester consist in finding the situations where the relationships between infinitesimal changes in its input and output variables degenerate. When such a situation occurs, the harvester will gain or lose one or more degrees of freedom, thus leading to a loss of control. As a consequence, such configurations are usually avoided when possible. Generally, the singular configurations of the harvester can be obtained by setting $\det(\mathbf{C}) = 0$, $\det(\mathbf{D}) = 0$, or both. The determinants of $\mathbf{C}$ and $\mathbf{D}$ can be expressed as follows:

$$\det(\mathbf{C}) = \left[\left(z - \frac{D}{2}\cos\theta - \frac{L}{2}\sin\theta\right)^2 + \left(y + \frac{L}{2}\cos\theta - \frac{D}{2}\sin\theta + \frac{L}{2}\right)^2\right]^{1/2} \cdot \left[\left(z - \frac{D}{2}\cos\theta - \frac{L}{2}\sin\theta\right)^2 + \left(y + \frac{L}{2}\cos\theta - \frac{D}{2}\sin\theta - \frac{L}{2}\right)^2 + L^2\right]^{1/2} \cdot \left[\left(z - \frac{D}{2}\cos\theta + \frac{L}{2}\sin\theta\right)^2 + \left(y - \frac{L}{2}\cos\theta - \frac{D}{2}\sin\theta - \frac{L}{2}\right)^2\right]^{1/2} = 0,$$

$$\det(\mathbf{D}) = \frac{L^2}{8}\left[(DL - 2Dy - 2Lz)\sin 2\theta + \left(2Ly - 2Dz - L^2\right)\cos 2\theta + \left(8z^2 + 2D^2\right)\cos\theta + (8yz - 4Lz + 2DL)\sin\theta + L^2 - 6Dz - 2Ly\right] = 0. \tag{15}$$

By examining (15), it is possible to extract the expressions corresponding to singular configurations. The following is a list of these expressions as well as their descriptions with respect to the mechanism's behaviors:

(i) $$\left[\left(z - \frac{D}{2}\cos\theta - \frac{L}{2}\sin\theta\right)^2 + \left(y + \frac{L}{2}\cos\theta - \frac{D}{2}\sin\theta + \frac{L}{2}\right)^2\right]^{1/2} = 0. \tag{16}$$

(a) The length of the linear generator joining nodes A_1 and B_1 is equal to zero. Node A_1 is thus coincident with node B_1. Moreover, node A_4 is also coincident with node B_2.

(b) The movement of the float is reduced to a rotation about the axis joining nodes A_1 and A_4. When this is the case, only one variable is needed to define the system. The harvester thus loses two degrees of freedom.

(c) Infinitesimal movements of node O' in a direction perpendicular to the line joining nodes A_1 and B_4 are possible without deforming the springs and the linear generators.

(d) External forces parallel to the line B_1B_4 are resisted by the harvester.

(ii) $$\left[\left(z - \frac{D}{2}\cos\theta + \frac{L}{2}\sin\theta\right)^2 + \left(y - \frac{L}{2}\cos\theta - \frac{D}{2}\sin\theta - \frac{L}{2}\right)^2\right]^{1/2} = 0. \tag{17}$$

(a) The length of the linear generator joining nodes A_3 and B_3 is equal to zero. Node A_2 is coincident with node B_4.

(b) The movement of the float is reduced to a rotation about the axis joining nodes A_2 and A_3. When this case occurs, only one variable can be used to describe the rotation of the float. The harvester thus loses two degrees of freedom.

(c) Infinitesimal movements of node O' in a direction perpendicular to the line joining nodes A_2 and B_1 are possible without deforming the springs and the linear generators.

(d) External forces parallel to the line B_1B_4 are resisted by the harvester.

(iii)
$$\begin{aligned}&(DL - 2Dy - 2Lz)\sin 2\theta \\ &+ (2Ly - 2Dz - L^2)\cos 2\theta + (8z^2 + 2D^2)\cos\theta \\ &+ (8yz - 4Lz + 2DL)\sin\theta + L^2 - 6Dz - 2Ly \\ &= 0.\end{aligned} \tag{18}$$

(a) Actually, it is impossible to extract the behaviors of the harvester from (18). This case corresponds to the boundaries of the input workspace and will be mapped in Section 5.2. Generally speaking, when this is the case, infinitesimal movements of the input variables along a direction perpendicular to a certain surface cannot be generated.

From (16) and (17), it can be seen that the singular configuration (i) corresponds to the situation where the length of the linear generator A_1B_1 is equal to zero while configuration (ii) corresponds to the situation where the length of the linear generator A_3B_3 is equal to zero. From an engineering point of view, the linear generators are generally limited to operate within a range of nonzero lengths. However, from the aspect of mechanism's analysis, the lengths of prismatic actuators can be set to be zero. This case belongs to one kind of the singular configurations of the proposed mechanism.

5. Workspaces

Since the input variables y, z, and θ are driven by water waves, the ranges of the input variables can be used to describe the strengths of the water waves. Moreover, the amount of the electricity produced by the harvester depends on the movements of the linear generators. The ranges of the output variables can be considered as an indicator of the efficiency of energy harvesting. In this section, the ranges of the input vectors are referred to as the input workspace while the ranges of the output vectors are referred to as the output workspace. The boundaries of the input and output workspaces usually correspond to singular configurations described in Section 4.2. From (16)–(18), it can be seen that the singular configurations are expressed in terms of the input variables. According to these expressions, the boundaries of the input workspace can be computed. Afterwards, these boundaries will be mapped from the input domain into the output domain in order to generate the output workspace.

5.1. Input Workspace. The input workspace of the harvester is a volume whose boundaries correspond to singular configurations discussed in Section 4.2. An example of such a workspace with $L = 1$ m and $D = 1$ m is shown in Figure 2.

In Figure 2, the surface corresponding to the singular configuration (iii) is identified by surface (iii). From this figure, it can be seen that the input workspace can be divided into three parts. The first part is defined by $-1 \le y \le 0$ and $0 \le \theta \le \pi/2$. It is bounded by surface (iii) and the planes

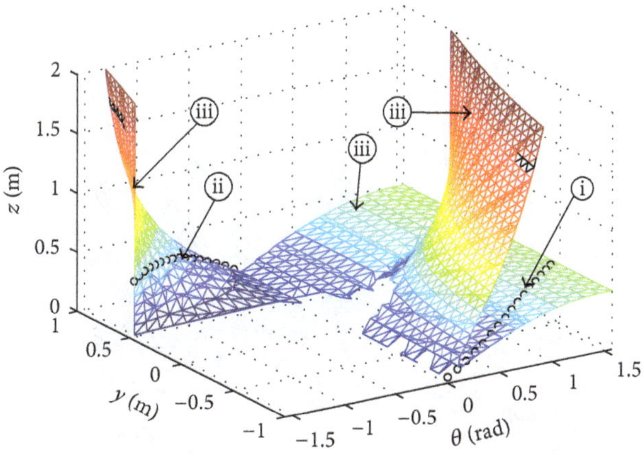

FIGURE 2: Input workspace of the harvester with $L = 1$ m and $D = 0.1$ m.

corresponding to $y = -1$, $\theta = \pi/2$, and $z = 2$. Moreover, the second part is defined by $0 \le y \le 1$ and $0 \le \theta \le \pi/2$. It is bounded by the planes corresponding to $\theta = \pi/2$, $y = 1$, and surface (iii). Finally, the third part is defined by $0 \le y \le 1$ and $-\pi/2 \le \theta \le 0$. It is bounded by the planes corresponding to $\theta = -\pi/2$, $y = -1$, and surface (iii). Furthermore, from Figure 2, it can also be observed that curves (i) and (ii) correspond to the singular configurations (i) and (ii), respectively. Since the harvester will be uncontrolled when it reaches a singular configuration, the boundaries of the input workspace and the singular curves (i) and (ii) should be avoided during the use of such a harvester.

5.2. Output Workspace. In order to obtain the output workspace, the singular configurations detailed in Section 4.2 should be rewritten in terms of the output variables firstly. From (16) and (17), it can be concluded that the singular configuration (i) in the output domain corresponds to $L_1 = 0$ while the singular configuration (ii) corresponds to $L_2 = 0$. Generally speaking, by substituting the solutions to the inverse kinematic problem into (18), an expression for singular configuration (iii) in terms of the output variables can be arrived at. However, this procedure is rather tedious. Here, Bezout's method [36] was used to derive the expression corresponding to singular configuration (iii) in the output domain due to its simplicity.

Equation (13) is firstly rewritten as

$$M_1\cos^2\theta + M_2\cos\theta + M_3 = 0, \tag{19}$$

where

$$\begin{aligned}M_1 &= 4(L^2 - L_2^2)L^2, \\ M_2 &= -\left[(L_1^2 - L_2^2)^2 + (L_2^2 - L_3^2)^2 - (L_1^2 - L_3^2)^2\right], \\ M_3 &= -4L^2(L^2 - L_2^2) - (L_1^2 - L_2^2)^2 - (L_2^2 - L_3^2)^2.\end{aligned} \tag{20}$$

Moreover, by substituting (12) into (18), the following equation is obtained:

$$N_1\cos^2\theta + N_2\cos\theta + N_3 = 0, \tag{21}$$

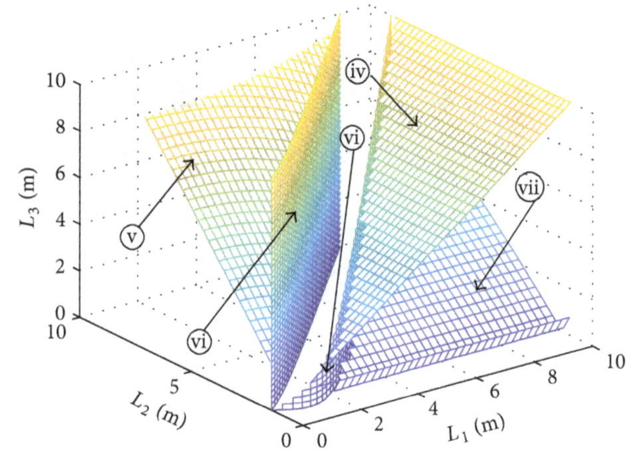

FIGURE 3: Output workspace of the water wave energy harvester with $L = 1$ m and $D = 0.1$ m.

where

$$N_1 = \left(L_1^2 - L_2^2\right)\left(L_2^2 - L_3^2\right),$$
$$N_2 = -\left(L_1^2 - L_2^2\right)^2 - \left(L_2^2 - L_3^2\right)^2, \quad (22)$$
$$N_3 = \left(L_2^2 - L_3^2\right)\left(L_2^2 - L_3^2\right).$$

It should be noted that (21) represents singular configuration (iii) expressed by L_1, L_2, L_3, and θ. Moreover, (19) is used to compute θ for the given values of L_1, L_2, and L_3. Generally, the solutions to θ obtained by solving (21) should satisfy (19). Furthermore, both (19) and (21) can be considered as two quadratics with respect to $\cos\theta$. According to Bezout's method, the condition that (19) and (21) have a comment root for $\cos\theta$ is as follows:

$$\begin{vmatrix} M_1 & M_2 \\ N_1 & N_2 \end{vmatrix} \begin{vmatrix} M_2 & M_3 \\ N_2 & N_3 \end{vmatrix} - \begin{vmatrix} M_1 & M_3 \\ N_1 & N_3 \end{vmatrix}^2 = 0. \quad (23)$$

Simplifying (23) yields

$$\left[\left(L_1^2 - L_3^2\right)\left(L_1^2 - 2L_2^2 + L_3^2\right)\right]^2$$
$$\cdot \left(4L^4 - 4L^2L_2^2 + L_1^4 - 2L_1^2L_2^2 + L_1^4\right) \quad (24)$$
$$\cdot \left(4L^4 - 4L^2L_2^2 + L_2^4 - 2L_2^2L_3^2 + L_3^4\right) = 0.$$

Equation (24) represents the surfaces corresponding to singular configuration (iii) in the output workspace. By plotting these surfaces, the output workspace of the harvester can be obtained. An example of such plots is shown in Figure 3 with $L = 1$ m and $D = 0.1$ m.

From Figure 3, it can be seen that the singular configuration (iii) determined by (24) corresponds to four surfaces (surfaces (iv)–(vii)) in the output workspace. Moreover, surfaces (iv), (v), (vi), and (vii) correspond to expressions $L_1 - L_3 = 0$, $L_1^2 - 2L_2^2 + L_3^2 = 0$, $4L^4 - 4L^2L_2^2 + L_1^4 - 2L_1^2L_2^2 + L_1^4 = 0$, and $4L^4 - 4L^2L_2^2 + L_2^4 - 2L_2^2L_3^2 + L_3^4 = 0$, respectively. It can also be observed that the output workspace of the harvester can be divided into two parts. The first part is bounded by surface (v), surface (vi), plane $L_1 = 0$, and plane $L_3 = 0$ while the second part is bounded by surfaces denoted by (iv), (vi), and (vii) and planes denoted by $L_2 = 0$ and $L_1 = 10$. This output workspace should be considered during the use and design of such a harvester.

It is noted that the forward and inverse kinematics, Jacobian matrix, and workspaces should be considered when such harvester is being designed. Moreover, when the harvester is put to use, the singular configurations should be avoided. The kinematics and Jacobian matrix are used to find the singular configurations.

6. Dynamic Analysis

The efficiency of the water wave harvesting is highly dependent on the dynamics of the harvester. Therefore, it is of utmost importance to research the dynamics of the harvester. In this section, the dynamic model of the harvester is developed. Furthermore, in order to compare the efficiency of a conventional heaving system with that of the proposed harvester, the dynamic model of the conventional heaving system is firstly introduced. Before introducing the dynamic models of the two systems, it is assumed that the linear water waves are applied on the two systems.

6.1. Dynamic Model of a Conventional Heaving System. A diagram of the conventional heaving wave energy harvester [37] composed of a float, a bar magnet, and a battery is shown in Figure 4. In order to compare the efficiency of the conventional heaving system with the proposed harvester, the floats of both systems are assumed to have the same size. Moreover, in this paper, the weight of the bar magnet was neglected.

According to [38], the motion equation of the float, driven by linear water waves, in a conventional heaving system is given by

$$\left(m + a_{wz}\right)\frac{d^2z}{dt^2} + \left(b_{rz} + b_{vz} + b_{pz}\right)\frac{dz}{dt}$$
$$+ \left(\rho g A_{wp} + Nk_s\right)z = F_{z0}\cos\left(\omega t + \alpha_z\right). \quad (25)$$

The coefficients in (25) are given as follows:

m is the mass of the float.

a_{wz} is the added mass.

b_{rz} is the damping coefficient.

b_{vz} is the viscous damping coefficient.

b_{pz} is the power take-off coefficient.

A_{wp} is the waterplane area when the body is at rest.

ρ is the density of seawater.

g is the acceleration due to gravity.

k_s is the spring constant of mooring lines and N is the number of lines (mooring restoring force).

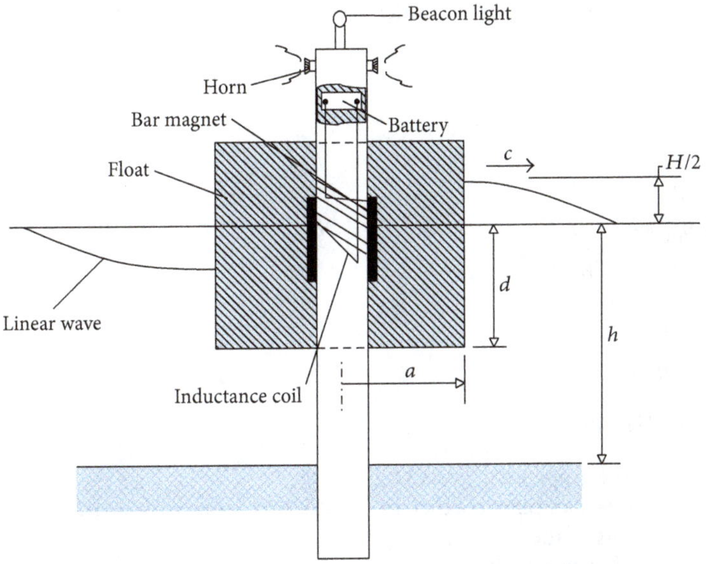

FIGURE 4: A conventional heaving wave energy harvester [37].

F_{z0} is the water-induced vertical force amplitude and $\omega = 2\pi/T$ is the circular wave frequency (T is the wave period).

α_z is the phase angle between the wave and force.

Finally, it should be noted that the computations of the above coefficients in (25) can be found in [38].

6.2. Dynamic Model of the Tensegrity-Based Water Wave Energy Harvester. As stated in Section 2, the harvester has three degrees of freedom. Therefore, three generalized coordinates, chosen as $\mathbf{q} = [q_1\ q_2\ q_3]^T = [y\ z\ \theta]^T$, are needed to develop the dynamic model.

In order to derive an appropriate dynamic model of the harvester, the following hypotheses are made:

(i) The links of the mechanism, except for the float, are massless.

(ii) The springs are massless.

(iii) There is no friction in the harvester's revolute, prismatic, and spherical joints.

The equations of motion of the harvester are developed using the Lagrangian approach; namely,

$$\frac{d}{dt}\frac{\partial T}{\partial \dot{\mathbf{q}}} - \frac{\partial T}{\partial \mathbf{q}} + \frac{\partial E}{\partial \mathbf{q}} = \mathbf{Q}_k, \tag{26}$$

where T and E are the kinetic and potential energies of the harvester and $\mathbf{Q}_k$ is the vector of nonconservative forces acting on the system. In [37], the translation of the float along Y axis is defined as surge, the translation of the float along Z axis is defined as heave, and the rotation of the float with respect to X axis is defined as pitch. The kinetic energy, due only to the surge, heave, and pitch movements of the float, can be expressed as

$$\mathbf{T} = \frac{1}{2}\dot{\mathbf{q}}^T\mathbf{M}\dot{\mathbf{q}}, \tag{27}$$

where

$$\mathbf{M} = \begin{bmatrix} m + a_{wy} & 0 & 0 \\ 0 & m + a_{wy} & 0 \\ 0 & 0 & I_y + A_w \end{bmatrix}. \tag{28}$$

I_y is the mass moment of inertia with respect to Y axis and A_w is added-mass moment of inertia due to pitching. The potential energies due to heaving and pitching motions of the top platform are described by McCormick [37] as

$$E_{pz} = \frac{1}{2}\rho g A_{wp} q_2^2,$$
$$E_{p\theta} = \frac{1}{2}C q_3^2, \tag{29}$$

where C is the restoring moment constant, defined for a bottom-flat body in terms of the draft. The total potential energy of the harvester becomes

$$\begin{aligned} E &= U + E_{pz} + E_{p\theta} \\ &= \frac{1}{2}\rho g A_{wp} q_2^2 + \frac{1}{2}C q_3^2 + K\left[\sqrt{\sigma_1^2 + \sigma_2^2} - l_0\right]^2 \\ &\quad + K\left[\sqrt{\sigma_3^2 + \sigma_4^2} - l_0\right]^2. \end{aligned} \tag{30}$$

The nonconservative forces, which correspond to the radiation damping force, viscous damping force, and water wave induced forces, can be expressed as

$$\mathbf{Q}_k = \begin{bmatrix} -b_{vy}\dot{q}_1 \\ F_{z0}\cos(\omega t) - (b_{rz} + b_{vz})\dot{q}_2 \\ M_{\theta 0}\cos(\omega t) - b_{r\theta}\dot{q}_3 \end{bmatrix}. \tag{31}$$

Substituting (27), (30), and (31) into (26), the dynamic model of the harvester can be rewritten as

$$\mathbf{M}\ddot{\mathbf{q}} + \mathbf{B}\dot{\mathbf{q}} + \mathbf{G} = \mathbf{F}, \tag{32}$$

where

$$\mathbf{B} = \begin{bmatrix} b_{vy} & 0 & 0 \\ 0 & b_{rz} + b_{vz} + b_{pz} & 0 \\ 0 & 0 & b_{r\theta} \end{bmatrix},$$
$$\mathbf{G} = \left[\frac{\partial E}{\partial q_1} \quad \frac{\partial E}{\partial q_2} \quad \frac{\partial E}{\partial q_3}\right]^T, \tag{33}$$
$$\mathbf{F} = \left[0 \quad F_{zo}\cos\omega t \quad M_{\theta 0}\sin\omega t\right]^T.$$

The elements of $\mathbf{G}$ are detailed in the Appendix. For (32), it should be noted that b_{vy} is viscous damping coefficient corresponding to the surge movements of the float while $b_{r\theta}$ is the radiation damping coefficient due to pitching motion. $M_{\theta 0}$ is the water-induced torque amplitude (applied on the float). The computations of b_{vy}, $b_{r\theta}$, $M_{\theta 0}$, and C can also be found in [37]. These computations are also not repeated here.

7. Energy Harvesting

In this section, two energy harvesting systems are researched, respectively. One is a conventional heaving system and the other is the tensegrity-based water wave harvester. Also, the powers of the two systems have been computed, respectively. The parameters of water waves are selected as $H = 0.2$ m, $T = 6$ s, and $h = 100$ m. H is the wave height measured from the trough to the crest while T is the wave period. h denotes the water depth. Moreover, the floats used in the two energy harvesting systems are supposed to have the same dimensions as $L = 1$ m, $D = 0.1$ m, and $d = 0.05$ m.

7.1. Conventional Heaving System. For a conventional heaving system, the motion of the float is expressed by (25). For the given water wave parameters and the dimensions of the float, the coefficients of (25) can be calculated according to [38]. The results are listed in Table 1.

Solving (25) yields the position and velocity of the float which are shown in Figure 5. The power of the heaving body is given by [37]

$$P_z(t) = F_z(t)\frac{dz(t)}{dt}, \tag{34}$$

where $F_z(t)$ is the wave introduced heaving force on the float.

The power for take-off, $P(t)$, is given by the difference between the available power ($P_z(t)$) and the power dissipated due to radiation ($P_{rz}(t)$) and viscous effects ($P_{vz}(t)$):

$$P(t) = P_z(t) - P_{rz}(t) - P_{vz}(t). \tag{35}$$

The average power for take-off over one period of time is given by

$$P_{\text{ave}} = \frac{1}{T}\int_T P(t)\,dt. \tag{36}$$

TABLE 1: Conventional heaving float coefficients.

Coefficient	Value	Unit
m	51.50	kg
a_{wz}	454.19	kg
b_{rz}	1065.50	N·s/m
b_{vz}	114.91	N·s/m
b_{pz}	0	N·s/m
A_{wp}	1	m^2
N	0	—
k_s	0	N/m
F_{z0}	2012.10	N
α_z	0	rad
ω_{nz}	4.47	Rad/s
b_{cz}	4518.60	N·s/m
Z_0	0.21	m

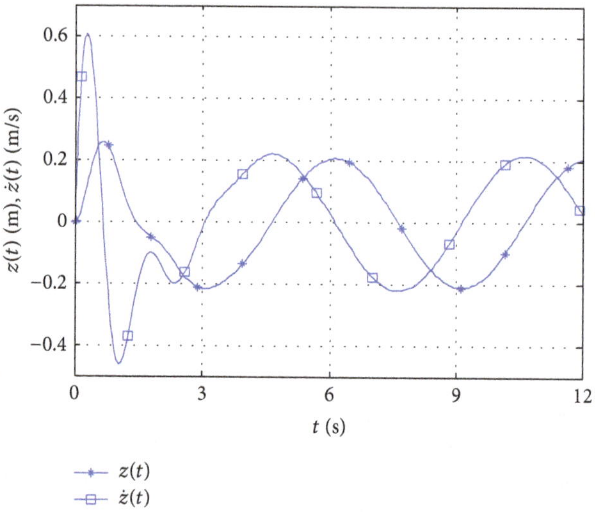

FIGURE 5: Motion of the conventional heaving system.

The water wave energy and power are [37]

$$E = \frac{\rho g^2 H^2 T^2 b}{16\pi}, \tag{37}$$

$$\mathbf{P} = \frac{\rho g^2 H^2 T b}{32\pi}\mathbf{i}. \tag{38}$$

Applying (36) over two wave periods of the function shown in Figure 6 gives an average power $P_{\text{ave}} = 0.154$ kW. Since the float's breadth is 1 m, then we can compare this result with the power contained in one meter of wave front. The maximum available power per meter of wave front is $P = 0.236$ kW (computed by (38)). Therefore, 65.17% of the wave energy can be harvested with electrical generators.

7.2. Tensegrity-Based Wave Energy Harvester. For the harvester considered here, it has infinitesimal mechanisms inherent of many tensegrity systems. This means that there are infinitesimal deformations of the mechanism that do not require any changes in the lengths of the harvester's

TABLE 2: Tensegrity-based harvester coefficients.

Coefficient	Value	Unit
m	51.50	kg
a_{wy}	2.02	kg
a_{wz}	454.19	kg
A_w	15.03	kg·m^2
I_x	4.33	Kg·m^2
C	849.58	N·m/rad
b_{rz}	1065.50	N·s/m
$b_{r\theta}$	88.79	N·m·s/rad
b_{vy}	114.91	N·s/m
b_{vz}	114.91	N·s/m
b_{pz}	0	N·s/m
A_{wp}	1	M^2
F_{z0}	2012.10	N
$M_{\theta 0}$	18.77	N·m
α_z	0	rad

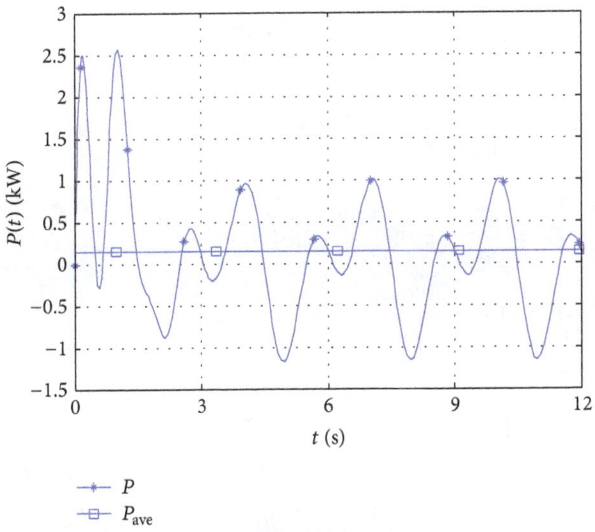

FIGURE 6: Power for take-off of the conventional heaving system.

components. It follows that some wave energy would not be harvested as the mechanism could deform some degree without the deformation being felt by the linear generators. However, since the deformations are infinitesimal, the effects of infinitesimal mechanisms are negligible.

Let the dimensions of the float in the tensegrity-based water wave harvester be the same as the conventional heaving float. The additional constant physical parameters are L_0 = 4 m and K = 10 N/m. Table 2 contains the values of the coefficients (computed according to [38]) for the equation of motion (see (32)).

The simulation is performed over two wave periods, that is, 12 seconds. Figures 7–9 show the position and velocity response of the float: surge, heave, and pitch.

Figure 10 shows the instantaneous power for take-off. The average power over two wave periods is P_{ave} = 0.186 kW. The

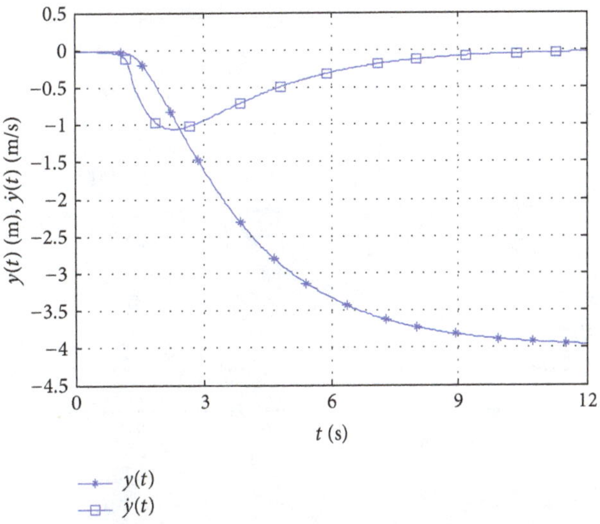

FIGURE 7: Surge motions of the tensegrity-based water wave energy harvester.

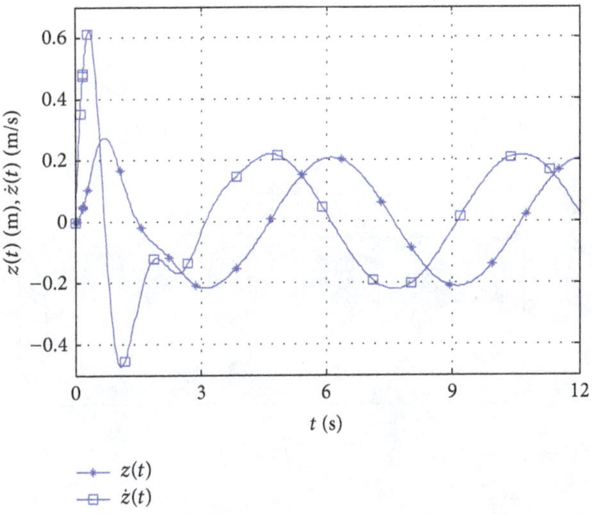

FIGURE 8: Heave motions of the tensegrity-based water wave energy harvester.

power contained in one meter of wave front is P = 0.236 kW (computed by (38)). Therefore, 78.76% of the available energy could be harvested by electrical generators. By comparing Figures 6 with 10, it is found that the proposed tensegrity-based harvester allows harvesting 13.59% more energy than a conventional heaving device under linear water wave conditions. For the conventional heaving device, the movement of the float is translation along the Z axis. It is proper to say that the conventional heaving device has one degree of freedom. However, the possible movements of the proposed harvester are rotations about the X axis and translations along the Y and Z axes (see Section 2). It is thus proper to say that the proposed harvester has three degrees of freedom. That is why the harvester can harvest more energy than a conventional device.

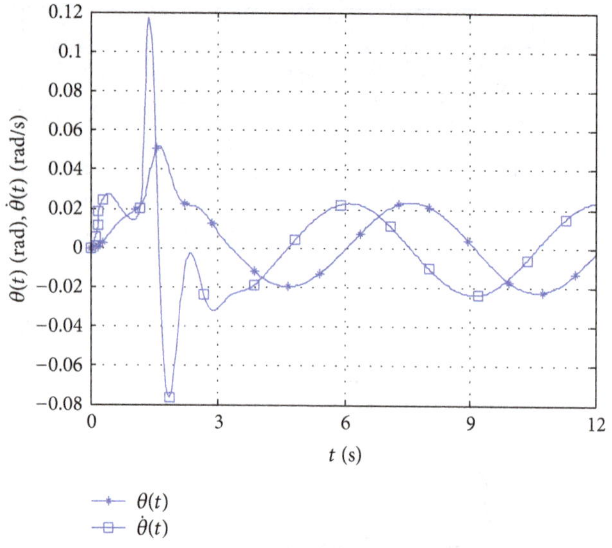

FIGURE 9: Pitch motions of the tensegrity-based water wave energy harvester.

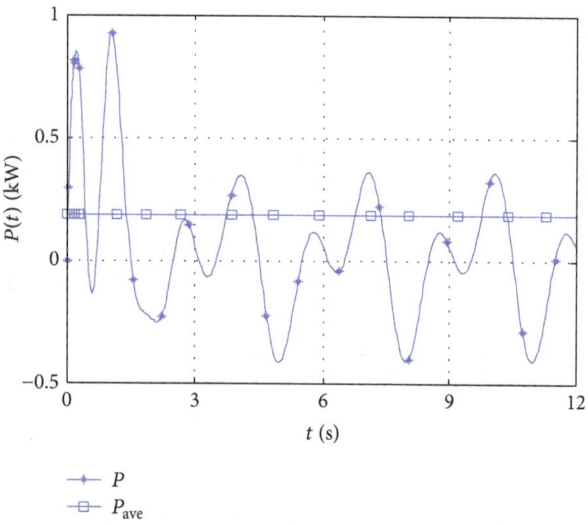

FIGURE 10: Tensegrity-based harvester: power for take-off.

8. Conclusion

A tensegrity-based water wave energy harvester was proposed in this work. The geometry of the harvester was described. The solutions to the direct and inverse kinematic problems were found by using a geometric method. The Jacobian matrix and singular configurations were subsequently computed. Then, the input and output workspaces were computed on the basis of the analysis of the obtained singular configurations. Afterwards, the dynamic analysis was performed considering the interaction with linear water waves, considering added mass, radiation damping, and viscous damping phenomena. It was shown that the proposed tensegrity-based water wave energy harvester allows harvesting 13.59% more energy than a conventional heaving system.

Appendix

Elements of G

The elements of **G** in (32) are as follows:

$$
\begin{aligned}
G_1 = \frac{\partial E}{\partial q_1} = 2K \Bigg\{ 1 & - L_0 \Bigg[\left(q_1 - \frac{L\cos q_3}{2} - \frac{D\sin q_3}{2} + \frac{L}{2} \right)^2 \\
& + \left(q_2 - \frac{D\cos q_3}{2} + \frac{L\sin q_3}{2} \right)^2 \Bigg]^{-1/2} \Bigg\} \cdot \Bigg(q_1 \\
& - \frac{L\cos q_3}{2} - \frac{D\sin q_3}{2} + \frac{L}{2} \Bigg) + 2K \Bigg(q_1 + \frac{L\cos q_3}{2} \\
& - \frac{D\sin q_3}{2} - \frac{L}{2} \Bigg) \cdot \Bigg\{ 1 \\
& - L_0 \Bigg[\left(q_1 + \frac{L\cos q_3}{2} - \frac{D\sin q_3}{2} - \frac{L}{2} \right)^2 \\
& + \left(q_2 - \frac{D\cos q_3}{2} - \frac{L\sin q_3}{2} \right)^2 \Bigg]^{-1/2} \Bigg\},
\end{aligned}
$$

$$
\begin{aligned}
G_2 = \frac{\partial E}{\partial q_2} = \rho g A_{wp} q_2 + 2K \Bigg(q_2 & - \frac{D\cos q_3}{2} \\
& + \frac{L\sin q_3}{2} \Bigg) \cdot \Bigg\{ 1 \\
& - L_0 \Bigg[\left(q_1 - \frac{L\cos q_3}{2} - \frac{D\sin q_3}{2} + \frac{L}{2} \right)^2 \\
& + \left(q_2 - \frac{D\cos q_3}{2} + \frac{L\sin q_3}{2} \right)^2 \Bigg]^{-1/2} \Bigg\} + 2K \Bigg\{ 1 \\
& - L_0 \Bigg[\left(q_1 + \frac{L\cos q_3}{2} - \frac{D\sin q_3}{2} - \frac{L}{2} \right)^2 \\
& + \left(q_2 - \frac{D\cos q_3}{2} - \frac{L\sin q_3}{2} \right)^2 \Bigg]^{-1/2} \Bigg\} \cdot \Bigg(q_2 \\
& - \frac{D\cos q_3}{2} - \frac{L\sin q_3}{2} \Bigg),
\end{aligned}
$$

$$
\begin{aligned}
G_3 = \frac{\partial E}{\partial q_3} = C q_3 + 2K \Bigg\{ 1 & - L_0 \Bigg[\left(q_1 - \frac{L\cos q_3}{2} - \frac{D\sin q_3}{2} + \frac{L}{2} \right)^2 \\
& + \left(q_2 - \frac{D\cos q_3}{2} + \frac{L\sin q_3}{2} \right)^2 \Bigg]^{-1/2} \Bigg\} \cdot \Bigg[\Bigg(q_1
\end{aligned}
$$

$$
\begin{aligned}
&- \frac{L\cos q_3}{2} - \frac{D\sin q_3}{2} + \frac{L}{2}\Big)\Big(\frac{L\sin q_3}{2} \\
&- \frac{D\cos q_3}{2}\Big) + \Big(q_2 - \frac{D\cos q_3}{2} + \frac{L\sin q_3}{2}\Big) \\
&\cdot\Big(\frac{D\sin q_3}{2} + \frac{L\cos q_3}{2}\Big)\Big] + 2K\Bigg\{1 \\
&- L_0\Big[\Big(q_1 + \frac{L\cos q_3}{2} - \frac{D\sin q_3}{2} - \frac{L}{2}\Big)^2 \\
&+ \Big(q_2 - \frac{D\cos q_3}{2} - \frac{L\sin q_3}{2}\Big)^2\Big]^{-1/2}\Bigg\}\cdot\Big[-\Big(q_1 \\
&+ \frac{L\cos q_3}{2} - \frac{D\sin q_3}{2} - \frac{L}{2}\Big)\Big(\frac{L\sin q_3}{2} \\
&+ \frac{D\cos q_3}{2}\Big) + \Big(q_2 - \frac{D\cos q_3}{2} - \frac{L\sin q_3}{2}\Big) \\
&\cdot\Big(\frac{D\sin q_3}{2} - \frac{L\cos q_3}{2}\Big)\Big].
\end{aligned}
\tag{A.1}
$$

Competing Interests

The authors declare that they have no competing interests.

Acknowledgments

This research is supported by the National Natural Science Foundation of China (no. 51375360).

References

[1] R. Motro, *Tensegrity: Structural Systems for the Future*, Kogen Page Science, Guildford, UK, 2003.

[2] R. E. Skelton and M. C. Oliveira, *Tensegrity Systems*, Springer, New York, NY, USA, 2009.

[3] C. R. Calladine, "Buckminster Fuller's 'Tensegrity' structures and Clerk Maxwell's rules for the construction of stiff frames," *International Journal of Solids and Structures*, vol. 14, no. 2, pp. 161–172, 1978.

[4] S. Pellegrino, "A class of tensegrity domes," *International Journal of Space Structures*, vol. 7, no. 2, pp. 127–142, 1992.

[5] F. Fu, "Structural behavior and design methods of Tensegrity domes," *Journal of Constructional Steel Research*, vol. 61, no. 1, pp. 23–25, 2005.

[6] L. Rhode-Barbarigos, N. B. Hadj Ali, R. Motro, and I. F. C. Smith, "Designing tensegrity modules for pedestrian bridges," *Engineering Structures*, vol. 32, no. 4, pp. 1158–1167, 2010.

[7] L. Rhode-Barbarigos, N. B. H. Ali, R. Motro, and I. F. C. Smith, "Design aspects of a deployable tensegrity-hollow-rope footbridge," *International Journal of Space Structures*, vol. 27, no. 2-3, pp. 81–95, 2012.

[8] S. Korkmaz, N. B. H. Ali, and I. F. C. Smith, "Configuration of control system for damage tolerance of a tensegrity bridge," *Advanced Engineering Informatics*, vol. 26, no. 1, pp. 145–155, 2012.

[9] R. E. Skelton, F. Fraternali, G. Carpentieri, and A. Micheletti, "Minimum mass design of tensegrity bridges with parametric architecture and multiscale complexity," *Mechanics Research Communications*, vol. 58, pp. 124–132, 2014.

[10] J. Duffy, J. Rooney, B. Knight, and C. D. Crane III, "Review of a family of self-deploying tensegrity structures with elastic ties," *Shock and Vibration Digest*, vol. 32, no. 2, pp. 100–106, 2000.

[11] A. Hanaor, "Double-layer tensegrity grids as deployable structures," *International Journal of Space Structures*, vol. 8, no. 1-2, pp. 135–143, 1993.

[12] G. Tibert, *Deployable tensegrity structures in space applications [Ph.D. thesis]*, Royal Institute of Technology, Stockholm, Sweden, 2002.

[13] N. Fazli and A. Abedian, "Design of tensegrity structures for supporting deployable mesh antennas," *Scientia Iranica*, vol. 18, no. 5, pp. 1078–1087, 2011.

[14] N. Vassart and R. Motro, "Multiparametered form finding method: application to tensegrity systems," *International Journal of Space Structures*, vol. 14, no. 2, pp. 147–154, 1999.

[15] K. Koohestani, "Form-finding of tensegrity structures via genetic algorithm," *International Journal of Solids and Structures*, vol. 49, no. 5, pp. 739–747, 2012.

[16] A. G. Tibert and S. Pellegrino, "Review of form-finding methods for tensegrity structures," *International Journal of Space Structures*, vol. 18, no. 4, pp. 209–223, 2003.

[17] S. H. Juan and J. M. M. Tur, "Tensegrity frameworks: static analysis review," *Mechanism and Machine Theory*, vol. 43, no. 7, pp. 859–881, 2008.

[18] D. E. Ingber, "Cellular tensegrity: defining new rules of biological design that govern the cytoskeleton," *Journal of Cell Science*, vol. 104, no. 3, pp. 613–627, 1993.

[19] D. E. Ingber, "The architecture of life," *Scientific American*, vol. 278, no. 1, pp. 48–57, 1998.

[20] D. E. Ingber, "Tensegrity I. Cell structure and hierarchical systems biology," *Journal of Cell Science*, vol. 116, no. 7, pp. 1157–1173, 2003.

[21] S. M. Levin, "The tensegrity-truss as a model for spinal mechanics: biotensegrity," *Journal of Mechanics in Medicine and Biology*, vol. 2, no. 3, pp. 375–388, 2002.

[22] I. J. Oppenheim and W. O. Williams, "Tensegrity prisms as adaptive structures," in *Proceedings of the ASME International Mechanical Engineering Congress and Exposition*, Dallas, Tex, USA, November 1997.

[23] C. Sultan, M. Corless, and R. E. Skelton, "Tensegrity flight simulator," *Journal of Guidance, Control, and Dynamics*, vol. 23, no. 6, pp. 1055–1064, 2000.

[24] C. Sultan, M. Corless, and R. E. Skelton, "Peak-to-peak control of an adaptive tensegrity space telescope," in *Smart Structures and Materials 1999: Mathematics and Control in Smart Structures*, vol. 3667 of *Proceedings of SPIE*, pp. 190–201, The International Society for Optical Engineering, Newport Beach, Calif, USA, March 1999.

[25] C. Paul, F. J. Valero-Cuevas, and H. Lipson, "Design and control of tensegrity robots for locomotion," *IEEE Transactions on Robotics*, vol. 22, no. 5, pp. 944–957, 2006.

[26] A. G. Rovira and J. M. M. Tur, "Control and simulation of a tensegrity-based mobile robot," *Robotics and Autonomous Systems*, vol. 57, no. 5, pp. 526–535, 2009.

[27] S. Hirai and R. Imuta, "Dynamic simulation of six-strut tensegrity rolling," in *Proceedings of the 2012 IEEE International Conference on Robotics and Biomimetics*, Guangzhou, China, 2012.

[28] M. Q. Marshall, *Analysis of tensegrity-based parallel platform devices [M.S. thesis]*, University of Florida, Gainesville, Fla, USA, 2003.

[29] S. M. M. Shekarforoush, M. Eghtesad, and M. Farid, "Kinematic and static analyses of statically balanced spatial tensegrity mechanism with active compliant components," *Journal of Intelligent & Robotic Systems*, vol. 71, no. 3-4, pp. 287–302, 2013.

[30] C. D. Crane III, J. Bayat, V. Vikas, and R. Roberts, "Kinematic analysis of a planar tensegrity mechanism with pre-stressed springs," in *Advances in Robot Kinematics: Analysis and Design*, J. Lenarčič and P. Wenger, Eds., pp. 419–427, Springer, Batz-sur-Mer, France, 2008.

[31] J. M. McCarthy, "21st century kinematics: synthesis, compliance, and tensegrity," *Journal of Mechanisms and Robotics*, vol. 3, no. 2, Article ID 020201, 2011.

[32] J. T. Scruggs and R. E. Skelton, "Regenerative tensegrity structures for energy harvesting applications," in *Proceedings of the 45th IEEE Conference on Decision and Control (CDC '06)*, pp. 2282–2287, IEEE, San Diego, Calif, USA, December 2006.

[33] M. R. Sunny, C. Sultan, and R. K. Kapania, "Optimal energy harvesting from a membrane attached to a tensegrity structure," *AIAA Journal*, vol. 52, no. 2, pp. 307–319, 2014.

[34] R. E. Vasquez, C. D. Crane III, and J. C. Correa, "Analysis of a planar tensegrity mechanism for ocean wave energy harvesting," *Journal of Mechanisms and Robotics*, vol. 6, no. 3, Article ID 031015, 2014.

[35] S. V. Sreenivasan, K. J. Waldron, and P. Nanua, "Closed-form direct displacement analysis of a 6-6 Stewart platform," *Mechanism and Machine Theory*, vol. 29, no. 6, pp. 855–864, 1994.

[36] C. D. Crane III and J. Duffy, *Kinematic Analysis of Robot Manipulators*, Cambridge University Press, New York, NY, USA, 1998.

[37] M. McCormick, *Ocean Wave Energy Conversion*, Dover, Mineola, NY, USA, 2007.

[38] R. E. Vasquez, *Analysis of a tensegrity system for ocean wave energy harvesting [Ph.D. thesis]*, University of Florida, Gainesville, Fla, USA, 2011.

4

Adaptive Neuro-Fuzzy Inference System based Path Planning for Excavator Arm

Nga Thi-Thuy Vu, Nam Phuong Tran, and Nam Hoai Nguyen

Hanoi University of Science and Technology, Vietnam

Correspondence should be addressed to Nga Thi-Thuy Vu; nga.vuthithuy@hust.edu.vn

Academic Editor: Huosheng Hu

This paper presents a scheme based on Adaptive Neuro-Fuzzy Inference Systems (ANFIS) to generate trajectory for excavator arm. Firstly, the trajectory is predesigned with some specific points in the work space to meet the requirements about the shape. Next, the inverse kinematic is used and optimization problems are solved to generate the via-points in the joint space. These via-points are used as training set for ANFIS to synthesis the smooth curve. In this scheme, the outcome trajectory satisfies the requirements about both shape and optimization problems. Moreover, the algorithm is simple in calculation as the numbers of via-points are large. Finally, the simulation is done for two cases to test the effect of ANFIS structure on the generated trajectory. The simulation results demonstrate that, by using suitable structure of ANFIS, the proposed scheme can build the smooth trajectory which has the good matching with desired trajectory even that the desired trajectory has the complicated shape.

1. Introduction

In the construction and mine fields, the excavator which is used to dig and transport of soil or coal is one of the important machines. The work environment of excavator is usually dangerous and harsh. Therefore, developing the automatic excavator system is the general trend. In the unmanual operation system, i.e., excavator system, the trajectory generation for the excavator base and arm is the hot spot because it determines the efficiencies of overall system.

In the real, the excavator arm is a three-degree of freedom (3DOF) manipulator robot. The trajectory planning can be done in both working space and joint space. In the working space, the trajectory is built for end-effectors in three-dimension reference frame so it is quite visual. However, the trajectory built in this space has to face with problems of inverse kinematic and manipulator redundancy [1]. Therefore, in the most case, the trajectory of the manipulator robot is planned in the joint space [2].

In the joint space, the trajectory is planned to meet some specific requirements such as time optimization, energy optimization, jerk optimization, obstacle avoidance, etc. In order to satisfy these conditions, the trajectory is usually predesigned with some via-points then the smooth curve is built using several interpolation such as polynomial, spline, Bezier, etc. In [3, 4] polynomial functions are used to generate the paths for robot arms. Reference [3] proposed a series of polynomials to create desired trajectory for robotic motion via a set of given point; they also addressed a problem of acceleration and jerk optimization. However, the main drawback of [3] is that numbers of parameter proportion to numbers of via-point, which leads to explosion of calculation when the numbers of given point are large. The problems of reducing vibration are solved in [4]; however, the generated trajectory is partial smooth. The Bezier Curve and modifier genetic algorithm are interested in [5] in order to create a path in dynamic field with avoiding obstacle and minimum path's length.

In recent year, neural networks and fuzzy systems which have ability to approximate functions and fit curves have been widely applied in the path planning field. These algorithms seem to be more flexible and potential than traditional one because the methods based on neural network and fuzzy system can create a path through many via-points without explosion of calculation. In [6–10], the shunting model technique is used to build neural network for path planning

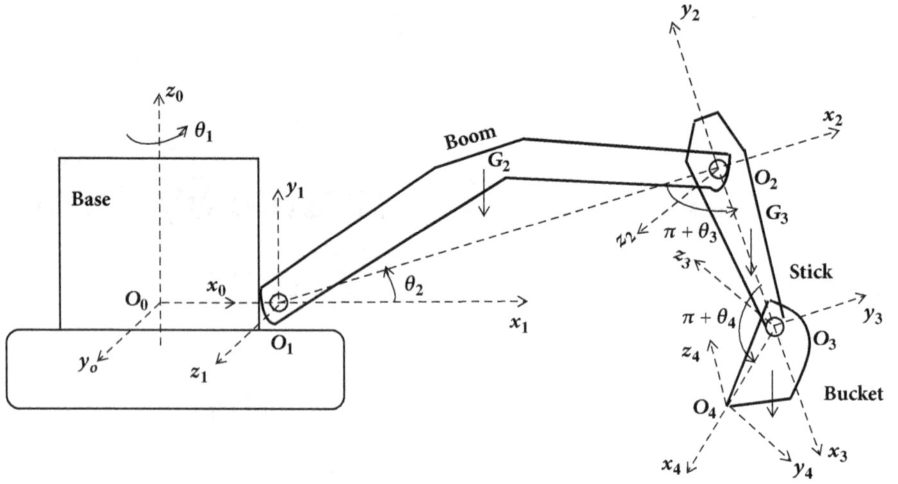

FIGURE 1: Block diagram of the excavator arm.

problems. In this method, the neural dynamics of each neuron is characterized by a shunting equation or simple additive equation [9]. The trajectories in [6]-[8] are generated for robots to avoid the static obstacles while in [9, 10] robots can work in dynamic environments with moving obstacles. The pulse-couple neural network is used in many application [11] and it is also applied into trajectory generation [12, 13]. This scheme can work in both static and dynamic environments but the complete information about working conditions is necessary. In the field of learning method, fuzzy system also is used to solve the path planning problems [14–17]. In [14] the fuzzy logic based on fuzzy sets algorithm is approached to plan the path for the robotic placement of fabrics on a work table. This fuzzy logic system is developed based on experimental data and it has ability to work with various materials and sizes, while optimal fuzzy scheme is introduced in [15] for path planning of manipulator robots. This is rule-based method which needs specific rules to generate the trajectory for robots and it can deal with moving obstacles. In order to generate a real-time and obstacle avoiding path for cushion robot, a fuzzy system which have capability to transform directly human knowledge in machine is utilized in [16]. Moreover, in [17] the fuzzy logic path planning algorithm is investigated to guarantee the safe motion with obstacle avoidance for mobile robot.

For the excavator, in order to meet the requirement of automatic trend, there are also some researches focusing on path planning topic. In [18, 19], the laser scanner, camera, and sensors are used to build 3D trajectory for automated excavator. This method gives the good result in the clean environment but the reliability of laser scanner and camera will reduce in the dusty environment. In [20], the current position of excavator arm is feedback to control system to predict the trajectory for next cycle. The neural network is used in [21] to determine the characteristic of the soil. From this result, in combination with the reaction force exerted on the bucket, the optimal trajectory is generated for excavator arm. In [22–24], the velocity and acceleration of bucket are used to build the path for excavator arm. The generated trajectory is optimal but velocity and acceleration are difficult to measure.

In this paper, an algorithm based on ANFIS is proposed to generate trajectory for excavator arm. Firstly, the trajectory is predesigned with some specific points in the work space to meet the requirements about the shape. Next, the inverse kinematic is used and optimization problems are solved to generate the via-points in the joint space. These via-points are used as training set for ANFIS to synthesis the smooth curve. In this scheme, the outcome trajectory satisfies the requirements about both shape and optimization problems. Moreover, the algorithm is simple in calculation as the numbers of via-points are large. Finally, the simulation is done for two cases to test the effect of ANFIS structure on the generated trajectory. The simulation results demonstrate that, by using suitable structure of ANFIS, the proposed scheme can build the smooth trajectory which has the good matching with desired trajectory even that the desired trajectory has the complicated shape.

2. Path Planning for Excavator Arm Based on ANFIS

2.1. Problem Description. Consider the excavator system as shown in Figure 1. It is assumed that the base is fixed and the arm of excavator operates in the $x_0O_0z_0$ plane.

To execute the digging task with satisfying technical constraints, the trajectory of excavator arm should go through some predesigned points. These points are selected from the desired shape, optimization criteria, constraints, etc. From given via-point, it is necessary to build the smooth curve for excavator to operate.

In order to minimize the time and jerk, the following optimization problem should be solvent [25]:

$$\begin{aligned}
&\textit{find} \quad \min J = k_T N \sum_{i=1}^{n-1} h_i + k_J \int_0^{t_f} (\dddot{q}(t))^2 \, dt \\
&\textit{subject to} \quad \left|\dot{q}_j(t)\right| \le VC_j, \quad j = 1, \cdots, N \\
&\qquad \left|\ddot{q}_j(t)\right| \le WC_j, \quad j = 1, \cdots, N \\
&\qquad \left|\dddot{q}_j(t)\right| \le JC_j, \quad j = 1, \cdots, N
\end{aligned} \tag{1}$$

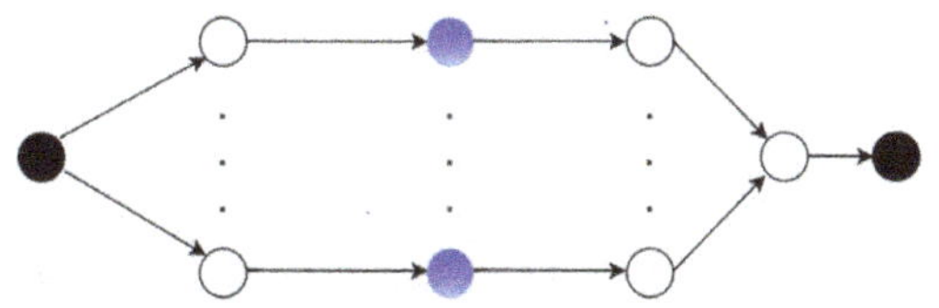

FIGURE 2: The ANFIS architecture.

where k_T and k_J are scalars, N are the numbers of joint, n are the numbers of via-points of, h_i is time interval between two via-points, $\dot{q}(t), \ddot{q}(t)$, and $\dddot{q}(t)$ are velocity, acceleration, and jerk of the jth joint, respectively, and VC_j, WC_j, and JC_j is the bound of velocity, acceleration, and jerk for the jth joint, respectively.

The objective function (1) can be expressed as follows [26]:

$$J = k_T N \sum_{i=1}^{n-1} h_i + k_J \sum_{j=1}^{N} \sum_{i=1}^{n-1} \left[\frac{\alpha_{i,j} - \alpha_{i,j}^2}{h_i} \right] \tag{2}$$

subject to the constraints

$$\begin{gathered} \max\left\{\left|\alpha_{j,1}\right|, \dots \left|\alpha_{j,n}\right| \le VC_j \\ \forall j = 1 \dots N,\ \forall i = 1 \dots n-1 \\ \max\left\{\left|\alpha_{j,1}\right|, \dots \left|\alpha_{j,n}\right| \le WC_j \quad \forall j = 1 \dots N \\ \left| \frac{\alpha_{j,i+1} - \alpha_{j,i}}{h_i} \right| \le JC_j \\ \forall j = 1 \dots N,\ \forall i = 1 \dots n-1 \end{gathered} \tag{3}$$

where $\alpha_{i,j}$ is the acceleration of the jth joint at the ith via-point.

Solve the optimal problem (2) by using Sequential Quadratic Programing technique (Optimization Toolbox of Matlab) to get the via-points.

After getting suitable via-points, the ANFIS is used to create the smooth trajectory for three joints.

2.2. ANFIS System Design and Training. In this research, ANFIS is utilized like a tool for curve fitting. The task of designing reference trajectory is to create a smooth-continuous path which passes some given points. Three ANFIS systems, which are based on the Sugeno model, are designed to plan paths for three joints. Each ANFIS system uses the same membership function for fuzzy sets, so we are going to analyze and represent one of the three ANFIS systems.

The input and output of each ANFIS system is time variable "t" and joint variable "theta", respectively. The ith if-then rule is as follows:

$$\text{Rule } i : \text{If } t \text{ is small then theta} = f_i(t) \tag{4}$$

The ANFIS architecture is shown as Figure 2. It consists of five layers.

(i) Layer 1: this layer performs a fuzzification process. The Gauss function is used as membership function in this study. It is defined as follows:

$$O_i^1 = \mu_i = e^{-(t-c_i)^2/2\sigma_i^2} \tag{5}$$

The parameters c_i and σ_i of first layer are typically referred as to the premise parameters.

(ii) Layer 2: this layer is fixed and nonadaptive. Its node has a function which multiplies the incoming signals from the outputs of the previous layer to obtain the firing strength of conditional clauses. If there exists only one input, then

$$O_i^2 = \alpha_i = \mu_i \tag{6}$$

(iii) Layer 3: this layer also has not included trainable parameters. The output of each node is the ratio of the ith rule's matching degree to the total of all rules' matching degree.

$$O_i^3 = \overline{\alpha_i} = \frac{\alpha_i}{\sum_i \alpha_i} \tag{7}$$

(iv) Layer 4: the parameters of this layer can be modified to adapt to training data. The function in each node of the layer is defined as

$$O_i^4 = \overline{\alpha_i} \times f_i = \overline{\alpha_i} \times (\mathrm{p}_i t_1 + r_i) \tag{8}$$

where p_i, r_i are referred as to consequent parameters.

(v) Layer 5: this layer has only one node. Its output is the sum of all outputs from the fourth layer.

$$O^5 = \sum_i \overline{\alpha_i} f_i \tag{9}$$

For training the ANFIS, it is able to apply the gradient method. But this method has slow convergence rate and tends to be trapped at local minima. To deal with this problem, [27] proposed a method, which is the combination of the gradient method and the least square estimator (LSE) method, namely, the hybrid algorithm. The training process is divided into two parts that is referred as forward-path and backward-path. In forward-path, premise parameters are kept unchanged, so the output of the ANFIS is a linear function of consequent parameters. Then, the least square error (LSE) method is applied to adjust these parameters. Next, the consequent parameters are fixed and premise parameters are updated based on the gradient algorithm. This hybrid algorithm is able to provide faster convergence and avoid the occurrence of local minima, because of the reduction in dimension of the search space. For these advantages of hybrid algorithm, we will use this method for the ANFIS training.

2.3. Path Planning Procedure. With the information in the previous parts, it is possible to generate the trajectory for excavator arm which satisfies some requirements about optimization and smooth. The sequence for this process has the following steps:

(i) Step 1: get the desired points based on shaped and optimal issues, then use inverse kinematics to obtain the via-points in the joint space as training sample.

(ii) Step 2: design ANFIS architecture.

(iii) Step 3: train ANFIS.

TABLE 1: Parameters of the ANFIS after training.

	c_i	σ_i	p_i	r_i
1st joint	2.4929	3.1575	-0.0260	0.6923
	2.8817	15.6040	0.0059	0.3450
	3.4474	20.7265	-0.0102	1.0390
2nd joint	1.2196	6.6245	-0.0359	-1.4945
	1.3634	8.1598	-0.0171	-1.7741
	0.9156	13.9925	0.0016	-2.0630
3rd joint	1.5545	4.9146	0.0681	0.7954
	4.0780	16.0198	0.0401	1.1314
	3.3035	20.4656	0.0327	0.2277

TABLE 2: Parameters of the ANFIS after training for the 1st joint (Case 2).

c_i	σ_i	p_i	r_i
1.1440	0.1237	-0.0235	0.6894
1.4447	3.2746	-0.0236	0.6910
1.6017	6.2853	0.0179	0.6860
1.7003	7.4553	0.0518	-0.1467
0.7788	9.6823	0.0055	0.3601
0.4393	12.8548	0.0059	0.3972
0.5287	15.8631	0.0434	-0.1261
0.5801	18.8813	0.0434	-0.1256
0.4406	22.3654	0	0.7859
1.0550	25.0155	0	0.7855

TABLE 3: Parameters of the ANFIS after training for the 2nd joint (Case 2).

c_i	σ_i	p_i	r_i
2.6620	0.5242	-0.0485	-1.5113
2.2012	4.6932	-0.0550	-1.4146
0.6739	10.0057	-0.0124	-1.8130
1.6379	15.2547	-0.0005	-2.0323
2.2804	19.8801	0.0034	-2.0965
2.0898	24.9830	0.0014	-2.0627

(iv) Step 4: use ANFIS to generate the trajectory for each joint.

In this work, the inverse dynamic calculation for excavator arm is based on the [28].

3. Simulation and Results

In order to verify the effectiveness of this scheme, the simulation is setup based on Optimization and Fuzzy toolboxes of Matlab. The constraints for optimal problem are similar as in [25]. The parameters for ANFIS are presented in detail in the following.

The simulation is done for two cases:

(i) *Case 1*. The numbers of rule are 3 for each joint.

(ii) *Case 2*. The numbers of rule are 6, 9, and 10 for the first, the second, and the third joint, respectively.

For Case 1, the parameters of the ANFIS for each joint after training are shown in the Table 1.

Simulation results for this case are shown in Figure 3.

In Figure 3, (a), (b), (c), and (d) are the matching errors of the first, the second, and the third joint and trajectory in the workspace, respectively. It can be seen from Figures 3(a), 3(b), and 3(c) that the matching error of each joint is quite small. The maximum absolute error is 0.015 rad for the first and the third joint, while this is about 0.006 rad for the second joint. In Figure 3(d), the desired trajectory and the approximated trajectory are presented. From this figure it is seen that the generated trajectory is quite close to the desired trajectory except the case of sudden change in the motion direction.

For *Case* 2, the numbers of rule for the first, the second, and the third joint are 10, 6, and 9, respectively. The parameters for each joint are given in Tables 2, 3, and 4. The results for this case are illustrated in Figure 4.

In the Figure 4, it is shown that the matching errors for all cases are insignificant, i.e., maximum absolute error for the first joint is 3e-3 rad, for the second joint is 5e-3 rad, and

TABLE 4: Parameters of the ANFIS after training for the 3rd joint (Case 2).

c_i	σ_i	p_i	r_i
1.9499	0.8483	0.0684	0.8396
2.3933	3.6964	0.0802	0.7339
0.4202	7.8501	0.0859	0.7016
0.8204	10.3098	0.0127	1.3822
0.5266	12.5082	-0.0044	1.6211
0.8842	16.2364	-0.0438	2.1717
0.5693	18.9473	-0.0471	2.2301
0.3972	22.3126	0.0001	1.2383
1.0464	25.0587	0	1.2392

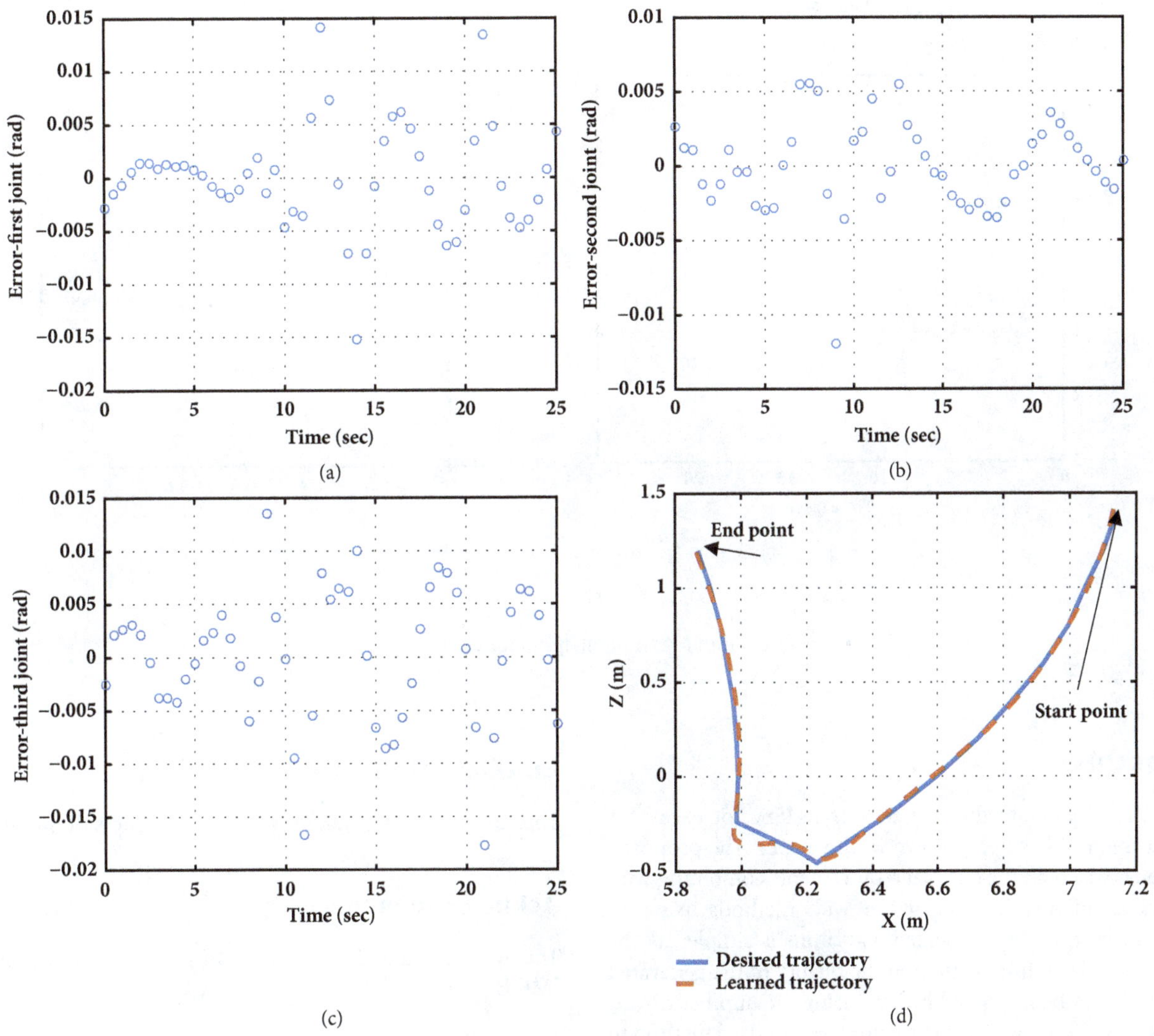

FIGURE 3: Training result for Case 1.

for the third joint is 4e-4 rad. The learned trajectory exactly follows the designed trajectory even under the condition of abrupt alteration of motion direction.

From the above simulation results, it is feasible to build the trajectory for excavator arm using an ANFIS. By choosing suitable structure and parameters of neural network as well as the numbers of fuzzy rule, the ANFIS can create the smooth trajectory which has the good matching with desired one despite the condition that the desired trajectory has a complicated shape.

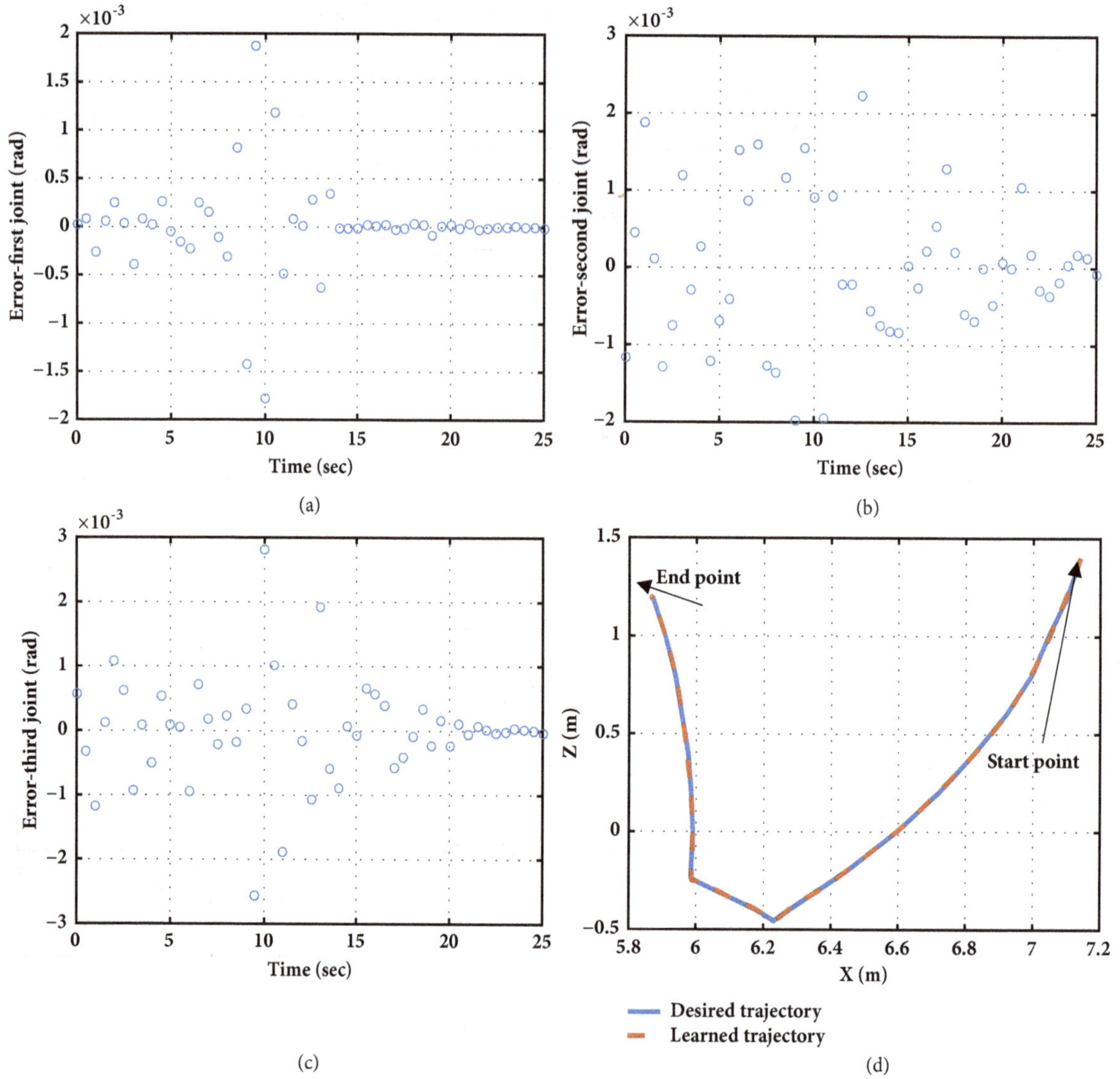

FIGURE 4: Training result for Case 2.

4. Conclusion

In order to generate the reference trajectory for excavator arm, a method has been shown in this paper. The proposed scheme is based on the optimal requirements combining with ANFIS technique. In comparison with methods using B-spline technique, the presented algorithm is simpler as the number of via-points is large, so the quality of the generated trajectory can be improved by increasing the number of via-points. Also, this characteristic helps the proposed method to deal with complicated shape trajectories. Finally, the simulation was shown for two cases to test the effect of the ANFIS structure on the generated trajectory. The simulation results demonstrated that, by using a suitable structure of the ANFIS, the proposed scheme can build the smooth trajectory which has the good matching with desired trajectory regardless of the fact that the desired trajectory has the complicated shape.

Conflicts of Interest

The authors declare that they have no conflicts of interest.

Acknowledgments

This work is funded by Ministry of Education and Training (MOET) under Grant no. B2018-BKA-70.

References

[1] J. Huang, P. Hu, K. Wu, and M. Zeng, "Optimal time-jerk trajectory planning for industrial robots," *Mechanism and Machine Theory*, vol. 121, pp. 530–544, 2018.

[2] A. Gasparetto, P. Boscariol, A. Lanzutti, and R. Vidoni, "Path planning and trajectory planning algorithms: a general overview," *Mechanisms and Machine Science*, vol. 29, pp. 3–27, 2015.

[3] Y. Guan, K. Yokoi, O. Stasse, and A. Kheddar, "On robotic trajectory planning using polynomial interpolations," in *Proceedings of the 2005 IEEE International Conference on Robotics and Biomimetics*, pp. 111–116, Shatin, China, July 2005.

[4] M. Dupac and P. Sewell, "Quick 3D trajectory planning for rotating extensible manipulators using piecewise polynomial interpolation," in *Proceedings of the Congress on Numerical Methods in Engineering*, Spain, 2017.

[5] M. Elhoseny, A. Tharwat, and A. E. Hassanien, "Bezier Curve Based Path Planning in a Dynamic Field using Modified Genetic Algorithm," *Journal of Computational Science*, vol. 25, pp. 339–350, 2018.

[6] S. X. Yang and M. Q.-H. Meng, "Real-time collision-free motion planning of mobile robots using neural dynamics based approaches," *IEEE Trans. Neural Netw*, vol. 14, no. 6, pp. 1541–1552, 2003.

[7] A. R. Willms and S. X. Yang, "Real-time robot path planning via a distance-propagating dynamic system with obstacle clearance," *IEEE Transaction on Systems, Man, and Cybernetics, Part B: Cybernetics*, vol. 38, no. 3, pp. 884–893, 2008.

[8] X. Yuan and S. X. Yang, "Multi-robot-based nanoassembly planning with automated path generation," *IEEE/ASME Transaction on Mechatronics*, vol. 12, no. 3, pp. 352–356, 2007.

[9] H. Li, S. X. Yang, and M. L. Seto, "Neural network based path planning for a multirobot system with moving obstacles," *IEEE Transaction on System, Man, and Cybernetics, Part C: Applications and Reviews*, vol. 39, no. 4, pp. 410–419, 2009.

[10] S. X. Yang and M. Meng, "Neural network approaches to dynamic collision-free trajectory generation," *IEEE Transaction on Systems, Man, and Cybernetics, Part B: Cybernetics*, vol. 31, no. 3, pp. 302–318, 2001.

[11] D. Wang and D. Terman, "Image segmentation based on oscillatory correlation," *Neural Computation*, vol. 9, no. 4, pp. 805–836, 1997.

[12] H. Qu, S. X. Yang, A. R. Willms, and Z. Yi, "Real-time robot path planning based on modified pulse-couple neural network model," *IEEE Trans. Neural Netw*, vol. 20, no. 11, pp. 1724–1739, 2009.

[13] W. Xueli, G. Yapei, and Z. Jianhua, "A novel algorithm for shortest path problem based on pulse couple neural network," in *Proceedings of the Chinese Control and Decision Conference*, pp. 2468–2473, 2015.

[14] G. T. Zoumponos and N. A. Aspragathos, "Fuzzy logic path planning for the robotic placement of fabrics on a work table," *Robotics and Computer-Integrated Manufacturing*, vol. 24, no. 2, pp. 174–186, 2008.

[15] C. Son, "Intelligent rule-based sequence planning algorithm with fuzzy optimization for robot manipulation tasks in partially dynamic environments," *Information Sciences*, vol. 342, pp. 209–221, 2016.

[16] P. Sun and Z. Yu, "Tracking control for a cushion robot based on fuzzy path planning with safe angular velocity," *Journal of Automatica Sinica*, vol. 4, no. 4, pp. 610–619, 2017.

[17] G. Zhou, N. Wang, X. Lu, and J. Ma, "Research on the fuzzy algorithm of path planning of mobile robot," in *Proceedings of the International Conference on Computer System, Electronics and Control*, China, 2017.

[18] H. Shao, H. Yamamoto, Y. Sakaida, T. Yamaguchi, Y. Yanagisawa, and A. Nozue, "Automatic excavation planning of hydraulic excavator," in *Proceedings of the International Conference on Intelligent Robotics and Applications*, 2008.

[19] A. Stentz, J. Bares, S. Singh, and P. Rowe, "A robotic excavator for autonomous truck loading," in *Proceedings of the IEEE/RJS Conf. on Intelligent Robots and Systems*, pp. 1885–1893, Victoria, Canada, 1998.

[20] Y. H. Zweiri, L. D. Seneviratne, and K. Althoefer, "Model-based automation for heavy duty mobile excavator," in *Proceedings of the IEEE/RSJ International Conference on Intelligent Robots and Systems*, vol. 3, pp. 2967–2972, October 2002.

[21] S. Lee, D. Hong, and H. Park, "Optimal path generation for excavator with neural networks based soil models," in *Proceedings of the IEEE International Conference on Multisensor Fusion and integration for intelligent Systems*, pp. 632–637, Seoul, Korea, 2008.

[22] F. Y. Wang and P. J. A. Lever, "On-Line trajectory planning for autonomous robotic excavation based on force/torque sensor measurements," in *Proceedings of the IEEE International Conference on Multisensor Fusion and Integration for Intelligent System*, pp. 371–378, Las Vegas, NV, USA, 1994.

[23] Z. Li, X. Li, S. Liu, and L. Jin, "A study on trajectory planning of hydraulic robotic excavator based on movement stability," in *Proceedings of the 13th International Conference on Ubiquitous Robots and Ambient Intelligence, URAI*, pp. 582–586, August 2016.

[24] Y. B. Kim, J. Ha, H. Kang, P. Y. Kim, J. Park, and F. C. Park, "Dynamically optimal trajectories for earthmoving excavators," *Automation in Construction*, vol. 35, pp. 568–578, 2013.

[25] A. Gasparetto and V. Zanotto, "A new method for smooth trajectory planning of robot manipulators," *Mechanism and Machine Theory*, vol. 42, no. 4, pp. 455–471, 2007.

[26] A. Gasparetto and V. Zanotto, "A technique for time-jerk optimal planning of robot trajectories," *Robotics and Computer-Integrated Manufacturing*, vol. 24, no. 3, pp. 415–426, 2008.

[27] J. S. R. Jang, "ANFIS: adaptive-network-based fuzzy inference system," *IEEE Transactions on Systems, Man, and Cybernetics*, vol. 23, no. 3, pp. 665–685, 1993.

[28] A. J. Koivo, M. Thoma, E. Kocaoglan, and J. Andrade-Cetto, "Modeling and control of excavator dynamics during digging operation," *Journal of Aerospace Engineering*, vol. 9, no. 1, pp. 10–18, 1996.

5

Service Arms with Unconventional Robotic Parameters for Intricate Workstations: Optimal Number and Dimensional Synthesis

Satwinder Singh and Ekta Singla

Mechanical Engineering Department, IIT Ropar, Rupnagar 140001, India

Correspondence should be addressed to Ekta Singla; ekta@iitrpr.ac.in

Academic Editor: Shahram Payandeh

A task-oriented design strategy is presented in this paper for service manipulators. The tasks are normally defined in the form of working locations where the end-effector can work while avoiding the obstacles. To acquire feasible solutions in cluttered environments, the robotic parameters (D-H parameters) are allowed to take unconventional values. This enhances the solution space and it is observed that, by inducing this flexibility, the required number of degrees of freedom for fulfilling a given task can be reduced. A bilevel optimization problem is formulated with the outer layer utilizing the binary search method for minimizing the number of degrees of freedom. To enlarge the applicability domain of the proposed strategy, the upper limit of the number of joints is kept more than six. These allowable redundant joints would help in providing solution for intricate workcells. For each iteration of the upper level, a constrained nonlinear problem is solved for dimensional synthesis of the manipulator. The methodology is demonstrated through a case study of a realistic environment of a cluttered server room. A 7-link service arm, synthesized using the proposed method, is able to fulfill two different tasks effectively.

1. Introduction

The tremendously increasing variety in the robotic services leads to *less repetitive tasks*. To provide solution to the resulting variations, adaptive design techniques and realization strategies have taken the attention of the researchers. Given a cluttered environment with fixed obstacles, as shown in Figure 1, tasks are generally defined as working locations for the robot end-effector. For such intricate workstations with narrow passages, manipulability needs to be induced in the required design to maneuver within the cluttered workcells. Focus of the work is to determine adaptive number of degrees of freedom (dof) for given work scenarios. There is no limitation of keeping degrees of freedom as six or less. To acquire even one connecting path in the given cluttered environments, a design with even larger dof may be used. Novelty of the work lies in enhancing the solution space through unconventional values of the robotic parameters.

In recent works, Yang and Chen [1] presented a study on optimizing the number of degrees of freedom for given tasks. The work is limited to specific conditions, without the description of any obstacles and/or workcells. In another work by Zhang and Wang [2], kinematic redundancy is utilized for avoidance of given obstacles. A fixed degrees-of-freedom manipulator is utilized in the work and no general strategy is provided for any given task and/or workcell. The techniques used to avoid collisions are the utilization of kinematic redundancy and/or having movable platform in most of the works. Modularity in robotic arms is worked upon by few researchers [3, 4] for providing a solution for changes in the environments. Normally, in the modular development strategies, a configuration of modular components is changed and then a systematic method is developed to compute its robotic parameters, thus formulating the kinematic equations. However, to design a service arm for *any given spatial workcell*, a general platform is required, with least input

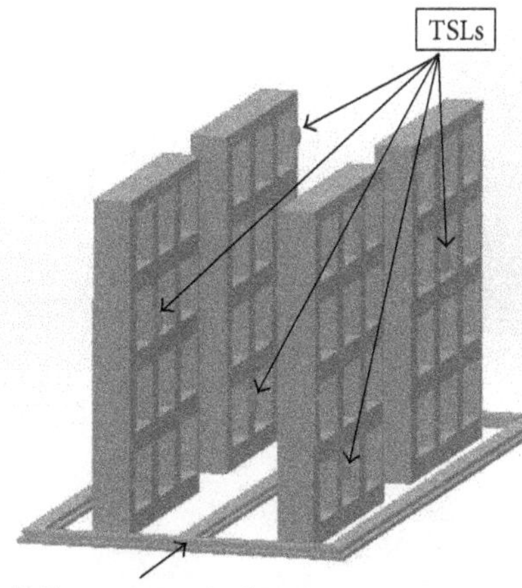

FIGURE 1: A server room environment with task space locations.

required from the user. Importance of the customized design of robotic arms has been discussed in recent works and the algorithms have been presented either with fixed degrees of freedom or with fixed or no environment [5, 6]. The field of task-based design needs to be further explored for algorithms adaptive to given workcells. This work provides one such solution to deal with given cluttered environment, through unconventional robotic parameters and by trading-off the requirement of number of degrees of freedom.

2. Problem Formulation

A nested optimization approach is proposed in this work with all the robotic parameters as design variables in the inner loop for dimensional synthesis and a unidirectional problem solving in number of dof in the outer loop. Important aspects related to the problem are the utilization of unconventional robotic parameters and varying number of degrees of freedom:

(1) Flexibility in the values of robotic parameters (D-H parameters in this work) leads to a larger search domain, which is required for a feasible solution in intricate workstations. For highly cluttered environments, this aspect is expected to play an important role. Related to this work, for a 6-link manipulator, Patel and Sobh [7] utilized a larger range of D-H parameters. The adaptable modules are fabricated to adapt the unconventional values of D-H parameters such as adaptable links [8] with adaptable connectors. The proposed methodology provides a general platform which can provide a robotic arm design corresponding to a given workstation, defined by a workspace model.

(2) The number of dof is not confined to any predefined value. Kinematic redundancy is kept acceptable in the varying number of dof. This is to gain the inherent advantages of a large number of dof, if required, according to the work scenario.

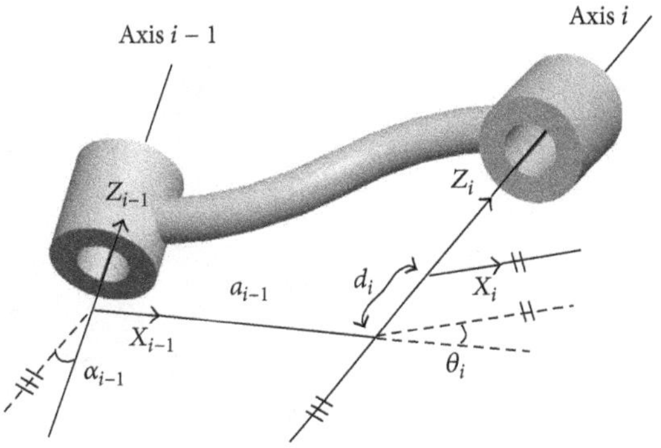

FIGURE 2: D-H representation.

The proposed strategy is illustrated through realistic workcells, with the changes in number of task space locations (TSLs).

2.1. Varying Number of Degrees of Freedom and Unconventional Robotic Parameters. For an n-linked serial manipulator, D-H convention is used to attach the reference frames to the serially joined links (refer to [9]). To define the relation between $(i-1)$th and ith links, $\forall 1 \le i \le n$, four parameters—twist angle (α_{i-1}), link length (a_{i-1}), joint offset (d_i), and joint angle θ_i—are associated.

As shown in Figure 2, the link parameters $((\alpha_{i-1}), (a_{i-1}))$ and joint offset (d_i) are fixed to provide a manipulator configuration. With n values of θ_i, this configuration will attain one particular posture. By varying the values of each (θ_i), the posture of this configuration can be changed; that is, position of the end-effector can be varied to reach the desired locations (TSLs)—say "N" in number.

A transformation matrix, using D-H parameters, defines transformation of frame i relative to frame $i-1$, represented as ${}^{i-1}_{i}T$, and is computed as

$$ {}^{i-1}_{i}T = \begin{bmatrix} C\theta_i & -S\theta_i & 0 & a_{i-1} \\ S\theta_i C\alpha_{i-1} & C\theta_i C\alpha_{i-1} & -S\alpha_{i-1} & -S\alpha_{i-1} d_i \\ S\theta_i S\alpha_{i-1} & C\theta_i S\alpha_{i-1} & -C\alpha_{i-1} & C\alpha_{i-1} d_i \\ 0 & 0 & 0 & 1 \end{bmatrix}. \quad (1) $$

For n dof, comprising $n+1$ frames, n number of transformation matrices are computed. The concatenated matrix provides the required transformation from frame-n corresponding to the end-effector to frame-0 attached to the base:

$$ {}^{0}_{n}T = {}^{0}_{1}T\,{}^{1}_{2}T\,{}^{2}_{2}T \cdots {}^{n-1}_{n}T. \quad (2) $$

The robotic parameters in manipulator kinematic equations are dependent on number of dof. For a single posture of an n-link manipulator, total $4n$ parameters are required to compute kinematic equations. Since the value of "n" is not defined *a priori*, the total number of design variables will vary with each change in dof n. This is handled through a nested optimization problem formulation.

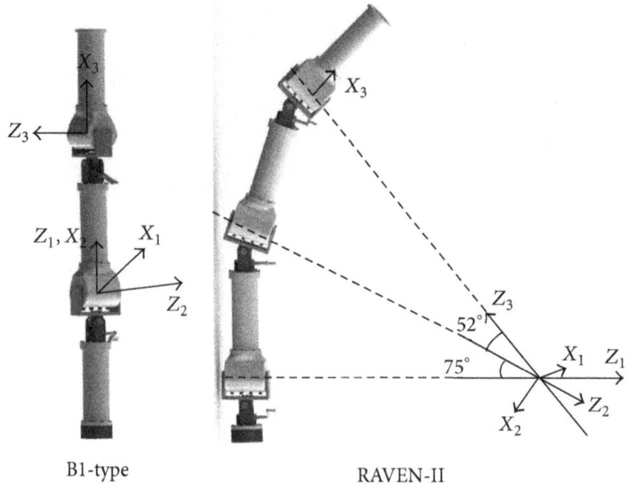

(a) Two links connected at conventional values of twist angle

(b) Illustration of unconventional twist angles, which provide flexibility to the design process

FIGURE 3: Unconventional robotic parameters.

Apart from the possibility of using redundant joints in the design, if and when required, the robotic parameters are also kept flexible in the proposed methodology. Figures 3(a) and 3(b) present the significance of conventional twist angles with normal values as 0 or $\pi/2$ and unconventional twist angles values. The latter provides a larger solution space and thus a possibility of getting solution even in highly constrained workcells.

2.2. Collision Avoidance. To work on a general problem with any given cluttered workspace, an obstacle avoidance strategy is required. It is important to check any collision among the robotic links and any of the environmental objects. In this work, emphasis is given on the thorough examination of any configuration in question; that is, the collision is not checked just for end-effector and/or for a few points on the robotic links. For this purpose, the solid model of the workspace is required in Stereolithographic (stl) format which is modelled in Solidworks Premium 2013 version. The stl file provides the connectivity information of modelled environment in triangulated mesh format. The forward kinematic procedure is used to model the robotic arm at every iteration. Each link is assumed to be a rectangular parallelepiped with square cross section of a prescribed width. For collision detection, the proximity query package is utilized. The obstacle avoidance approach computes the minimum distance between the two solid models, represented in their triangulated form. In case there is collision between the two objects, the package furnishes the data about the colliding pairs. The collided pair may belong to the robot links or robot obstacles. The strategy computes a positive minimum distance "D" between the models which is utilized in formation of corresponding constraints.

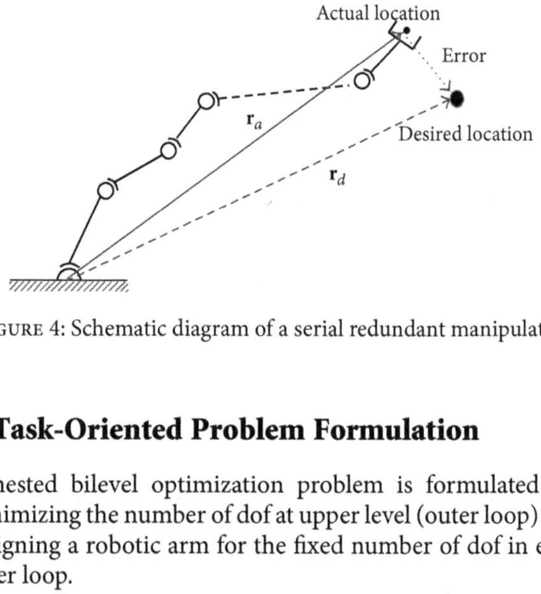

FIGURE 4: Schematic diagram of a serial redundant manipulator.

3. Task-Oriented Problem Formulation

A nested bilevel optimization problem is formulated for minimizing the number of dof at upper level (outer loop) and designing a robotic arm for the fixed number of dof in each inner loop.

3.1. Reachability at Working Locations: Objective Function. For each current value of "n" in the outer loop, robotic parameters are synthesized for *reachability* at the required TSLs. Reachability is measured as a squared error function of *Euclidean distances*, that is, distance between the end-effector position of the manipulator and the specified TSL as

$$P_{\text{error}} = \left\|\mathbf{r}_d - \mathbf{r}_a\right\|^2. \tag{3}$$

A schematic diagram of the end-effector position of a redundant manipulator while reaching at the desired task locations is shown in Figure 4. For an n-link manipulator, working for N TSLs—with the actual position of the end-effector corresponding to the jth TSL as vector $\mathbf{r}_{ja}$ and the desired TSL position as vector $\mathbf{r}_{jd}$—the error square sum for N number of TSLs is computed as

$$P_{j_{\text{error}}} = \sum_{j=1}^{N} \left[\left\| \mathbf{r}_{d_j} - \mathbf{r}_{a_j} \right\|^2 \right]. \tag{4}$$

In Cartesian coordinates, (3) can be expressed as

$$P_{\text{error}} = (x_d - x_a)^2 + (y_d - y_a)^2 + (z_d - z_a)^2. \tag{5}$$

Finally, the cumulative error for all the N TSLs can be written as

$$P_{\text{error}} = \sum_{j=1}^{N} \left[\left(x_{d_j} - x_{a_j}\right)^2 + \left(y_{d_j} - y_{a_j}\right)^2 + \left(z_{d_j} - z_{a_j}\right)^2 \right]. \tag{6}$$

To include the orientation of a frame attached to the end-effector of the manipulator relative to the base frame, Euler angle conventions have been used. With α_{td}, β_{td}, and γ_{td} representing the Euler angles corresponding to a particular configuration of the manipulator under consideration, the objective function can be revised as

$$\begin{aligned} P_{\text{error}} = \sum_{t=1}^{N} &\left[(x_{td} - x_{ta})^2 + (y_{td} - y_{ta})^2 + (z_{td} - z_{ta})^2 \right] \\ &+ (\alpha_{td} - \alpha_{ta})^2 + (\beta_{td} - \beta_{ta})^2 + (\gamma_{td} - \gamma_{ta})^2. \end{aligned} \tag{7}$$

Corresponding actual orientation parameters (α_{ta}, β_{ta}, and γ_{ta}) can be derived from the concatenative rotation matrix of the end-effector with respect to the base frame.

The actual coordinates of the manipulator end-effector are given by the forward kinematics using D-H parameters $(a_{i-1}, \alpha_{i-1}, d_i, \theta_i)$. The transformation matrices are multiplied together to form an arm matrix. The first three elements of the global transformation matrix give the current location of the end-effector in Cartesian form, that is, (x_{ia}, y_{ia}, z_{ia}). It is worth mentioning here that the base point coordinates can also be considered as design variables, in case the application allows a flexibility in the installation point of the robot:

$$T_{\text{tool}}^{\text{base}} = T_{\text{translation}} T_1^0 T_2^1 T_3^2 \cdots T_n^{n-1}, \tag{8}$$

where $T_{\text{translation}}$ is the coordinates of the base point.

3.2. Unconventional Robotic Parameters: Design Variables. The global transformation matrix is calculated to determine the actual position in Cartesian coordinates. The arm matrix is the function of the number of dof and D-H parameters and all are considered as design variables for the formulated problem. Out of all the D-H parameters, link parameters are fixed for each TSL, while the joint variables are varying to provide jth configuration. The change in joint variables is required to reach at different TSLs for each particular robotic posture. With the number of TSLs as "N", total "$N + n$" joint variables are required. The joint variables are expressed as

$$\begin{gathered} \theta_{11}, \theta_{12}, \ldots, \theta_{1n}; \\ \theta_{21}, \theta_{22}, \ldots, \theta_{2n}; \\ \vdots \\ \theta_{N1}, \theta_{N2}, \ldots, \theta_{Nn}. \end{gathered} \tag{9}$$

The total number of design variables for n-link manipulator reaching "N" TSLs is, therefore, $(3 + N) * n$. Since the number of dof "n" is also a design variable, it leads to an inherent challenge of handling varying number of design variables in each iteration. Therefore, a bilevel problem is formulated to fix the number of dof at the upper level. The dimensional synthesis variable vector is expressed as $\mathbf{x}$ and it is further a function of "n".

3.3. Constraints Handling. The minimization of the objective function is subject to the constraints due to the limits on the design variables and due to the environmental obstacles.

3.3.1. Limiting Values of D-H Parameters: Inequality Constraints . All the D-H parameters and degrees of freedom are design variables in the above formulated problem. The limiting bound/range on all the design variables are imposed as constraints for the optimization method. The limits on D-H parameters for the ith link can be written as

$$\begin{aligned} &a_{il} \le a_i \le a_{iu}, \\ &\alpha_{il} \le \alpha_i \le \alpha_{iu}, \\ &d_{il} \le d_i \le d_{iu} \quad \text{for all joints as revolute}, \\ &\theta_{ji_l} \le \theta_{ji} \le \theta_{ji_u} \quad \forall j, \end{aligned} \tag{10}$$

where $i = 1, 2, \ldots, n$, $j = 1, 2, \ldots, N$, and l in suffix represents the lower bound, while u represents the upper. Each inequality in (10) gives rise to a pair of constraints in the form

$$\begin{aligned} g_{i1}(\mathbf{x}) &= a_{il} - a_i \le 0, \\ g_{i2}(\mathbf{x}) &= a_i - a_{iu} \le 0. \end{aligned} \tag{11}$$

The bounds on the number of degrees of freedom are also discussed later.

3.3.2. Cluttered Workstations: Inequality Constraints. To compute the minimum distance between the robot and the workspace or among the robot links, the workspace and each link of the manipulator are presented in a triangulated mesh format, as discussed in Section 2.2. A function $D(\mathbf{x})$ is defined as the minimum distance between two nearest triangle pairs. The minimum distance of separation (positive in nature), reported by the package, is returned as D. The function D gets a negative value representing the *overall* intersection, when there is a collision between objects (robot-obstacle or link-link).

The inequality constraint including both cases is represented as

$$g_j(\mathbf{x}) = -D_j \leq 0, \quad j = 1, 2, \ldots, N, \tag{12}$$

where N is the number of TSLs. Thus, the jth constraint gets active only when there is an intersection of the manipulator with any obstacle and/or among the links, at the posture corresponding to the jth TSL.

All these constraints, along with the objective function, constitute the problem in ($\mathbf{x}$). As discussed earlier, this constrained optimization problem is solved by using augmented Lagrangian method.

4. Binary Search Method: Outer Loop

The binary search method is applied at the outer layer of the formulated bilevel optimization problem. The technique is used to locate the target value in a sorted array [10]. In this work, targeted value is the minimum number of dof at which the manipulator can be designed for reachability at desired TSLs. The range of dof is the array under consideration from which the middle element will be selected and every element of the array is an integer.

It is used to find out the optimal number of degrees of freedom n. Total number of design variables are dependent on dof as discussed in Section 3.2, that is, $[(3+N) * n]$, where N is the number of TSLs. "n" is an integer value and possesses a finite range of 3–12 in this work.

To initialize the method, the array is split in half and the middle element is selected as an input to the inner loop; that is,

$$n_{\text{mid}}^{t} = \frac{n_u - n_l}{2}, \tag{13}$$

where n_u and n_l are the upper and the lower limit of the number of dof and t represents the iteration number. In the inner loop, a nonlinear optimization problem has been formulated and augmented Lagrangian method is used to solve the highly constrained problem. n_{mid}^{t} is the value at which design vector $\mathbf{x}$ is to be determined in inner loop optimization.

In case the solution does not exist at n_{mid}^{t}, it becomes the new n_l as

$$n_{\text{mid}}^{t+1} = \frac{n_u - n_{\text{mid}}^{t}}{2}. \tag{14}$$

However, if there exists a design solution at n_{mid}^{t}, then it becomes new n_u and

$$n_{\text{mid}}^{t+1} = \frac{n_{\text{mid}}^{t} - n_l}{2}. \tag{15}$$

The iterative process terminates with $n_u - n_l = 1$.

5. Augmented Lagrangian Method: Inner Loop

This constrained optimization method uses the combination of duality and penalty aspects. A penalty function is induced in the objective function to check the constraint violation. The Hessian of Lagrangian can be ill-conditioned in some cases due to which it affects the rate of convergence. The dual method can be applied only on convex functions. So, in augmented Lagrangian method, a moderate penalty is applied to augment the objective function into its convex function [11].

Suppose $f(\mathbf{x})$ is the objective function to be minimized subject to the inequality constraints

$$g_i(\mathbf{x}) \geq 0 \quad i = 1, 2, \ldots, I, \tag{16}$$

and equality constraints

$$h_j(\mathbf{x}) = 0 \quad j = 1, 2, \ldots, J, \tag{17}$$

where I and J are the number of inequality and equality constraints, respectively. The augmented Lagrangian function is then expressed as

$$\begin{aligned}\mathcal{F}(\mathbf{x}) = f(\mathbf{x}) &+ \sum_{i=1}^{I} \left[\mu_i g_i(\mathbf{x})\right] \\ &+ \frac{1}{2} R \sum_{i=1}^{I} \left[\max\left(0, g_i(\mathbf{x})\right)\right]^2 + \sum_{j=1}^{J} \left[\lambda_j h_j(\mathbf{x})\right] \\ &+ \frac{1}{2} R \sum_{j=1}^{J} \left[\max\left(0, h_j(\mathbf{x})\right)\right]^2,\end{aligned} \tag{18}$$

where μ_i and λ_j are the Lagrange multipliers corresponding to ith inequality and jth equality constraints, respectively, whereas R is the penalty parameter.

6. Methodology

For a given workcell (in triangulated mesh format) with the tasks defined as the working locations (in Euclidean space), the problem formulated in the previous sections facilitates the solution of the problem with minimum number of dof. The complete methodology is summarized through the following steps.

Upper Level Start

(1) Define n_u and n_l for binary search algorithm.

(2) Compute the n_{mid} value and update it to lower level.

Lower Level Start

(1) Formulate the objective function $\mathcal{F}(\mathbf{x})$ referring to (18).

(2) Specify the constraints due to parameter bounds and obstacles as mentioned in Section 3.3.

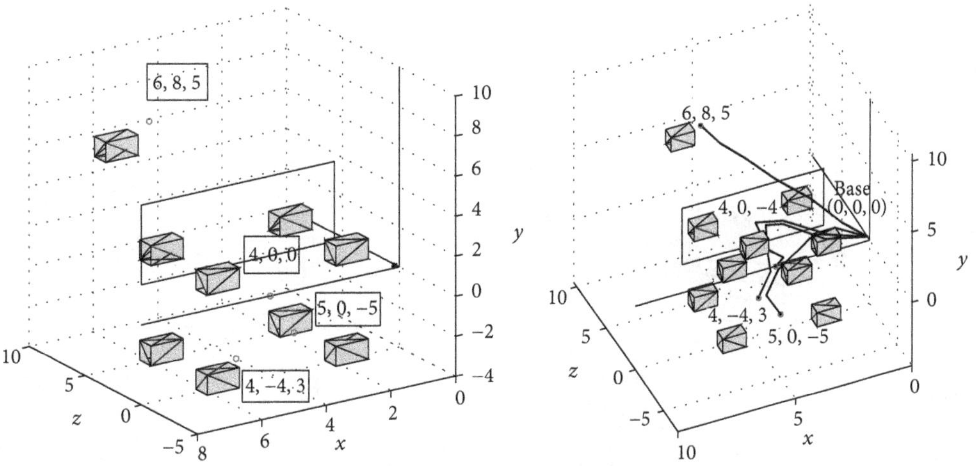

(a) A 10-obstacle environment with 4 TSLs

(b) A skeletal view of the 6-link manipulator in the environment.

FIGURE 5: An exemplary case study for strategy validation.

(3) Apply the augmented Lagrangian method to solve the formulated NLP problem. This involves updating the Lagrange multipliers as discussed in previous section.

(4) Update the solution status to upper level.

Lower Level Ends

(1) Based on the solution status at lower level, update n_{mid} according to (14) and (15).

(2) Check termination criteria. If the condition mentioned after (15) is fulfilled, the corresponding inner loop solution $\mathbf{x}$ is the required feasible solution with minimum number of dof.

Upper Level Ends. It is possible that, on termination, the objective function is not zero, which signifies that no solution is possible within the prescribed range of dof. The range of dof can also be expanded to increase the redundancy. The bounds on the link lengths can be relaxed, if allowed. The range of degrees of freedom is 3 to 12 in this work. Only revolute joint has been taken in the design process which means joint offset will be fixed parameter in all the cases.

7. Results and Discussion

The problem, discussed and formulated in the previous sections, is implemented in C++. For the execution of the code, the information of workspace environment in the triangulated mesh format, the task space locations in the environment, base position of the manipulator, and the limits on the D-H parameters (design variables) are required as the input data.

Using the proposed strategy, manipulator design for several environments has been synthesized and some of the results are presented in this section.

TABLE 1: Six-link manipulator design.

S. number	α_i	a_i	d_i	θ_{i1}	θ_{i2}	θ_{i3}	θ_{i4}
1	0	0	0.4	0.74	0.32	0.18	0.55
2	0.97	2.20	−0.4	0.43	0.48	0.69	−0.87
3	0.64	1.99	−0.37	0.03	0.9	0.94	−0.38
4	0.55	1.99	−0.19	−0.14	1.44	0.77	0.54
5	0.77	1.49	−0.24	−0.14	1.26	0.5	0.92
6	0.99	1.49	−0.41	0.07	0.6	0.38	0.70

7.1. Example 1: Strategy Validation

Case 1. An exemplary workspace is taken into consideration for the validation of the strategy. It has 10 blocks scattered in the space, as shown in Figure 5(a). The required manipulator needs to work at 4 TSLs $[(6,8,6), (4,0,0), (4,-4,3)$, and $(5,0,-5)]$, while avoiding all the blocks. The base of the manipulator is fixed at (0,0,0). The design of an 8-link manipulator has been reported in the work of Singla et al. [5] for the same environment and TSLs. The number of dof was reported fixed in the work.

By using the proposed strategy given in this chapter, the 6-link manipulator can reach all the given TSLs. In the first iteration, the degrees of freedom have been taken as 8 at the outer loop and initialize the augmented Lagrangian method in the inner loop for reachability. The 8-link design outcome is expected because, with same input, it has been reported earlier.

The n_{mid}-update is done according to the algorithm discussed in Section 4. The number of dof is updated to 6 at the outer loop and this updated value is checked for the success in the inner loop. Table 1 presents the robotic parameters and Figure 5(b) shows the corresponding skeletal view of the manipulator in the workspace.

Binary search will update the new value of dof that is 5 in this case. At number of dof 5, the method fails to converge.

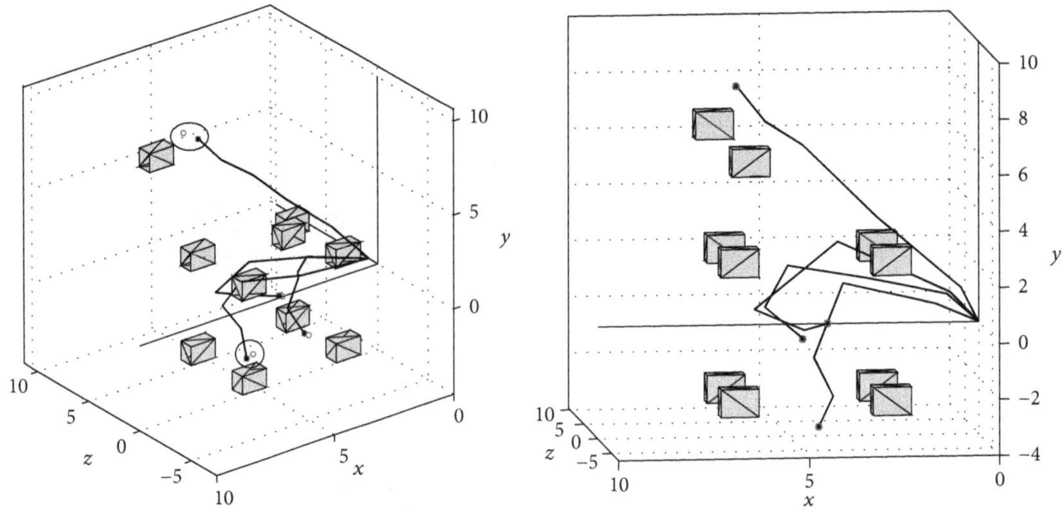

(a) A 5-link design failed to reach

(b) A 5-link manipulator design

FIGURE 6: A 5-link design by varying the limits of the design variables.

A positive error value that is 1.0139 is left, which signifies that no solution exists at dof 5. The method will terminate here with a 6-link design as an output. The case showcases the importance of varying number of links.

Case 2. It is quite possible to achieve the required set of design variables with 5 dof, if variable bounds are relaxed. For 5-dof design, the link length limits in the failed case are $[3, 3, 2, 1.5, 1.5]$ which are changed to $[4, 4, 4, 2.5, 2.5]$. With the changed limits, the reachability is achieved as shown in Figure 6(b). With the initial limits, robot arm was nearly outstretched to its full length and yet not able to reach one of the TSLs, as shown in Figure 6(a). Therefore, the variations in the link lengths are tried and successfully delivered the result.

7.2. Example 2: Comparison with Conventional Values of D-H Parameters. This example is included to compare the design results having conventional values of twist angles with the proposed flexibility in twist angles. The environment and the working locations are kept the same as used in the previous case study. In this example, the outer loop is not active and the number of dof is fixed manually at different levels. The problem starts with a 6-link configuration in which the link lengths and the joint offsets can vary and all the twist angles are taken as either perpendicular or parallel. The values of all the twist angles are taken the same as for PUMA configuration. This example is an attempt to compare whether a manipulator with conventional twist angles that is 0 or $\pm\pi/2$ achieves the objective with same number of dof. If not, then how many dof are required to reach? Hence, the problem started with minimum number of dof which has been calculated in the previous example. It is observed that, with least variation in twist angles, no solution is obtained with number of dof 6, 7, and 9. Twist angle can vary along with other variables from dof 6 onwards because there is no set conventional value of twist angle in redundant manipulators. In this example, with first 6-dof puma configuration, a 9-link manipulator is the design outcome. Corresponding D-H parameters are shown in Table 2 and the configurations are presented in Figure 7. There is an important aspect that with increase in solution space degrees of freedom can be decreased for reaching desired locations.

TABLE 2: Nine-link manipulator design with conventional twist angles.

S. number	α_i	a_i	d_i	θ_{i1}	θ_{i2}	θ_{i3}	θ_{i4}
1	0	0	−0.51	0.41	1.39	3.00	−2.32
2	−1.48	3.06	−0.75	0.77	−2.89	2.01	−1.01
3	−0.01	0.61	−0.23	0.69	2.80	1.74	0.00
4	−1.49	1.99	0.99	0.39	1.34	−0.61	−0.50
5	−1.51	1.79	0.86	0.34	−2.17	−2.66	2.85
6	1.55	1.95	−0.51	0.39	−1.15	−2.99	−1.10
7	−0.01	1.91	−0.28	0.44	1.15	−2.50	0.91
8	0.09	1.81	0.32	0.49	−0.76	−2.98	−0.72
9	0	1.89	0.27	0.50	0.011	−1.13	−0.01

7.3. Example 3: Realistic Environment of a Server Room. In this case study, a server room environment has been taken into consideration, as shown in Figure 8(a). The manipulator has to reach 3 TSLs, $(185, 15, 80)$, $(36, 25, 80)$, and $(180, 30, 43)$, as shown in Figure 8(b) and the base point of the manipulator is taken at $(110, 8, 30)$. Minimum 6-link manipulator is required to reach all these TSLs. The corresponding design is presented in Table 3 and Figure 8(c) presents the visualization for desired working postures.

In the first case study, it has been shown that, with less number of dof and varying the limits of variable, it is possible to compute the solution set of parameters. Now, in the present scenario, number of TSLs increased from 3 to 5 and placed in such a way that 6-link design failed to reach which means strategy will automatically select the higher number

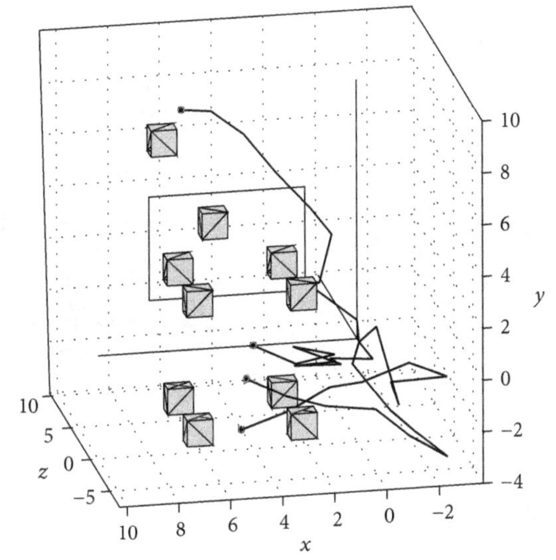

FIGURE 7: A 9-link manipulator design with the first 6 links in PUMA configuration.

(a) Server room model

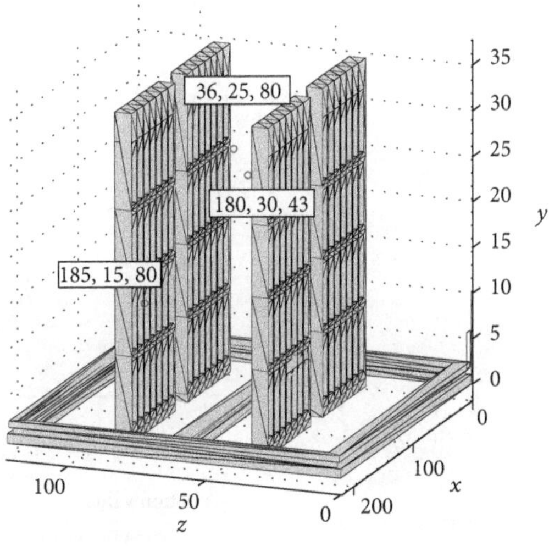

(b) Work environment and the TSLs

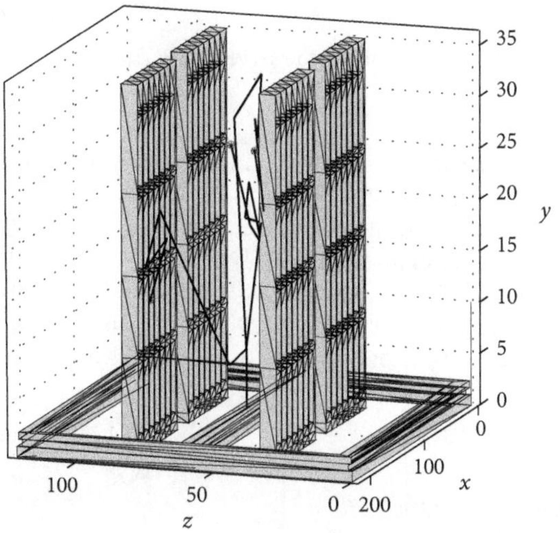

(c) Skeletal view of the 6-link manipulator in the server room

FIGURE 8: A server room case study.

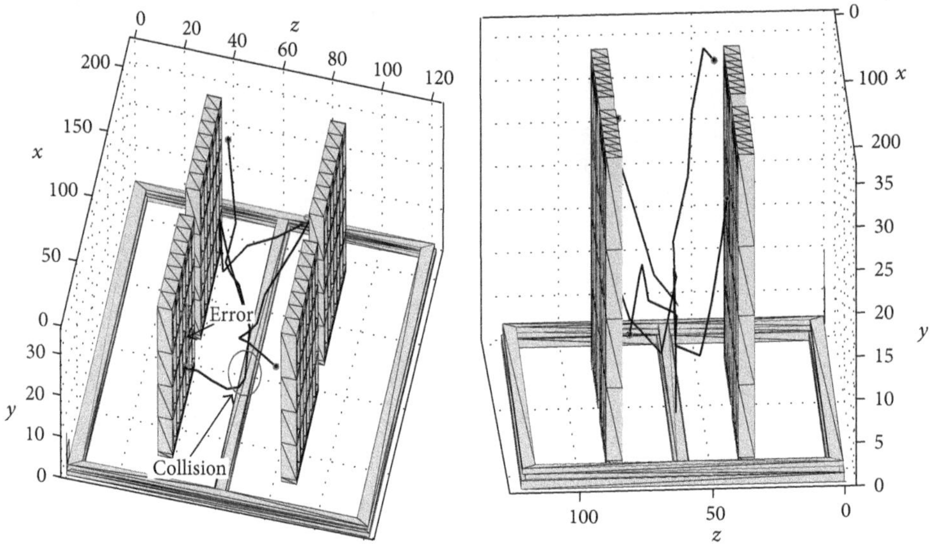

(a) Seven-link manipulator in the server room, intermediate iteration number 5

(b) Seven-link design in the server room: final solution

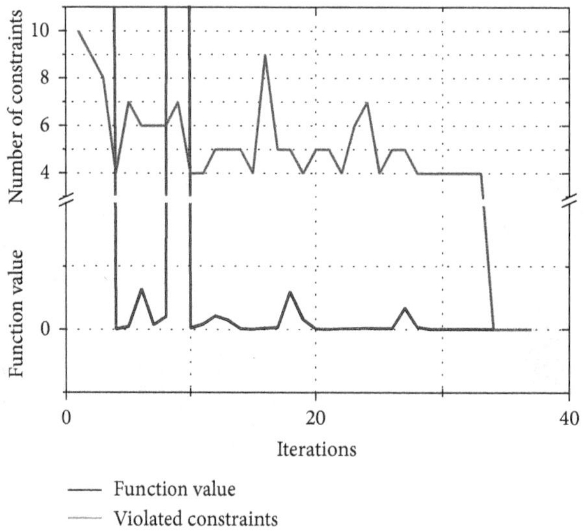

(c) Number of violated constraints and the objective function evaluation

FIGURE 9: Analysis: the server room case study.

of dof. A 7-link manipulator design is selected to reach 5 TSLs [(185, 15, 80), (36, 25, 80), (180, 30, 43), (36, 35, 43), and (120, 35, 82)] with the same base point as shown in Figure 9(a) showing the violation of constraints in an intermediate iteration. The method will give the solution when there is no collision and all the parameters are converged according to the termination criteria. Figure 9(b) shows the model which is an outcome of a fully converged problem. The corresponding D-H parameters are shown in Table 4.

TABLE 3: Six-link manipulator design for server room environment.

S. number	α_i	a_i	d_i	θ_{i1}	θ_{i2}	θ_{i3}
1	0	0	0.09	0.13	1.9	−2.8
2	2.5	20.03	0.09	2.8	1.25	−2.09
3	0.72	20.3	0.08	2.7	0.47	−2.7
4	2.05	20.2	0.06	−0.44	−0.71	0.38
5	2.06	20.3	0.06	−0.75	−0.38	0.032
6	1.75	20.3	0.071	−0.30	0.58	0.08

7.4. Example 4: Synthesis for Both Position and Orientation. This example presents a manipulator synthesis problem for both position and orientation of the end-effector, that is, for the entire arm-wrist combination. The workspace consists of an enclosed environment including a table and a big box, representing a cupboard, as shown in Figure 10(a). The figure includes two TSLs along with the desired approaching directions of the end-effector. The prescribed positions for these locations are (−30, 50, 60) and (60, 40, 70) with $(\pi/2, 0, -\pi/2)$

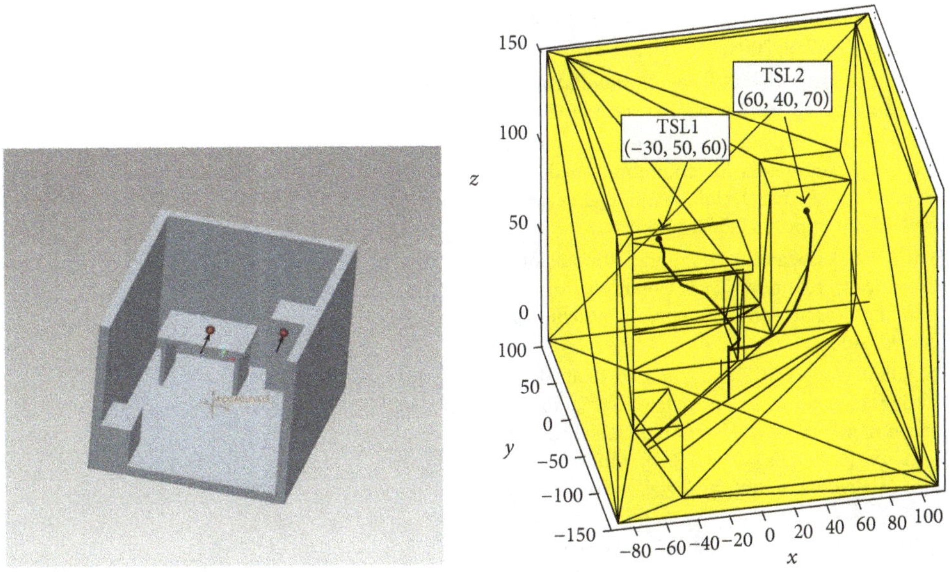

(a) The workspace (b) The resulting configurations

FIGURE 10: Seven-link manipulator in a room environment with both position and orientation prescribed.

TABLE 4: Seven-link manipulator design for server room environment.

S. number	α_i	a_i	θ_{i1}	θ_{i2}	θ_{i3}	θ_{i4}	θ_{i5}
1	0	0	−2.72	0.81	−0.26	−0.35	1.89
2	2.49	15.99	0.00	−0.58	0.08	−2.9	−0.89
3	1.05	14.99	0.58	0.50	−0.36	0.53	0.56
4	−0.28	13.99	0.19	−0.07	−0.42	−0.06	2.80
5	1.19	12.97	0.26	−0.82	0.09	0.01	1.79
6	1.33	11.97	−0.04	0.45	0.04	−0.01	−0.43
7	0.73	10.98	0.11	0.06	−0.66	−0.00	−0.25

TABLE 5: Synthesis result: tasks included both specified position and orientation.

S. number	a_i	α_i	d_i	θ_{1i}	θ_{2i}
1	17.4	0.70	−2.1	0.17	0.93
2	17.2	0.11	−1.4	0.31	1.41
3	15.8	1.16	−0.08	0.50	0.53
4	15.8	1.40	1.8	0.10	−0.55
5	16.1	1.01	1.07	−0.43	0.36
6	16.6	0.95	2.6	−0.25	0.90
7	18.7	1.00	3.1	0.76	−0.08

as the three orientation angles, the same for both locations. A 7-link manipulator is synthesized for the situation, with fixed base point locations as $(0, -30, 30)$. Table 5 contains the synthesis solution and pictorial view of the corresponding configurations is presented in Figure 10(b).

8. Conclusion

In this work, the *variation* in the working environments for a service robotic arm is dealt with through a general methodology for a constrained workcell. For this purpose, least constraints are applied at the number of dof and over the values of the robotic parameters. A bilevel optimization problem is formulated, involving the minimization of the number of degrees of freedom required to work in the given environment and for the given tasks (positions and orientations). Binary search algorithm is implemented for the outer layer of the optimization problem, which solves the unidirection problem in number of dof. In each of the outer iterations, a nonlinear optimization problem is solved for dimensional synthesis. The proposed methodology is validated through the reduction of the number of dof required for an environment, reported in earlier works. The importance of *flexibility* in robotic parameters is illustrated through two different cases in a realistic workcell of a computer server room where a robotic assistance is required for maintenance.

Competing Interests

The authors declare that they have no competing interests.

Acknowledgments

Authors gratefully acknowledge the financial support of Department of Science and Technology (DST), Government of India, for this work.

References

[1] G. Yang and I.-M. Chen, "Task-based optimization of modular robot configurations: minimized degree-of-freedom approach," *Mechanism and Machine Theory*, vol. 35, no. 4, pp. 517–540, 2000.

[2] Y. Zhang and J. Wang, "Obstacle avoidance for kinematically redundant manipulators using a dual neural network," *IEEE*

Transactions on Systems, Man, and Cybernetics, Part B: Cybernetics, vol. 34, no. 1, pp. 752–759, 2004.

[3] I.-M. Chen, "Modular robots," in *Handbook of Manufacturing Engineering and Technology*, pp. 2129–2168, Springer, 2015.

[4] Z. M. Bi, W. J. Zhang, I.-M. Chen, and S. Y. T. Lang, "Automated geneartion of the D–H parameters for configuration design of modular manipulators," *Robotics and Computer-Integrated Manufacturing*, vol. 23, no. 5, pp. 553–562, 2007.

[5] E. Singla, S. Tripathi, V. Rakesh, and B. Dasgupta, "Dimensional synthesis of kinematically redundant serial manipulators for cluttered environments," *Robotics and Autonomous Systems*, vol. 58, no. 5, pp. 585–595, 2010.

[6] S. Rubrecht, E. Singla, V. Padois, P. Bidaud, and M. de Broissia, "Evolutionary design of a robotic manipulator for a highly constrained environment," in *New Horizons in Evolutionary Robotics*, S. Doncieux, N. Bredèche, and J.-B. Mouret, Eds., vol. 341 of *Studies in Computational Intelligence*, pp. 109–121, Springer, 2011.

[7] S. Patel and T. Sobh, "Task based synthesis of serial manipulators," *Journal of Advanced Research*, vol. 6, no. 3, pp. 479–492, 2015.

[8] M. Brandstötter, A. Angerer, and M. Hofbaur, "The curved manipulator (cuma-type arm): realization of a serial manipulator with general structure in modular design," in *Proceedings of the 14th IFToMM World Congress*, pp. 403–409, October 2015.

[9] J. Craig, *Introduction to Robotics: Mechanics and Control*, Addison-Wesley Longman Publishing, Boston, Mass, USA, 1989.

[10] D. E. Knuth, "Optimum binary search trees," *Acta Informatica*, vol. 1, no. 1, pp. 14–25, 1971.

[11] K. Deb, *Optimization for Engineering Design: Algorithms and Examples*, PHI Learning, 2012.

6

Method for Effectively Utilizing Node Energy of WSN for Coal Mine Robot

Xiliang Ma[1,2] and Ruiqing Mao[1]

[1]*Xuzhou Institute of Technology, Xuzhou, Jiangsu 221111, China*
[2]*School of Mechanical and Electrical Engineering, China University of Mining and Technology, Xuzhou, Jiangsu 221116, China*

Correspondence should be addressed to Xiliang Ma; xlma818@sina.com

Academic Editor: L. Fortuna

As a special application scenario, the data collected by wireless sensor networks of coal mine robot is from vital and dangerous environment. Therefore, the nodes need to work as long as possible. In order to efficiently utilize the node energy of wireless sensor network, this paper proposes a self-organizing routing method for wireless sensor networks based on Q-learning. The method takes many factors into account, such as the hop number, distance, residual energy, and node communication loss and energy. Each node of the wireless sensor networks is mapped into an Agent. Periodic training is carried out to optimize the route choice. Each Agent chooses the optimal path for data transmission according to the calculated Q evaluation value. Simulation results show that the self-organizing sensor networks using Q-learning can balance the energy consumption of the nodes and prolong the lifetime of the networks.

1. Introduction

Coal mine environment is often threatened by toxic gases and high temperature, so coal mine robots [1] often replace human to enter the pit to carry on the detection or the rescue tasks. Coal mine robots need to sense the states of themselves, as well as the environment of the underground. So a variety of sensors need to be deployed outside coal mine robots, such as the sensors to detect temperature, humidity, wind speed, wind direction, wind pressure, light intensity, coal dust and toxic gases, distance, the speed of the robot's left and right tracks, and pitch angle. There are several disadvantages with wired sensors. First, cables for power and signal connections are required, and wiring is a piece of tedious work. Second, temporary data collection cannot be performed and the system expansion is inconvenient., Third, the robot shells need to provide threading holes and location holes, which will reduce the stiffness and strength of the robot shells and reduce the explosion-proof performance of the coal mine robots. Vibration or collision during robots' motion affects the connection and joint quality of the cable. This will affect the reliability of the data detection and even cause coal mine robots not to work properly in the underground. Wireless sensor network technologies can make up for these deficiencies.

Different from the general wireless sensor networks (WSNs) [2], a wireless sensor network deployed on a coal mine robot consists of a number of small sensor nodes. The distance between the nodes is relatively small. As coordinators and data collectors, these nodes transmit the data to the sink nodes and monitor the environment and the state parameters of the robot in real time. Each node has different data types and must meet the requirements of explosion-proof or intrinsic safety. So the nodes are usually powered by batteries. The data is collected by the nodes if from vital and dangerous environments. Therefore, the nodes need to work as long as possible. As a special application scenario, how to efficiently utilize the node energy of coal mine robot WSN is a difficult problem. Node energy consumption is mostly generated by sensor communications. In order to prolong the lifetime of the network, this paper designs a routing mechanism based on reinforcement learning and applies it to wireless body area networks. This paper also studies the energy efficiency, which further improves the energy efficiency and prolongs the network life cycle.

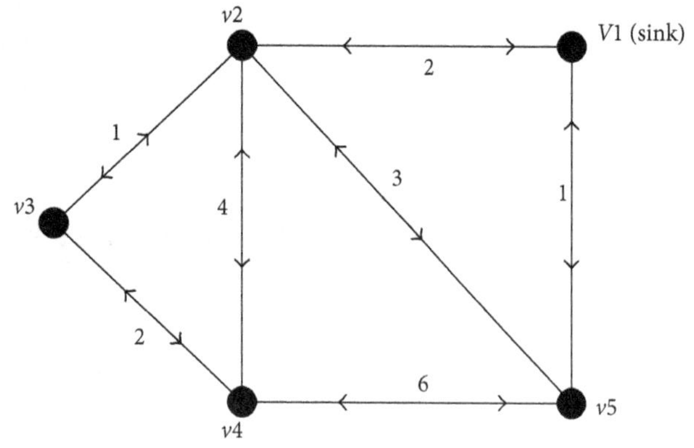

FIGURE 1: Model of wireless sensor network for coal mine robots.

2. Wireless Sensor Networks for Coal Mine Robot

2.1. A Wireless Sensor Network Model for Coal Mine Robot. The coal mine robot wireless sensor network has a single layer structure, which can be represented by a directed graph $G(V, A)$. The graph is shown in Figure 1, where V represents the sensor node set, V_1 is a sink node, and the wireless sensor network is wired to the coal mine robot control system by the sink node. Each node is fixed relative to the coal mine robot. The nodes transmit messages to the sink node in a direct or multihop manner. A represents a set of directed arcs, each of which represents the connection of two nodes in V that can communicate directly. The label beside an edge in the graph represents the corresponding path weight w_{ij}. The path with a smaller weight means that the channel has a better quality. The nodes on this path have more residual energy and the hop consumes less energy.

2.2. Node Energy Model. The energy consumption of wireless sensor network nodes is divided into 3 parts: perception, communication, and data processing. In this paper, we mainly consider the energy consumption of data communication, that is, the energy consumed by sending and receiving data. This paper refers to the energy model in LEACH [3], as shown in Figure 2. The energy consumption of the sending node includes two parts: the processing circuit and the power amplifier. The energy consumption of the receiving node is only used to process the circuit [4]. Suppose the distance between the receiving point and the sending point is d and the threshold of distance is d_0. When $d < d_0$, the free-space path loss model is adopted. If $d \geq d_0$, multipath fading model [5] is adopted.

In Figure 2, it can be seen that the energy consumption of nodes comes mainly from the process of transmitting and receiving data, which is closely related to the transmission distance. In Figure 2, the meaning of each symbol is as follows: E_{elec} is node energy consumption for sending and receiving 1-bit data. ε_{amp} is the power coefficient of the received data; n is the path decay index, whose typical value is 2 or 4. d is communication distance between two nodes; k is the number of bytes sent; E_{PA} is the amount of energy consumed by processing received data.

In short distance free-space communication, the transmission power is directly proportional to the transmission distance. And the transmission power can be simply expressed as $E = \lambda d$, in which λ is a fixed parameter. The energy consumption will be four times as much as the original energy when the distance is doubled. Therefore, when the transmission distance is large, the energy consumption will be greatly reduced by adding forwarding nodes. As shown in Figure 2, a power amplifier involves the path decay index n. When the communication distance between nodes $d < d_0$, the path attenuation index n is set to 2 and the energy consumed between two nodes is proportional to the square of their distance. When $d \geq d_0$, n is set to 4. The communication distance between nodes is large. The energy consumed by sending k-bit data to a node at intervals of N is shown in

$$E_{\text{Tx}}(d,k) = \begin{cases} kE_{\text{elec}} + k\varepsilon_{\text{amp}}d^2, & d < d_0 \\ kE_{\text{elec}} + k\varepsilon_{\text{amp}}d^4, & d \geq d_0. \end{cases} \tag{1}$$

The energy consumed by receiving k-bit data is shown in

$$E_{\text{Rx}}(k) = kE_{\text{elec}} + kE_{\text{PA}}. \tag{2}$$

In WSN, the energy consumption of nodes is very sensitive to the number of hops [6, 7]. Therefore, we define the energy function of the number of hops as follows: $E_h = e^{\text{hop}(i,s)/H}$, where $\text{hop}(i, s)$ represents the hop count of the next hop from Agent j to sink. The hop count from Agent i to sink is H. Assuming that Agent i selects Agent j as the next hop, the energy consumption for transmitting k-bit packets between Agent i and Agent j is

$$E_{i,j} = \left(2kE_{\text{elec}} + k\varepsilon_{\text{amp}}d^{\theta}\right)e^{\text{hop}(i,s)/H} \quad \theta \in [2,4]. \tag{3}$$

3. Energy Saving Methods

In order to prolong the lifetime of wireless sensor networks for coal mine robots, it is necessary to balance the energy consumption of all nodes. At the same time, the nodes with less remaining energy should be used as few as possible. In this paper, the Q-learning algorithm of reinforcement learning is applied to the wireless sensor networks of coal mine robots. This algorithm takes into account the distance between nodes, hops, communication energy consumption, and the residual energy consumption of nodes.

3.1. Agent Reinforcement Learning. Broadly speaking, an intelligent Agent senses the environment and performs an action, reinforcement learning, which is derived from animal learning and stochastic approximation [8]. It is an unsupervised machine learning technique. It can use the uncertain environment reward to find optimal behavior sequence and realize on-line learning in dynamic environment. Agents are independent, cooperative, and self-learning. They do not require human control. Agent nodes perceive the change of the environment and get the maximum reward value and get

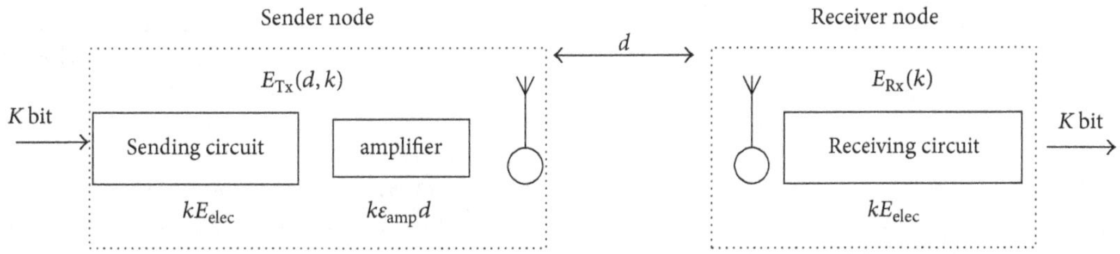

FIGURE 2: Energy consumption model of wireless sensor nodes in coal mine robot.

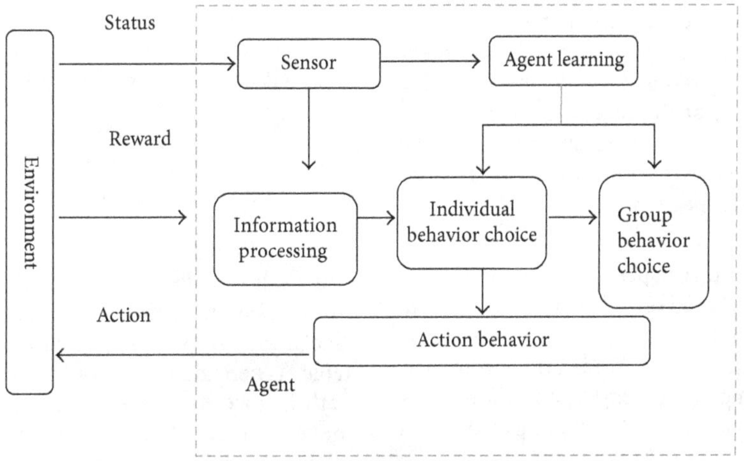

FIGURE 3: Basic principles of Agent's reinforcement learning.

an optimal strategy for corresponding decision actions. The principle of Agent's enhanced learning is shown in Figure 3. As can be seen from the diagram, Agent needs to interact with the environment. First, Agent senses information from complex environments. Then, Agent processes information, improves the performance, chooses a behavior, and makes group behavior choices. According to its individual and group behavior choices, Agent makes a decision, selects an action, and influences the environment.

3.2. Q-Learning Algorithm. Q-learning [9] is the most widely used algorithm in reinforcement learning algorithms. It is an unsupervised learning method whose input is a feedback from constantly changing and complex environment. Q-learning can be modeled by Markov decision process (MDP) [10]. MDP can be defined as a four-tuple (S, A, P, R), where S represents the set of states, A represents a set of actions, P represents the state transition function of the environment, and R is the reward function of the environment.

In MDP, the state transition function P and the reward function R are only related to the current state and actions and not related to the previous states and actions. The purpose of Agent's reinforcement learning is to learn a strategy [11] $\pi : S \rightarrow A$. Agent takes an action a_t according to the current status S_t; that is, $\pi(S_t) = a_t$. By following an arbitrary policy, the cumulative value $V^{\pi}(S_t)$ obtained from any initial state S_t is

$$V^{\pi}(S_t) = r_t + \gamma r_{t+1} + \gamma^2 r_{t+2} + \cdots = r_t + \gamma V^{\pi}(S_{t+1}) = \sum_{i=0}^{\infty} \gamma^i r_{t+i}. \quad (4)$$

In formula (4), r represents the return value, and γ represents the discount factor which reflects the relative proportion of the delay return to the immediate return. The goal of Agent is to learn a strategy to make $V^{\pi}(s_t)$ maximum; this strategy is called an optimal policy and is represented by π^*:

$$\pi^* = \arg\max_{\pi} V^{\pi}(S_t), \quad \forall s. \quad (5)$$

Q-learning is a model independent reinforcement learning algorithm. It learns the optimal strategy directly. Q-learning neither needs a prior knowledge model about the state transfer function P and the return function R nor needs to learn these models in the course of learning. According to the characteristics of the coal mine robots' wireless sensor networks, a model-free Q-learning algorithm is proposed in this paper. The algorithm is simple, fast, and easy to use.

The single step Q value updating formula of the Q-learning algorithm is

$$\begin{aligned} Q_{t+1}(s_t,\alpha_t) &= (1-\alpha)\,Q_t(s_t,\alpha_t) \\ &\quad + \alpha\left[r_{t+1} + \gamma\max Q_t\left(s_{t+1},\alpha'\right) - Q_t(s,\alpha_t)\right]. \end{aligned} \tag{6}$$

The contents of the Q table are constantly updated through formula (6) in the learning process [12]. In the formula, $Q_t(s_t,\alpha_t)$ is the function of status-action pair (s_t,α_t) at moment t; α and γ are learning rates and discount factors, respectively. s_t is the state of the environment at moment t. r_{t+1} is the return value of the state given by the environment at moment $t+1$, $\max Q_t(s_{t+1},\alpha')$ is the maximum Q value of the environment state s_{t+1} at the moment $t+1$, and α' is any action taken by Agent in the environment state. The Q-learning algorithm converges to an optimal solution by iteratively searching the state space.

3.3. QLSORP Algorithm. In this paper, a routing algorithm for wireless sensor networks based on Q-learning is designed. The algorithm is applied to the routing protocol of wireless sensor networks for coal mine robots. The routing protocol is called QLSORP (Q-Learning-based Self-Organization Routing Protocol). Q-learning is applied to the routing algorithms in wireless sensor networks.

The process of finding the optimal path in wireless sensor networks is equivalent to a Markov process. In each step, an instantaneous reward value is generated. The prerequisite for implementing the routing algorithm is that the return value R, the status s, and the action a must be determined. The return value is the ratio of the residual energy of Agent nodes to the energy consumed by communication. With R_j representing the current remaining energy of the node, according to formula (3), the return value of the Agent node j can be calculated as follows:

$$V^{\pi}(j) = \frac{R^j}{E_{i,j}} = \frac{R^j}{\left(2kE_{\text{elec}} + k\varepsilon_{\text{amp}}d^{\theta}\right)e^{\text{hop}(i,s)/H}} \tag{7}$$

$$\theta \in [2,4].$$

In the formula, the return value takes into account the remaining energy in the network and the energy consumption required for node communication. Then, a key point is found to balance the energy loss of nodes. State changes with time. Action refers to the process in which Agent nodes choose a path as the optimal one. Considering the communication energy consumption and the remaining energy, the Agent nodes on the routing path continuously send learning packets to the neighbor nodes to obtain the desired return value. The nodes will select the path with the highest Q value as the optimal one. Thus, the whole energy of the network is balanced and the network life cycle is prolonged under the premise of guaranteeing the efficiency of data transmission. The algorithm is described as follows.

Step 1. The sink node sends a learning evaluation message to its neighboring nodes in the same cycle. The wireless sensor network is initialized, and all sensor nodes are started. Each sensor node records the number of its neighboring nodes and energy consumption to its neighboring nodes. At the same time, the energy consumption threshold of each sensor node is set and the return value of the node is set to 0.

Step 2. Define a set D to store the information of the Agent nodes that have been learned during this cycle to prevent infinite loops in the path creation process. The source node periodically sends learning information to its neighboring nodes and determines whether the neighboring nodes exist or not in the set D; at the same time, the return value to each neighboring node is calculated. The neighboring node with a high return value is selected as the next hop routing node.

Step 3. Repeat Step 2. The neighboring node of the selected node is calculated and then probes the route of the next hop. The sink nodes are found successively.

Step 4. When the return value of other nodes is received, the sink node updates the Q value table by formula (6) according to the Q-learning mechanism in the reinforcement learning algorithm. The path is selected according to the Q value in the table. When the energy of the sensor nodes in the selected path is lower than the set threshold or the sensor node no longer has any effect, a message is sent to the source node along the opposite direction of the selected path. Then the source sensor node gives up choosing this path. Instead, the path with the second largest Q value will be chosen to transmit the information.

Step 5. The source node selects the path with large Q value to transmit the information stably to the sink node. The residual energy information of each sensor node is updated simultaneously.

Step 6. The sink node periodically sends learning message. According to the message, the source node probes the path, selects the path, and sends the message to the sink node. The value of Q varies with the return value of the node. The Q table is updated and stored in the sink node.

Figure 4 is the flow chart of the QLSORP algorithm.

4. Simulation Experiment Analysis

4.1. Experiment General Situation. Network simulation is a basic method for wireless sensor network research. The wireless sensor network of coal mine robot mainly involves temperature, humidity, wind speed, wind direction, light intensity, coal dust, oxygen, gas, distance sensor, track speed, and pitching angle of robot. Communication between wireless sensor nodes and between nodes and sink nodes constitutes a wireless sensor network. The topology is shown in Figure 5. Simulation analysis is carried out to verify the efficiency of routing algorithm based on Q-learning for wireless sensor networks. It is assumed that the transmission

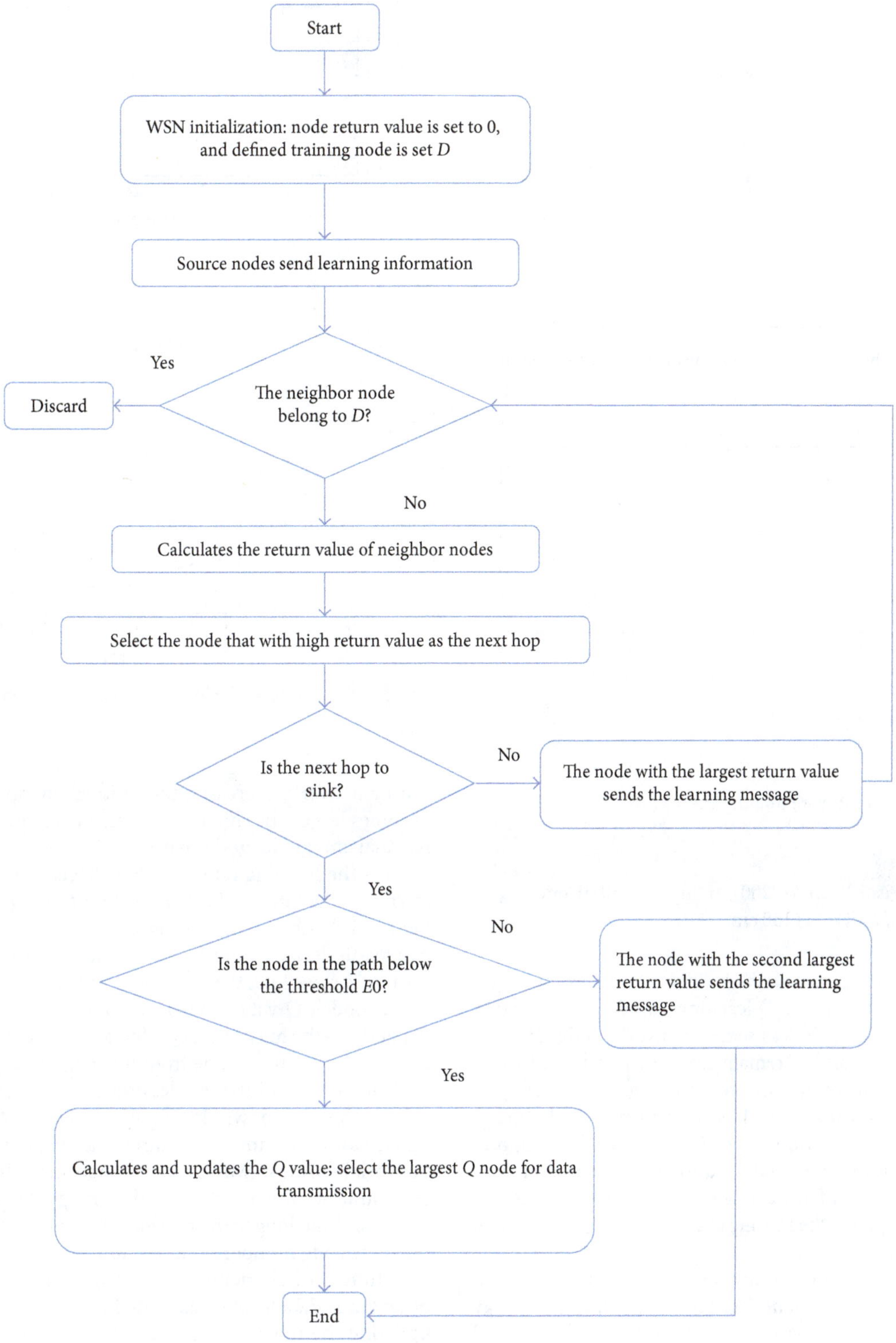

FIGURE 4: Flow chart of the algorithm.

radius of the sensor node is 20–60 cm. The transmitted packet size is 10 K bits. The initial energy of the nodes is 5 J. The nodes send data to the sink node in a single hop or multihop. The sink node is wired to the control system of the coal mine robot, whose energy is always constant. The movement of the nodes is not considered because the wireless sensor network is fixed after the arrangement of the sensor and the robot body. The QLSORP algorithm was developed and implemented in the environment of MATLAB 2016; other simulation parameters are set as follows: the size of the

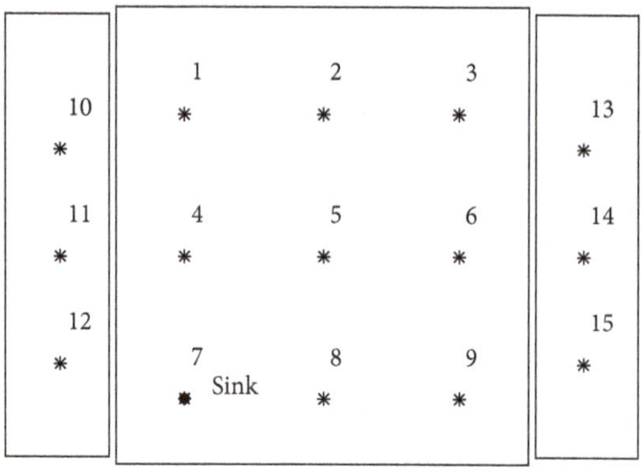

FIGURE 5: Node topology of wireless sensor network for coal mine robot.

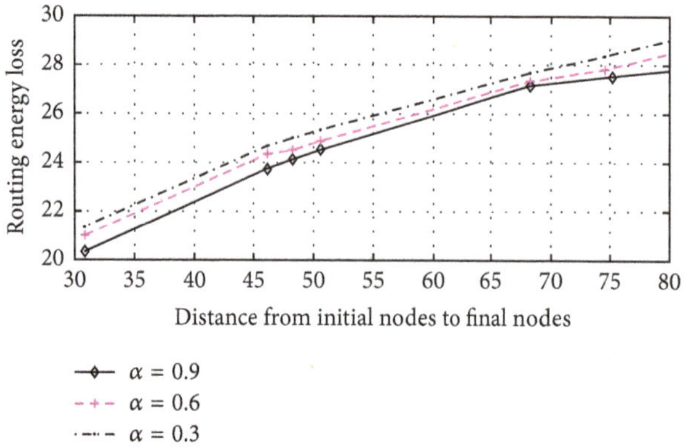

FIGURE 6: Routing energy loss.

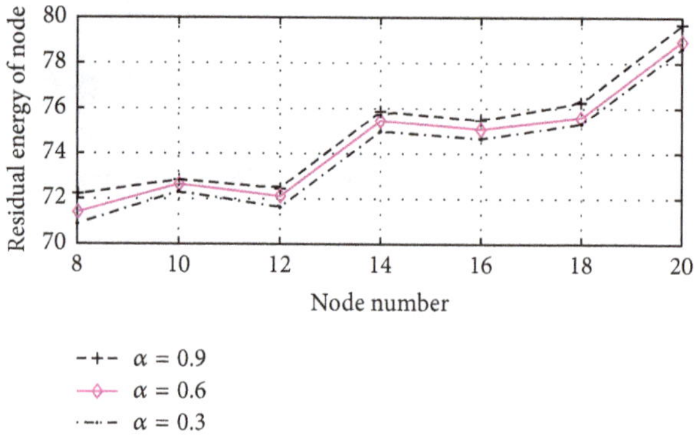

FIGURE 7: Residual energy of node.

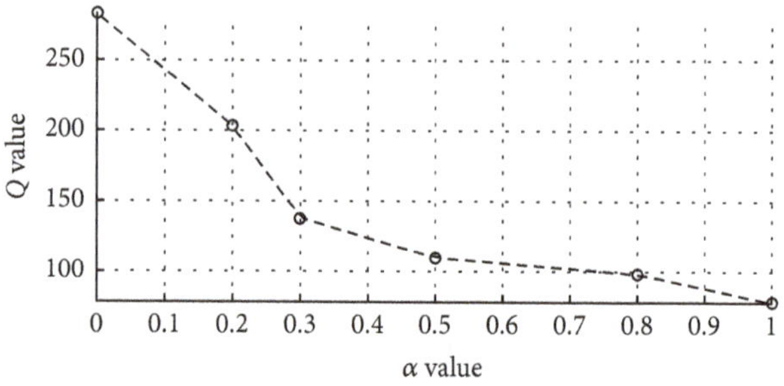

FIGURE 8: The impact of alpha values on Q.

network area is 120 cm $*$ 200 cm, E_{elec} is 50 nJ/bit, E_{PA} is 100 pJ/bit/cm2, and $d0$ is 100 cm.

4.2. Experimental Results and Analysis. From formula (6), it can be seen that in the Q-learning based self-organizing routing method for wireless sensor networks (QLSOP), the two parameters γ and α remain unknown and need to be preset. In order to reduce the computational complexity of the routing algorithm, $\gamma = 1$ is used to reduce the delay. α is desirable for any number of 0-1, such as 0.9, 0.6, and 0.3. The simulations are conducted to verify the influence of parameter α on the lifetime of wireless sensor networks, the choice of node paths, the routing energy loss, and the residual energy of nodes.

Figure 6 shows the routing loss of the source nodes at different distances to sink nodes. It is obvious that the energy loss of the nodes routing to the sink nodes is proportional to the distance between them. The energy consumed by routing increases as the distance between the nodes and the sink nodes increases. α is the learning rate. Its size has a certain influence on the loss of routing energy. The smaller the learning rate is, the more the loss of routing is.

The number of nodes also affects the routing energy consumption of the source node routing to the sink node. The residual energy of the node is obtained by simulation, as shown in Figure 7. When the number of nodes in wireless sensor networks increases, the residual energy in the nodes becomes larger. The higher the learning factor, the more the residual energy the nodes have.

α is the learning rate, its value affects convergence of Q-learning. The effects of different α on Q value are shown in Figure 8. With the increasing of α, the value of Q in the fixed path decreases gradually. Formula (6) shows that when α value is close to 0, the value of Q depends on the preserved Agent node's Q value. When α value is close to 1, Q value depends on the maximum Q value of the Agent's neighboring node and the return value from the neighboring node.

The lifetime of the wireless sensor network for the coal mine robot is shown in Figure 9. Figure 9(a) is a 3D surface for node number, communication path distance, and network lifetime in wireless sensor networks. The shorter the communication path, the less the energy consumed by the nodes and the longer the network lifetime. In addition, with the increasing number of nodes in wireless sensor networks, the number of alternative routes through which the source node sends data to the sink node increases. By continuously updating the return value, the optimal path is selected and the consumption of the routing energy is reduced, which results in a longer network lifetime. Figure 9(b) shows the comparison between the DSR algorithm and the QLSORP algorithm proposed in this paper. The curve at the bottom of the graph represents the network lifetime of the DSR under the same simulation conditions. Its performance is lower than that of QLSORP. The reason is that the QLSORP algorithm balances the energy between the nodes of wireless sensor networks. From Figures 9(a) and 9(b) we can see that when

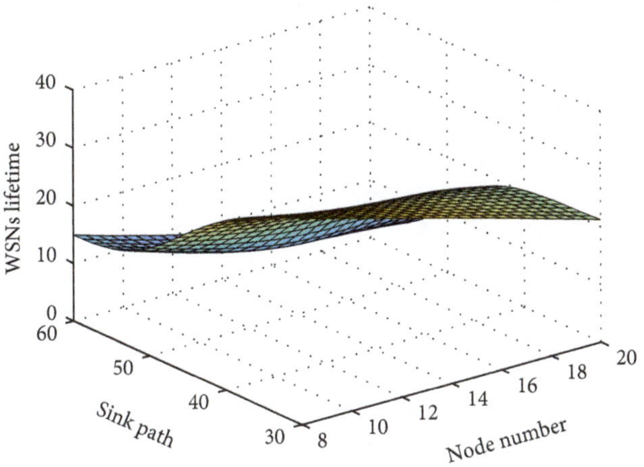

(a) Network lifetime under node number and path distance

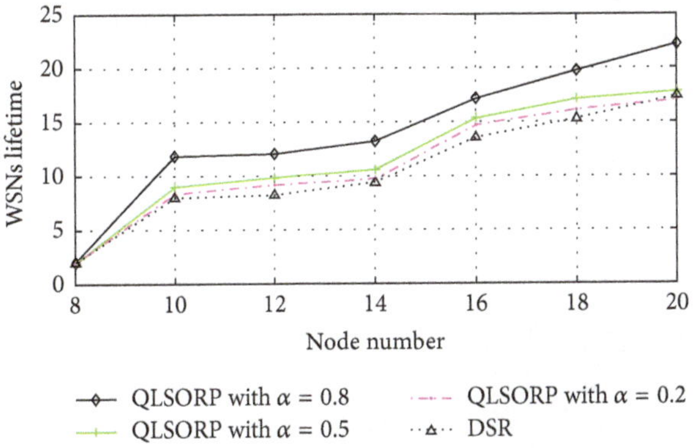

(b) Comparison of network lifetime between QLSORP and DSR

FIGURE 9: Network lifetime.

the number of nodes is near 18, with the increase of α, the lifetime of WSNs also increases accordingly. It is noteworthy that the optimal path is invariable, regardless of the value of α.

5. Conclusions

This paper proposes a QLSORP routing algorithm for wireless sensor network in coal mine robot based on Q-learning. The wireless sensor network nodes are regarded as intelligent Agent nodes. A self-organizing routing method based on Q-learning for wireless sensor networks is designed. It takes into account the hop number, distance, residual energy, communication loss, energy, and so on. By calculating the return value of neighboring nodes, Q value of the path is updated constantly. The updated Q value table is kept in the sink node. The sink node searches the optimal path by comparing Q values and selects the path with the largest Q value for data transmission. When the node in the path is accidentally dead or the node energy is exhausted, the path with the second largest Q value is selected for data transmission. This balances the energy between nodes. Compared with the DSR algorithm, the QLSORP algorithm can significantly reduce the energy consumption of the network. The algorithm can prevent the premature death of some nodes in the network. The coal mine robot wireless sensor network node energy can be utilized effectively. The lifetime of wireless sensor networks is prolonged.

This paper mainly studies the effectiveness of the routing algorithm based on Q-learning to balance the node energy of the WSN and improve the lifetime of the whole network. The robustness of WSN has not been considered, such as the impact of environmental noise interference on WSN, or the coverage hole problem of WSN [13] caused by the failure of some nodes or the exhaustion of energy. These problems all affect the practical effect of WSN. The next work will further study the topology self-cure algorithm [14] for node failure or energy exhaustion and improve the robustness to environmental noise [15] and other disturbances.

Conflicts of Interest

The authors declare that there are no conflicts of interest regarding the publication of this paper.

References

[1] X. Ma and H. Zhu, "Gas concentration prediction based on the measured data of a coal mine rescue robot," *Journal of Robotics*, vol. 2016, pp. 1-10, 2016.

[2] A. E. Forster and A. L. Murphy, "Exploiting reinforcement learning for multiple sink routing in WSN," in *Proceedings of the IEEE Internatonal Conference on Mobile Adhoc and Sensor Systems, MASS*, Italy, October 2007.

[3] P. Ciciriello, L. Mottola, and G. P. Picco, "Efficient routing from multiple sources to multiple sinks in wireless sensor networks," *Wireless Sensor Networks*, vol. 4373, pp. 34–50, 2007.

[4] M. Meng, X. Wu, B.-S. Jeong, S. Lee, and Y.-K. Lee, "Energy efficient routing in multiple sink sensor networks," in *Proceedings of the 5th International Conference on Computational Science and Its Applications (ICCSA '07)*, pp. 561-566, IEEE, August 2007.

[5] J.-M. Dricot, S. Van Roy, G. Ferrari, F. Horlin, and P. De Doncker, "Impact of the environment and the topology on the performance of hierarchical body area networks," *EURASIP Journal on Wireless Communications and Networking*, vol. 2011, no. 1, pp. 1–17, 2011.

[6] X. Yan, Y. Zhang, H. Tang, and S. Li, "An ETBG optimization algorithm based on analytic hierarchy process in WSS," in *Proceedings of the 2nd International Conference on Computer Science and Electronics Engineering (ICCSEE 2013)*, China, March 2013.

[7] R. R. Yager, "Weighted maximum entropy OWA aggregation with applications to decision making under risk," *IEEE Transactions on Systems, Man, and Cybernetics: Systems*, vol. 39, no. 3, pp. 183–189, 2009.

[8] J. K. Goyal and K. S. Nagla, "A new approach of path planning for mobile robots," in *Proceedings of the 3rd International Conference on Advances in Computing, Communications and Informatics (ICACCI '14)*, pp. 863–867, IEEE, New Delhi, India, September 2014.

[9] X. Shi, Q. Ran, M. Fan, H. Yu, and L. Wang, "Dynamic weighted DV-Distance algorithm for wireless sensor networks," *Chinese Journal of Scientific Instrument*, vol. 34, no. 9, pp. 1975–1981, 2013.

[10] S. L. Zheng, H. Che, Y. Q. Fan, D. Hu, and X. Xiao, "Agent-based mobile sink routing algorithm in wireless sensor networks,"

Journal of Electronic Measurement and Instrument, vol. 27, no. 2, pp. 127–134, 2013.

[11] H. Zhao, Z. H. Hu, and Y. Wwn, "ACS based differentiated service routing algorithm in wireless sensor network," *Journal on Communications*, vol. 34, no. 10, pp. 106–115, 2013.

[12] J. Elias, "Optimal design of energy-efficient and cost-effective wireless body area networks," *Ad Hoc Networks*, vol. 13, pp. 560–574, 2014.

[13] L. H. Yan and Y. Y. HE, "Robust approach foe holes recovery of wireless sensor networks," *Computer Science*, vol. 44, no. 2, pp. 123–128, 2017.

[14] L. F. Liu, J. G. Wu, Z. Q. Zou, H. Chen, and L. Niu, "opology self-cure algorithm aiming at node failure problem in wireless sensor networks," *Journal of Southeast University*, vol. 39, no. 4, pp. 695–699, 2009.

[15] A. Fortuna, L. Frasca, M. Iachello, M., and V. T. Pham, "Robustness to noise in synchronization of network motifs: Experimental results," *An Interdisciplinary Journal of Nonlinear Science*, vol. 22, no. 4, Article ID 043106, 2012.

7

A Low-Cost Model Vehicle Testbed with Accurate Positioning for Autonomous Driving

Benjamin Vedder,[1] **Jonny Vinter**,[1] **and Magnus Jonsson**[2]

[1] *Department of Electronics, RISE Research Institutes of Sweden, Sweden*
[2] *School of Information Technology, Halmstad University, Sweden*

Correspondence should be addressed to Benjamin Vedder; benjamin.vedder@ri.se

Academic Editor: Huosheng Hu

Accurate positioning is a requirement for many applications, including safety-critical autonomous vehicles. To reduce cost and at the same time improving accuracy for positioning of autonomous vehicles, new methods, tools, and research platforms are needed. We have created a low-cost testbed consisting of electronics and software that can be fitted on model vehicles allowing them to follow trajectories autonomously with a position accuracy of around 3 cm outdoors. The position of the vehicles is derived from sensor fusion between Real-Time Kinematic Satellite Navigation (RTK-SN), odometry, and inertial measurement and performs well within a 10 km radius from a base station. Trajectories to be followed can be edited with a custom GUI, where also several model vehicles can be controlled and visualized in real time. All software and Printed Circuit Boards (PCBs) for our testbed are available as open source to make customization and development possible. Our testbed can be used for research within autonomous driving, for carrying test equipment, and other applications where low cost and accurate positioning and navigation are required.

1. Introduction

It is common to use model cars for automotive research, and several studies are published in that field. For example, research within vehicle platooning has been carried out on model cars equipped with floor marking and distance sensors [1, 2], with the goal of developing Model Predictive Control (MPC) algorithms for controlling the distance between adjacent vehicles. Model cars have also been used to develop obstacle avoidance algorithms for mobile robots [3] and in student projects to teach them about autonomous driving [4]. These projects are specific and aimed at certain tasks, and the hardware and software are not available to replicate for use within other areas.

The need for a generic model vehicle platform for education and research within autonomous driving is recognized in the community, and several attempts at answering that need have been made. A project named MOPED [5] provides a model car that has three Raspberry Pi single board computers [6] connected over ethernet to simulate part of the complexity of a modern full scale car. Two of the computers on the MOPED model car run AUTOSAR [7] while the third one runs the default Raspbian Linux distribution (https://www.raspberrypi.org/downloads/raspbian/) that comes with the Raspberry Pi. The reason for using AUTOSAR is to represent a software stack similar to the one on a full scale car; however it requires software tools that are not available as open source or freeware for developing the AUTOSAR portions of the software. Further, the MOPED model car only provides low level control functions for the motor and steering servo, meaning that the users have to implement trajectory following and positioning algorithms themselves, as well as equipping the car with the necessary sensors. Gulliver [8] is another initiative that addresses the need of a miniature vehicle platform in research and education environments. In their publication, the authors describe high level algorithms for handling different traffic scenarios that can be used given access to a model vehicle that can follow trajectories. While these projects are an aid in education and research about autonomous driving, a significant amount of work is still required from the researchers or students to get a self-driving model car up and running. Note that, with self-driving in this context, we refer to the ability to follow a predefined trajectory accurately and repeatedly, which has been one of the goals in our work as explained below.

Our work focuses on providing a hands-on hardware and software testbed that can be used to build self-driving model vehicles with minimal effort. The goal is to provide the possibility to get a model vehicle up and running and follow a custom trajectory autonomously with 5 cm or better accuracy in just one day of work, given that the user has a background within electronics. To achieve this, we have developed low-cost hardware and both embedded and desktop software controlling model vehicles with Ackerman steering that can be modeled with a bicycle model [9]. Our testbed also has support for Hardware-in-the-Loop (HIL) testing [10, 11] by simulating parts of the vehicle dynamics, which is useful during development and automatic testing. Thus, the contribution of this work is to answer to what extent the aforementioned goal can be achieved with low-cost hardware, provide help for other researchers who want to implement their own self-driving model car, and answer what performance and accuracy can be expected. All software and hardware design for our testbed are available on github (https://github.com/vedderb/rise_sdvp) for making it possible to study and extend our platform. To our knowledge there is nothing available today that can fulfill the goal that we have for our tested, and as expressed in the aforementioned studies [5, 8] there is a need for that in the education and research communities.

In addition to usage as a research and education platform, our testbed can be used in measurement and data collection applications. For example, we have used it to pull a trailer with different radar units to be characterized around a variety of radar targets. This way the radars can be moved along a predefined trajectory around the targets, while the measurements they take are logged together with accurate position stamps. Another possible application that we have been considering is using model cars based on our testbed to carry light sensors and map the light intensity of artificial lighting in outdoor environments. The open source nature of the software in our testbed and the versatile visualization tools that are part of it make this a relatively simple task.

The remainder of this paper is organized as follows: in Section 2 we describe the architecture of our testbed and in Section 3 we describe how positioning is performed. Section 4 describes our trajectory following approach and in Section 5 we present our conclusions from this work.

2. Architecture Overview

Our testbed consists of a control Printed Circuit Board (PCB) we have developed that can be connected to a VESC (https://vesc-project.com) open-source motor controller over Controller Area Network- (CAN-) bus. Together with a battery and a Global Navigation Satellite System (GNSS) antenna they can be connected to the Permanent Magnet Synchronous Motor (PMSM) and steering servo suitable for a model car. The control PCB together with a

Figure 1: Our custom control PCB and our VESC 5 kW motor controller next to a car key for size comparison.

Figure 2: The 1:6 scale model car in our testbed.

5 kW VESC motor controller is shown in Figure 1 together with a car key for size comparison. Another unit of the same control PCB configured as a communication interface can be connected to a laptop computer running RControlStation, which is the monitoring and control Graphical User Interface (GUI) for our testbed. The stationary control PCB can also be connected to a GNSS antenna and act as a Real-Time Kinematic Satellite Navigation (RTK-SN) base station for the testbed, eliminating the need for an external base station. This gives a minimal stand-alone configuration of our testbed, which is able to follow trajectories autonomously. The cost of this configuration, excluding the laptop cost, is in the range of €900 to €2000 depending on the choice of model car, battery size, VESC version, and GNSS antennas. The schematics and hardware layout of our control PCB, the control PCB firmware and the RControlStation software, and the VESC firmware and configuration software are all available on github (https://github.com/vedderb).

While the aforementioned minimal configuration is sufficient for getting everything running and carrying out experiments, a Raspberry Pi (https://www.raspberrypi.org/) single board computer can be added for remote debugging, for video streaming, and for providing WiFi or 4G cellular connectivity. Our github repository also contains a command line utility for the Raspberry Pi that among other things provides a TCP/UDP to USB bridge for communication with RControlStation over WiFi or 4G. Figure 2 shows a photo of one of our model cars and Figure 3 shows a block diagram

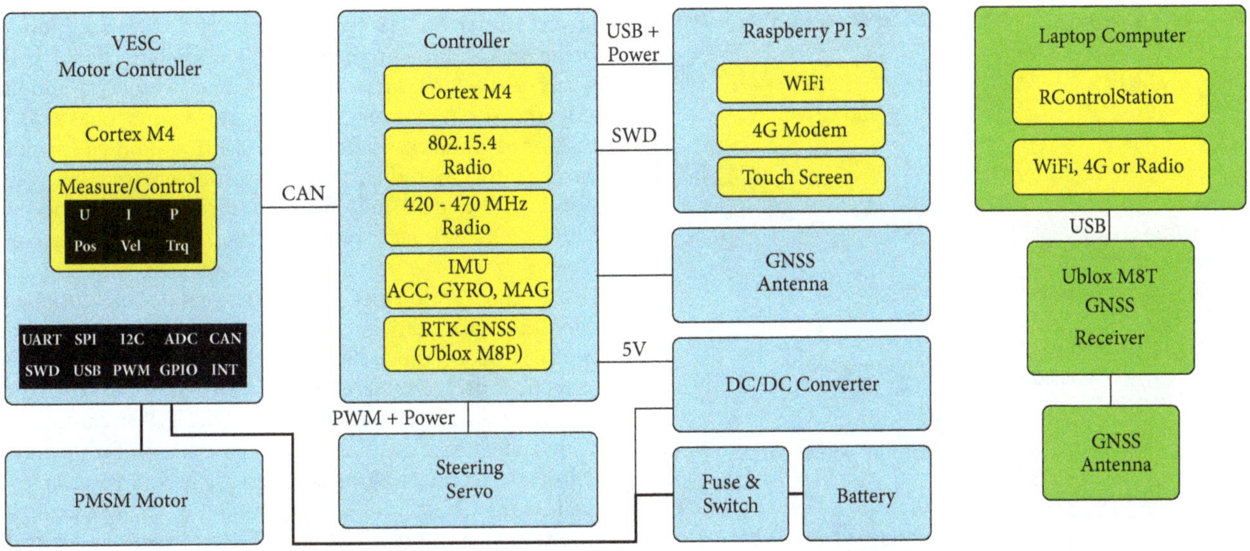

FIGURE 3: A block diagram of the configuration of our model car. A laptop computer for control and monitoring while acting as a RTK-SN base station is also shown.

of its configuration. Note that it also has a touch screen for the Raspberry Pi for showing network connections and other useful information, such as the battery charge level. Going through the block diagram in Figure 3, our model car, shown in Figure 2, has the following components:

(i) A lithium-ion battery with 10 cells in series and 3 Ah, providing up to 6 hours of power depending on the driving speed.

(ii) An integrated fuse and power switch, between the battery and the rest of the circuit.

(iii) A DC/DC converter that provides a 5V rail.

(iv) Our custom controller PCB, powered from the 5V rail.

(v) Our VESC motor controller, powered from the battery and connected to the controller over CAN-bus.

(vi) A PMSM motor, connected to the VESC motor controller.

(vii) A steering servo and a GNSS antenna, connected to the controller.

(viii) A Raspberry Pi single board computer, connected to the controller over USB. It is powered backwards through the USB ports from the controller. The Serial Wire Debug (SWD) port of the controller is also connected to the Raspberry Pi, so that the controller can be programmed and debugged remotely.

(ix) Outside the car, there is a laptop computer connected over TCP to the Raspberry Pi using the WiFi or 4G cellular connection. Our RControlStation software runs on the laptop computer and utilizes the connection to control and monitor the model car.

(x) The laptop is connected to Ublox M8T RTK-SN receiver with an antenna mounted on a tripod to act as a base station for the model car, enabling high precision positioning (see Section 3 for more details). RControlStation handles the setup of the Ublox M8T, as well as forwarding of the required correction data for RTK-SN to the model car.

The control PCB is based on an ARM Cortex M4 microcontroller and runs the ChibiOS (http://www.chibios.org) real-time operating system. The microcontroller carries out sensor fusion for position estimation (see Section 3), the trajectory following algorithm (see Section 4), and all other functionalities of the model vehicle, meaning that the connection to the laptop is only required for monitoring and sending high level control commands. In addition to the microcontroller, the control PCB contains an Inertial Measurement Unit (IMU), a CAN transceiver, two radios, DC/DC converters, an Ublox M8P GNSS receiver, and various connection ports for possible extensions.

After the control PCB, the other essential part of the electronics and software on the model car is the VESC motor controller. The VESC is also developed by us, partly in parallel with our testbed, but this paper only describes it briefly. It can drive the motor of the model car with high efficiency over a wide dynamic range using the state-of-the-art motor control technique Field Oriented Control (FOC) [12] with Space Vector Modulation (SVM) [13]. The VESC is able to operate from zero speed with high torque without position sensors on the motor by taking advantage of a nonlinear observer [14] and effective startup algorithms. Further, it provides position and speed feedback from the motor over CAN-bus as well as closed loop speed control, which is essential for the positioning system of the model car to work accurately as described in Section 3. Another essential functionality for our testbed that the VESC provides is automatic identification of all motor parameters necessary for sensorless FOC, which are rarely available in the datasheet of inexpensive model car motors. All the configuration and parameter detection of the VESC can be performed with

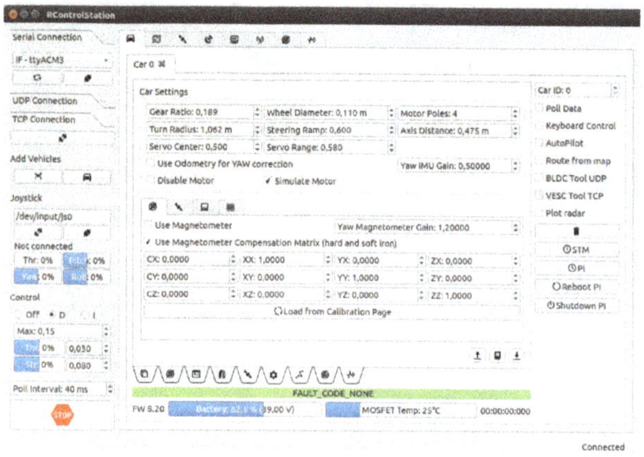

FIGURE 4: A screenshot of RControlStation, where parameters for the model car can be edited.

the accompanied VESCTool GUI, which among other things provides configuration wizards to get all settings right.

In the same way as the VESC motor controller can be configured to run with almost any PMSM without writing code and/or using expensive lab equipment for motor parameter identification, the control software can easily be configured for different model car configurations. Figure 4 shows a screenshot of RControlStation where several of the model car parameters can be edited and stored in the control PCB, such as the wheel diameter, gear ratio, and turning radius. This enables users of our testbed to configure the hardware and software to work with any model vehicle size and configuration from easy-to-use GUIs, which is one of the main objectives of our testbed. RControlStation also provides many debugging and plotting tools aiding with setting up model vehicles and performing experiments, such as magnetometer calibration and data visualization. For example, all graphs and map plots in this paper are generated in vector format using RControlStation, without requiring any additional code.

To develop and evaluate the trajectory following algorithm, as well as aiding with development of automatic test case generation with HIL, the control PCB supports a simulation mode where tests can be performed with just a USB connection to it without using the model car. This mode is implemented by simulating the behavior of the motor and mechanics with inertia and drag, and by updating the position of the vehicle with only dead reckoning from the simulated motor feedback, and heading calculated from the commanded steering angle (see Section 3 for details about the position estimation). The rate of movement of the commanded steering angle is limited to capture the behavior of a realistic steering servo. This simplifies and speeds up testing and development, while capturing many important aspects of the real-time hardware.

3. Positioning

Accurate positioning is an essential part of our testbed. Our main position source is RTK-SN, which is based on consumer GNSS technology with the addition of carrier-phase measurements of the satellite signals. The carrier-phase measurements together with conventional code measurements on both the rover (the object to be positioned) and a base station with a known position within a 10 km radius from the rover are required. This means that a data link between the rover and the base station with a data rate of around 100 bytes/second, depending on the number of satellites, is required. The data stream with code and carrier-phase measurements from the base station, together with information about the position of the base station is usually referred to as correction data and sent using a format such as RTCM3 (http://www.navipedia.net/index.php/RTK_Standards). RControlStation can either act as a base station for the model vehicles by connecting an appropriate GNSS receiver to it (such as the Ublox M8T), or act by connecting to an existing base station using the TCP or NTRIP (https://en.wikipedia.org/wiki/Networked_Transport_of_RTCM_via_Internet_Protocol) protocol. The correction data is then sent from RControlStation to the vehicles using the communication link of choice (4G, WiFi, or radio) using the RTCM3 format. The position accuracy of the rover relative to the base station with RTK-SN is around 1 cm under optimal conditions.

Traditionally RTK-SN has been an expensive technology, but recently less expensive solutions have become available [15]. In previous work we have compared the performance of high and low cost RTK-SN systems [16] and came to the conclusion that, besides the longer initial convergence time and low update rates of the low cost systems, the performance is similar. We have also studied how low cost RTK-SN performs in urban environments [17] using the same control PCB as presented here, and came to the conclusion that it performs well even when the view of the sky is poor given that multiple satellite constellations and/or sensor fusion with dead reckoning is used.

3.1. Challenges. The main challenge with using inexpensive RTK-SN equipment with our testbed is the low position update rate, as well as latency and jitter between updates. The Ublox M8P RTK-SN receiver provides a position update rate of 5 Hz, which is too low for accurate positioning and control at speeds of up to 80 km/h, which our model cars are capable of. Figure 5 shows the measured age of consecutive position updates from the Ublox M8P. As can be seen, the samples are between 95 and 135 ms old and have jitter of up to 40 ms. If the model car moves at 10 m/s, which is less than half of the maximum speed, a latency of 100 ms causes a position error of 1 m. At the same speed, the jitter between consecutive position samples can cause errors of up to 0.5 m. Having a low update rate and high latency on one or more of the sensors on the system is a common problem within robotics, especially when low cost hardware is involved [18], and there are various methods to handle that.

To deal with the low update rate, latency, and jitter of the RTK-SN position samples, dead reckoning from sensor fusion between the odometry data from the VESC motor controller and samples from the IMU is combined with the

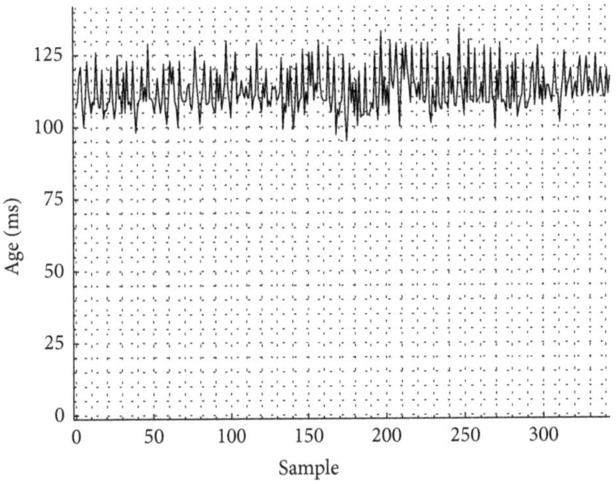

FIGURE 5: Delay jitter of the RTK-SN samples over time.

RTK-SN position samples to get a total update rate of 100 Hz without latency. We have come up with the method below of performing the combination. Note that the RTK-SN position in **(3)** and **(4)** first is moved to the center of the vehicle based on the current estimation of the heading angle and offset of the GNSS antenna from the vehicle center.

(1) The IMU is sampled at 1 kHz and fed to a quaternion-based orientation filter as proposed by Madgwick et al. [19], which provides the roll, pitch, and heading of the model vehicle. When the magnetometer of the IMU is configured to be included, we have improved the position filter by adding tilt compensation for estimating the heading [20] as well as ellipsoid fitting for hard and soft iron compensation of the magnetometer [21].

(2) The motor position is sampled from the VESC motor controller at 100 Hz and combined with the estimated heading from the position filter to calculate a relative dead reckoning position. This position is stored in a 100 samples long FIFO buffer together with the current GNSS time stamp derived from the pulse per second (PPS) signal from the Ublox M8P receiver.

(3) When RTK-SN position samples are received, they are compared to the dead reckoning position sample from the FIFO buffer from **(2)** with the closest time stamp to the sample, which will be 5 ms away in the worst case. The difference between the closest FIFO buffer sample and the RTK-SN sample is then used to correct the current position by moving it in the direction of the difference as

$$p_{xy} = p_{xy\_old} + d_{xy}G_{stat} + d_{xy}G_{dyn}d_p \quad (1)$$

where p_{xy} is a vector with the new xy-position of the vehicle, p_{xy_old} is the previous position of the vehicle, d_{xy} is the difference vector between the latest RTK-SN sample and the closest position in time from the FIFO buffer, G_{stat} is a scalar configurable static gain, G_{dyn} is a scalar configurable dynamic gain, and d_p is a scalar of how far the vehicle has moved since the previous RTK-SN update. Noise between consecutive RTK-SN samples is rejected by gradually moving the current position using this method instead of moving it the full difference at once for every sample.

(4) When the magnetometer is not used to provide an absolute heading reference, the heading of the orientation filter in **(1)** is updated every time a RTK-SN sample is received by first computing an RTK-SN heading as

$$\phi_{rtk} = -a\tan 2\left(y_{rtk} - y_{prtk}, x_{rtk} - x_{prtk}\right) \quad (2)$$

where (x_{rtk}, y_{rtk}) is the latest RTK-SN position sample and (x_{prtk}, y_{prtk}) previous RTK-SN position sample. After that a heading difference is calculated as $\phi_d = \phi_{rtk} - \phi_{FIFO}$, where ϕ_{FIFO} is the heading of the sample from the FIFO buffer from **(2)** closest to the average time between the latest and previous RTK-SN samples. The heading in the current position is then updated by rotating it in the angular direction of ϕ_d with a step proportional to how far the vehicle has moved between the latest and previous RTK-SN samples and the heading correction gain. Scaling by the distance moved is used because consecutive RTK-SN samples do not provide any heading information when the vehicle is stationary.

The rationale of this approach is that the dead reckoning position is accurate over short distances but drifts as the distance increases. The time delay of the RTK-SN samples is short enough to not cause significant degradation of the dead reckoning, meaning that the current position can be calculated accurately at a high rate by using an old absolute position and the relative movements that have occurred since then. Notice that the heading estimation is critical for the dead reckoning to perform well, and more details about that can be found in our previous work [17].

3.2. Performance Evaluation. To evaluate the performance of the latency and jitter compensation described above, we have driven the model car along a trajectory with rapid accelerations and decelerations and measured the difference between the RTK-SN position and the dead reckoning position for each sample with and without the FIFO delay compensation. Figure 6 shows the difference without compensation, and Figure 7 shows the difference with compensation. As can be seen, without the compensation the difference is stable at constant low speed but goes to 1.1 m during the accelerations and decelerations. With the compensation enabled, the difference is limited to 10 cm during the acceleration and to 20 cm during the deceleration, due to wheel slippage. The reason that the difference in Figure 6 is low during constant speed without compensation is that the position converges to an invalid value, which will be apparent when the speed changes rapidly, whereas when the compensation is enabled the position will be closer to the true position at all times. It can also be noted that during the constant speed shown in Figure 6 there is over 10 cm position jitter while Figure 7 shows less than 3 cm jitter; this is due to the delay jitter between the RTK-SN samples shown in Figure 5.

To obtain an estimate about the absolute accuracy and repeatability of the position, we have downloaded a 20 m radius circle trajectory to the model car with a constant speed of 4 km/h. While the car was following that trajectory lap after lap on artificial grass, it made visible traces that allowed us to

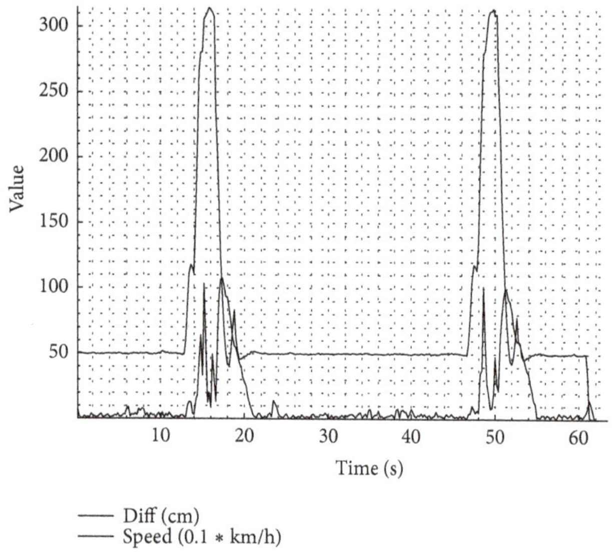

FIGURE 6: The speed and difference between the RTK-SN and the dead reckoning position during hard acceleration and deceleration. Delay compensation for the RTK-SN samples is disabled.

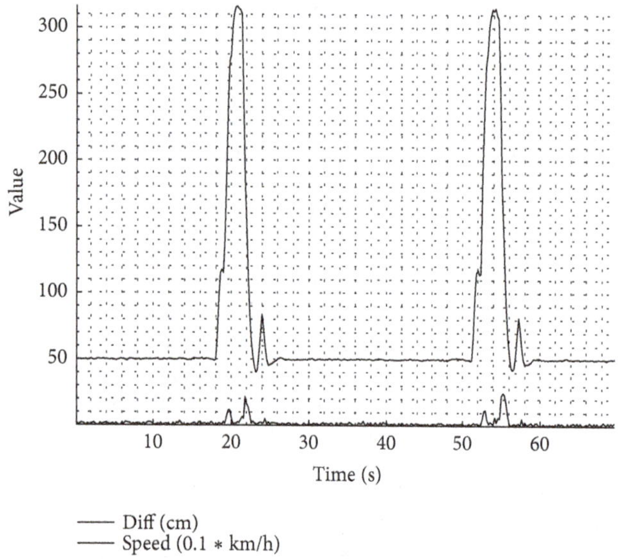

FIGURE 7: The speed and difference between the RTK-SN and the dead reckoning position during hard acceleration and deceleration. Delay compensation for the RTK-SN samples is enabled.

visually inspect the deviation of the car tires from the traces, as seen in Figure 8. As far as we could observe, the tires of the car stayed within the traces with less than half the tire width, or 2.5 cm, for the entire experiment. We also measured the diameter of the circle on the grass, and it had the correct diameter as accurately as we were able to measure with our equipment. The position difference between the RTK-SN samples and the closest samples in the dead reckoning FIFO stayed below 3 cm for the entire experiment, giving us confidence that our model vehicle can estimate its absolute position with 3 cm accuracy.

FIGURE 8: Our 1:6 model car repeatedly following a circular trajectory on artificial grass. The traces from previous laps are visible and can be used to estimate the lateral positioning and control repeatability.

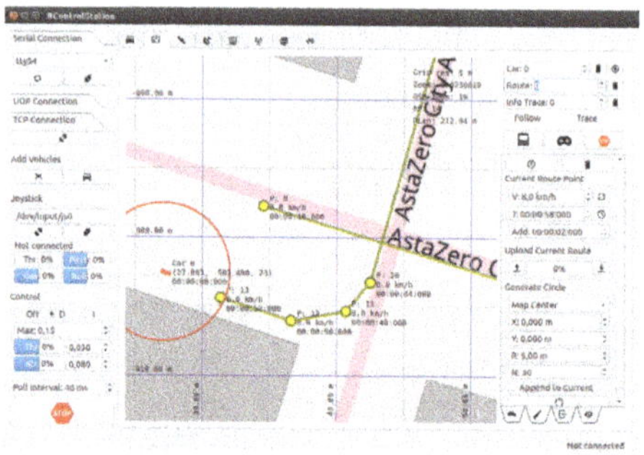

FIGURE 9: A screenshot of the map page in RControlStation, where trajectories can be edited.

4. Trajectory Following

An important aspect of our testbed is the ability to edit and follow trajectories. We define a trajectory as a list of points where each point has a xy-positions and a speed or time stamp, depending on the mode of operation. Our testbed support three trajectory following modes: **(1) speed based**, where the model vehicle adjusts its speed proportional to the set speed and relative distances to the two closest trajectory points; **(2) absolute time**, where the model vehicle adjusts its speed such that it reaches the trajactory points at absolute RTCM (https://en.wikipedia.org/wiki/Coordinated_Universal_Time) times (derived from the GNSS receiver clock); and **(3) relative time**, where the model vehicle adjusts its speed such that it reaches the trajectory points at times relative to when the start command was issued. Creating trajectories is most intuitive using the speed mode, but the time-based modes are necessary to synchronize multiple vehicles in a scenario. There is also a synchronization command available that can be sent to the model vehicles in real time, so that they continuously update their speed to reach a trajectory point specified in the command at the time specified in the command, based on the distance left along the trajectory to that point.

RControlStation allows users to graphically edit trajectories with an overlay of OpenStreetMap [22], as shown in the screenshot in Figure 9. We chose OpenStreetMap because it is accompanied by a complete set of tools for map creation and rendering, available as open source software. This makes it easy to update and create artificial maps, in e.g., test areas, and render them on a server. The transforms for converting OpenStreetMap rendered map tiles with Web Mercator (https://en.wikipedia.org/wiki/Web_Mercator) projection to the coordinate system of our testbed for viewing in RControlStation are well documented on their wiki page (https://wiki.openstreetmap.org/wiki/Main_Page). This was a significant aid in implementing the map rending functionality of RControlStation.

4.1. Lateral Control. Lateral control along the trajectories of the model vehicles in our testbed is performed using the pure pursuit algorithm, which is a common method within robotics [23]. Essentially it works in the following way: **(1)** draw a circle around the vehicle with a radius of the chosen look-ahead distance; **(2)** calculate the point where that circle intersects the trajectory; if multiple points are found pick the one furthest ahead on the trajectory; and **(3)** adjust the steering angle of the vehicle such that it follows an arc that intersects with that point. When visualized, it looks like the vehicle follows a point that moves away from it along the trajectory. More details about the pure pursuit algorithm can be found in the literature [23]. We have also added support for the case when the vehicle is further away from the trajectory than the look-ahead distance, in which case it will follow the closest point on the trajectory until the circle around the vehicle intersects with the trajectory, after which the algorithm is carried out as usual. This is useful when sharp turns or oscillations cause the vehicle to lose the trajectory, or when it is started far away from the trajectory. Also, if the model vehicle is to be used with automatic test case generation, it is helpful to have the ability to drive to the closest point on a recovery trajectory and drive back to the initial position along it.

Improvements to the pure pursuit algorithm commonly found in literature are interpolation of the trajectory to find points between trajectory points [23] and to use an adaptive look-ahead distance [24]. Both of these improvements have been employed in our testbed to increase the trajectory following performance. We have also implemented one to our knowledge unique improvement to the pure pursuit algorithm, which is adding gain to the steering angle calculation when the point to be followed is far away. The steering angle is corrected as

$$A_{corr} = \begin{cases} A_{st}(1 + 0.2D) & \text{if } D < 20\,\text{m} \\ 5A_{st} & \text{otherwise} \end{cases} \tag{3}$$

where A_{corr} is the corrected steering angle, A_{st} is the steering angle that leads to the goal point along a circle that tangents with the car, and D is the distance to the goal point. The gain helps when the vehicle starts far away from the trajectory heading away from it, where it without the gain would follow an arc longer than necessary to reach the trajectory. Figure 10 shows the arc the vehicle follows without the distance gain and Figure 11 shows the arc that it follows with the gain. The starting position is the same in both figures, namely, where the red line starts. As can be seen, the arc in Figure 11 is significantly shorter. Note that the look-ahead distance of the vehicle in the figures is illustrated with a red circle around it and that the point it is aiming for is drawn on the trajectory.

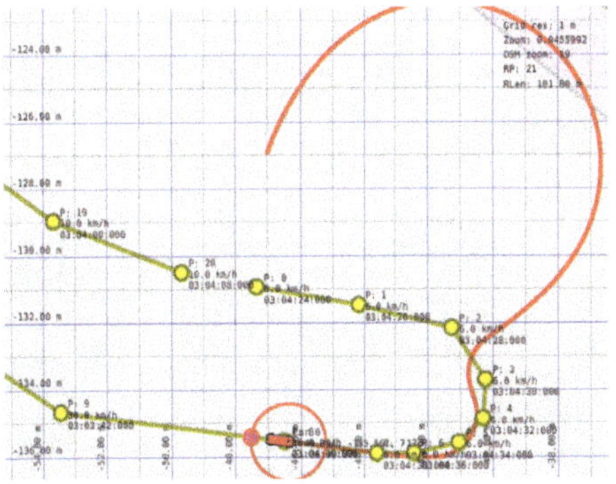

FIGURE 10: The red trace shows how the model car navigates back to the trajectory with angle distance gain disabled.

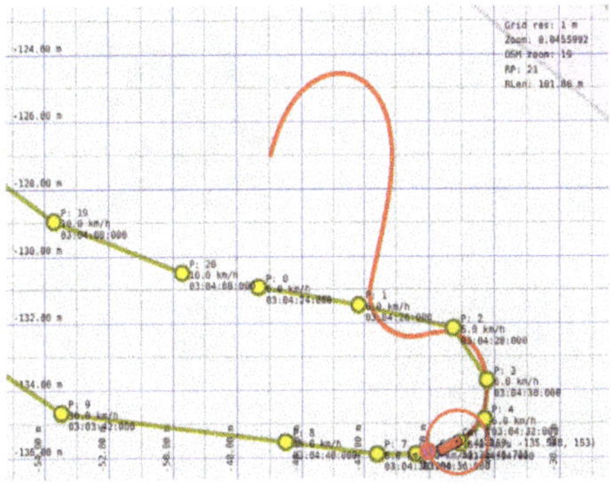

FIGURE 11: The red trace shows how the model car navigates back to the trajectory with angle distance gain enabled.

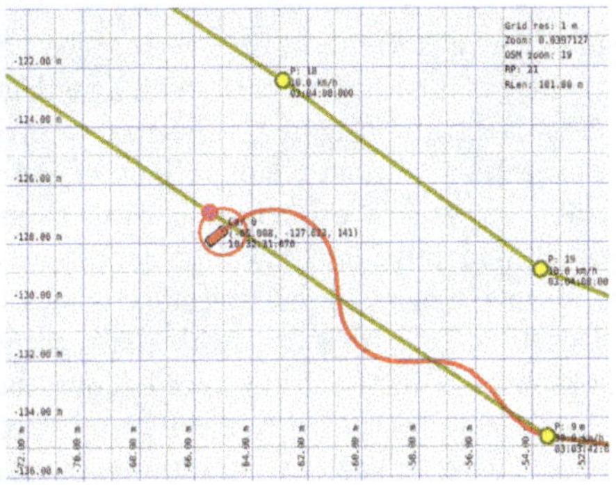

FIGURE 12: The red trace shows how the model car oscillates at 30 km/h with too small fixed look-ahead distance.

Using the simulation mode of the control PCB and a trajectory that has variable speeds and sharp turns, we have set up an experiment to evaluate the performance of the lateral control with different settings and improvement strategies of the pure pursuit algorithm. First we disabled adaptive look-ahead distance and found a static distance that is long enough to follow the trajectory in a stable manner. As an unstable example, Figure 12 shows how the car behaves when the

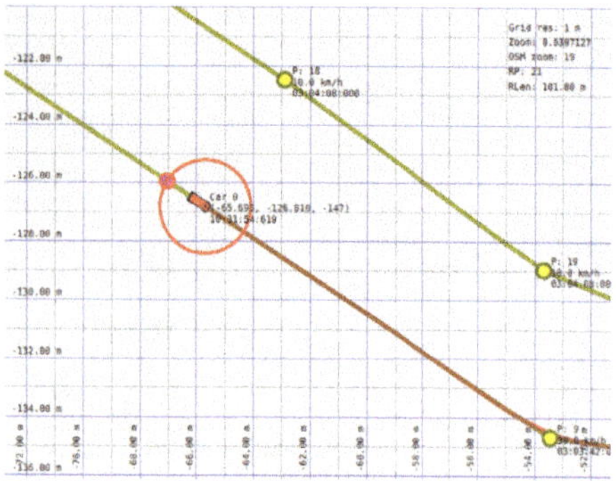

FIGURE 13: The red trace shows how the model car follows the trajectory at 30 km/h with sufficient look-ahead distance.

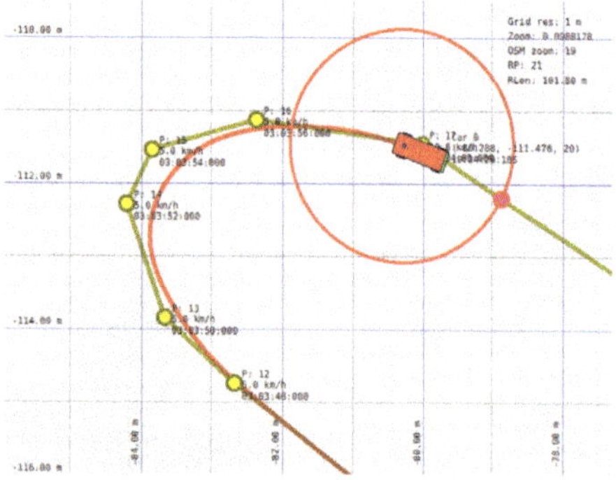

FIGURE 14: The red trace shows how the model car followed the trajectory with an arc without adaptive look-ahead distance enabled.

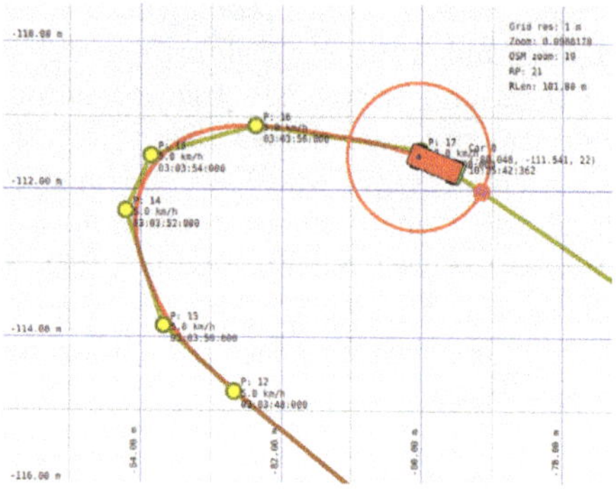

FIGURE 15: The red trace shows how the model car followed the trajectory with an arc with adaptive look-ahead distance enabled.

FIGURE 16: The model car pulling a trailer with radar to be characterized.

look-ahead distance is too short on a high speed part of the trajectory, whereas the same scenario with a sufficient look-ahead distance is shown in Figure 13. The maximum deviation from that trajectory without adaptive look-ahead distance during a stable lap was 40 cm, whereas it was 13 cm for the same track with adaptive look-ahead distance enabled. The difference in deviation from the trajectory comes from the low-speed parts of that trajectory with sharp turns, where a short look-ahead distance at low speed allows the vehicle to follow the trajectory tightly while still being stable due to the low speed. Figure 14 shows a tight turn without adaptive look-ahead distance and Figure 15 shows the same turn with adaptive look-ahead distance. For reference, the adaptive look-ahead distance is calculated based on the speed of the vehicle as

$$d = d_{base}\left(1 + |v| * 0.05\right)^2 \tag{4}$$

where d is the calculated adaptive look-ahead distance, v is the speed in m/s, and d_{base} is the look-ahead distance when the speed is 0. This equation was derived experimentally.

5. Conclusion

In this paper we have presented a novel testbed for research and development within the areas of autonomous driving and accurate positioning. We were able to achieve high position accuracy and low latency using low-cost hardware by fusing data from multiple sensors (Accelerometer, Gyroscope, odometry, and RTK-SN) to take advantages of their individual strengths. For trajectory following we have implemented the well-known pure pursuit algorithm along with two common and one unique improvements as described in Section 4.

Model vehicles based on our testbed can follow trajectories autonomously, and all tools necessary for visualization and trajectory generation are provided. Our testbed is scalable to a wide range of model vehicles and can be set up in just one day of work given familiarity with the testbed. The custom hardware and all involved software are open source for easy development and extension. We have also performed a wide range of tests to ensure high performance and reliability in various situations. To our knowledge there is no similar testbed available today, and there is a significant need for it in the research and education communities.

In addition to use within research and education, our testbed can be used in data acquisition applications where sensors need to be moved accurately according to a defined path, while data is stored together with position and speed. For example, Figure 16 shows a photo of our model car pulling a trailer with radar to be characterized along a predefined path around different targets. The open source nature of our testbed provides a solid base for such applications and enables further functionality to be implemented without the burden of implementing the positioning and navigation functionality.

Glossary

CAN:	Controller area network. A communication bus standard for industrial and automotive applications.
Dead reckoning:	The process of estimating the current position by advancing a previously determined position using inertial or speed measurements. Dead reckoning generally relies on measuring derivatives, meaning that the error will drift without an upper bound if no absolute measurements, e.g., GNSS, are used
FOC:	Field Oriented Control (see [12])
GNSS:	Global Navigation Satellite System
GUI:	Graphical User Interface
HIL:	Hardware-In-the-Loop
IMU:	Inertial Measurement Unit. Achip that can measure acceleration and angular velocity with three axes each
MPC:	Model predictive control (see [2])
Odometry:	Tracking the rotation of the wheels. In our case this is done by tracking the rotation of the motor
PCB:	Printed circuit board
PMSM:	Permanent magnet synchronous motor. A three-phase motor without brushes, often referred to as BLDC motor
PPS:	Pulse per second. A pulse that is emitted with accurate timing in the beginning of each second, synchronized to GNSS time
RTK-SN:	Real-time kinematic satellite navigation (see [15, 16])
SVM:	Space vector modulation (see [13])
SWD:	Serial wire debug. A programming and debugging interface.

Data Availability

As our testbed is open source, including all PCB designs, embedded software, and desktop software, the experiments presented in this paper can be replicated by constructing a model car using the resources available at https://github.com/vedderb/rise_sdvp. The data, plots, and trajectories, as well as additional data and plots from the experiments presented in this paper, can be found here https://github.com/vedderb/rise_sdvp/tree/master/Misc/Test%20Data/paper_2018-06. To replicate the experiments and collect similar data, as well as to generate similar plots, the trajectories in the routes directory from the previous link can be used together with the appropriate terminal commands from the car terminal of RControlStation. RControlStation is part of the open source software, and all terminal commands can be listed by typing help in the car terminal.

Conflicts of Interest

The authors declare that there are no conflicts of interest regarding the publication of this paper.

Acknowledgments

This research has been funded through EISIGS (grants from the Knowledge Foundation) and by VINNOVA via the FFI Projects Chronos steps 1 and 2.

References

[1] F. C. Braescu and C. F. Caruntu, "Prototype model car design for vehicle platooning," in *Proceedings of the 2nd International Conference on Optimization of Electrical and Electronic Equipment, OPTIM 2017 and Intl Aegean Conference on Electrical Machines and Power Electronics, ACEMP 2017*, pp. 953–958, Brasov, Romania, May 2017.

[2] F. C. Braescu, "Basic control algorithms for vehicle platooning prototype model car," in *Proceedings of the 21st International Conference on System Theory, Control and Computing, ICSTCC 2017*, pp. 180–185, Sinaia, Romania, October 2017.

[3] A. Fenesan, D. Szöcs, T. Pana, and W-H. Chen, "Building an electric model vehicle and implementing an obstacle avoidance algorithm," in *Proceedings of the 2012 International Conference and Exposition on Electrical and Power Engineering (EPE)*, pp. 49–53, Iasi, Romania, October 2012.

[4] F. Bormann, E. Braune, and M. Spitzner, "The C2000 autonomous model car," in *Proceedings of the 4th European DSP Education and Research Conference, EDERC 2010*, pp. 200–204, France, December 2010.

[5] J. Axelsson, A. Kobetski, Z. Ni, S. Zhang, and E. Johansson, "MOPED: A mobile open platform for experimental design of cyber-physical systems," in *Proceedings of the 40th Euromicro Conference Series on Software Engineering and Advanced Applications, SEAA 2014*, pp. 423–430, Italy, August 2014.

[6] G. Halfacree and E. Upton, *Raspberry Pi User Guide*, Wiley, 1st edition, 2012.

[7] R. Svenningsson, R. Johansson, T. Arts, and U. Norell, "Formal Methods Based Acceptance Testing for AUTOSAR Exchangeability," *SAE International Journal of Passenger Cars - Electronic and Electrical Systems*, vol. 5, no. 1, pp. 209–213, 2012.

[8] M. Pahlavan, M. Papatriantafilou, and E. M. Schiller, "Gulliver: A Test-Bed for Developing, Demonstrating and Prototyping Vehicular Systems," in *Proceedings of the 2012 IEEE Vehicular Technology Conference (VTC 2012-Spring)*, pp. 1-2, Yokohama, Japan, May 2012.

[9] K. Hulme, E. Kasprzak, K. English, D. Moore-Russo, and K. Lewis, "Experiential learning in vehicle dynamics education via motion simulation and interactive gaming," *International Journal of Computer Games Technology*, vol. 2009, Article ID 952524, 15 pages, 2009.

[10] A. Soltani and F. Assadian, "A hardware-in-the-loop facility for integrated vehicle dynamics control system design and validation," *IFAC-PapersOnLine*, vol. 49, no. 21, pp. 32–38, 2016.

[11] A. Mouzakitis, D. Copp, R. Parker, and K. Burnham, "Hardwarefor testing automotive ECU diagnostic software," *SAGE Measurement and Control Journal*, vol. 42, no. 8, pp. 238–245, 2009.

[12] J. P. John, S. S. Kumar, and B. Jaya, "Space vector modulation based field oriented control scheme for brushless DC motors," in *Proceedings of the 2011 International Conference on Emerging Trends in Electrical and Computer Technology, ICETECT 2011*, pp. 346–351, India, March 2011.

[13] M. Gaballah and M. El-Bardini, "Low cost digital signal generation for driving space vector PWM inverter," *Ain Shams Engineering Journal*, vol. 4, no. 4, pp. 763–774, 2013.

[14] J. Lee, J. Hong, K. Nam, R. Ortega, L. Praly, and A. Astolfi, "Sensorless control of surface-mount permanent-magnet synchronous motors based on a nonlinear observer," *IEEE Transactions on Power Electronics*, vol. 25, no. 2, pp. 290–297, 2010.

[15] T. Takasu and A. Yasuda, "Development of the low-cost RTK-GPS receiver with an open source program package RTKLIB," in *Proceedings of the International Symposium on GPS/GNSS*, pp. 4–6, Jeju, Korea, 2009.

[16] M. Skoglund, T. Petig, B. Vedder, H. Eriksson, and E. M. Schiller, "Static and dynamic performance evaluation of low-cost RTK GPS receivers," in *Proceedings of the 2016 IEEE Intelligent Vehicles Symposium, IV 2016*, pp. 16–19, Sweden, June 2016.

[17] B. Vedder, J. Vinter, and M. Jonsson, "Accurate positioning of bicycles for improved safety," in *Proceedings of the 2018 IEEE International Conference on Consumer Electronics (ICCE)*, pp. 1–6, Las Vegas, Nev, USA, January 2018.

[18] M. Bošnak, D. Matko, and S. Blažič, "Quadrocopter hovering using position-estimation information from inertial sensors and a high-delay video system," *Journal of Intelligent Robotic Systems*, vol. 67, no. 1, pp. 43–60, 2012.

[19] S. O. H. Madgwick, A. J. L. Harrison, and R. Vaidyanathan, "Estimation of IMU and MARG orientation using a gradient descent algorithm," in *Proceedings of the IEEE International Conference on Rehabilitation Robotics (ICORR '11)*, pp. 1–7, July 2011.

[20] T. Ozyagcilar, "Implementing a tilt-compensated ecompass using accelerometer and magnetometer sensors, An4248," Freescale Semiconductor, Application Note, 2015.

[21] A. Vitali, "Ellipsoid or sphere fitting for sensor calibration, Dt0059," ST Microelectronics, Design Tip, 2016.

[22] M. Haklay and P. Weber, "Openstreetmap: user-generated street maps," *IEEE Pervasive Computing*, vol. 7, no. 4, pp. 12–18, 2008.

[23] H. Ohta, N. Akai, E. Takeuchi, S. Kato, and M. Edahiro, "Pure pursuit revisited: Field testing of autonomous vehicles in urban areas," in *Proceedings of the 4th IEEE International Conference on Cyber-Physical Systems, Networks, and Applications, CPSNA 2016*, pp. 7–12, Japan, October 2016.

[24] M. Park, S. Lee, and W. Han, "Development of lateral control system for autonomous vehicle based on adaptive pure pursuit algorithm," in *Proceedings of the 2014 14th International Conference on Control, Automation and Systems (ICCAS)*, pp. 1443–1447, Gyeonggi-do, South Korea, October 2014.

8

A Bioinspired Gait Transition Model for a Hexapod Robot

Qing Chang[1] and Fanghua Mei[2]

[1]School of Mechanical Engineering, Tianjin University of Commerce, Tianjin, 300134, China
[2]School of Automation Science and Electrical Engineering, Beijing University of Aeronautics and Astronautics, Beijing 100191, China

Correspondence should be addressed to Qing Chang; changbit@163.com

Academic Editor: Keigo Watanabe

Inspired by the analysis of the ant locomotion observed by the high-speed camera, an ant-like gait transition model for the hexapod robot is proposed in this paper. The model which consists of the central neural system (CNS), neural network (NN), and central pattern generators (CPGs) can produce the rhythmic signals for different gaits and can realize the transition of these gait automatically and smoothly according to the change of terrain. The proposed model suggests the neural mechanisms of the ant gait transition and can improve the environmental adaptability of the hexapod robot. The numerical simulation and corresponding physical experiment are implemented in this paper to verify the proposed method.

1. Introduction

There is no doubt that the legged locomotion can improve the adaptability of the animals to the various kinds of terrains. To find out how to generate the coordinate and flexible gait, the leg mechanism and neural control system of the legged animals were systematic analyzed [1–4].

The legged robots that aim at walking in the complex environment include biped robots [5], quadruped robots [6], and hexapod robots [7]. On the one hand, the mechanism of the robot leg is becoming much more complicated so as to realize proper gaits in different terrain [8]. On the other hand, the robot's control system tends to get some inspiration from the animal's locomotion neural system. It is found that the rhythmic locomotion of many animals is controlled by a series of the central pattern generator (GPG) and then the CPG is applied in the bioinspired robot control [9]. Many CPG models were used in hexapod robots to achieve desired locomotion. Rostro proposed a CPG model which is built of spiking neurons and can produce rhythmic signals for hexapod robots to realize walking, jogging, and running, but the rhythmic signals of different locomotion are produced separately [10]. Zhong proposed a CPG model for a novel hexapod robot whose legs can radially be free distributed around the robot body, the relevant parameters which decide the locomotion of robot are tuned according to the processed feedback information [11]. Arena provides a multitemplates approach to cellular nonlinear networks (MTA-CNN), and the templates of GPG can be reorganized by changing the synaptic connections, but this adjustment is rather complex and cannot be done online [12]. Ren proposed a CPG model with multiple chaotic central pattern generators with learning for the hexapod robot called AMOSII, and the model can change the connections of the generators to plan the locomotion for the malfunction compensation, but the connections must decide before locomotion [13]. The proposed CPG models above can produce kinds of gaits for hexapod robots, but how to achieve the transition between the different gaits is rarely considered by the previous works. For instance, Arena uses different templates to define gait, but every gait pattern is related to a predefined template, so the hexapod robot can only perform some predefined gait pattern.

In this paper, inspired by the analysis of the ant locomotion observed by the high-speed camera, an ant-like gait transition model for the hexapod robot is proposed. The bionic gait transition model which can realize a smooth transition among regular gait consists of reconfigurable CPGs, the central neural system (CNS), and neural network (NN). CNS is responsible for acquiring environment information and determining the basic locomotion parameters such as velocity and direction. The NN is a three-layer perceptron

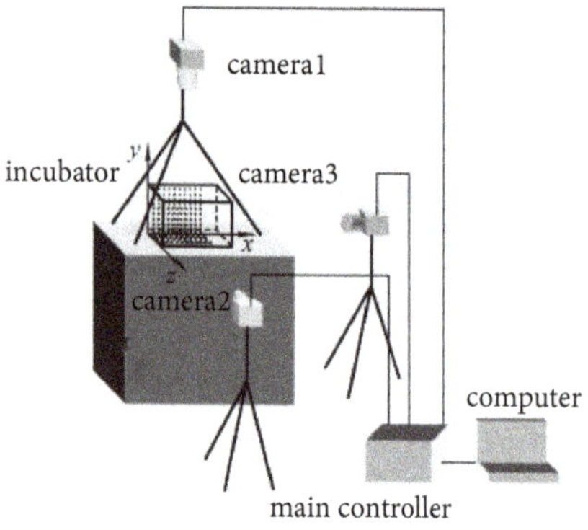

Figure 1: Schematic of the experimental system.

neural network which can adjust the gait transition factors f_p (probing) and f_s (sliding) real-time based on the velocity (V) and surface roughness of ground (S) acquired from CNS. The CPGs can reconfigure themselves according to the gait transition factors to realize a smooth gait transition among sliding gait, probing gait, and regular gait. The three parts mentioned above contribute to the desired gait transition that adapts to the variable environment. To verify the proposed model, numerical simulation and corresponding physical experiment were implemented in this paper.

2. Observation and Analysis of Ant Gait

2.1. Materials and Methods. The observation experiments were performed with several ants which were similar in size (11.3mm~11.8 mm). The ant was put inside of a transparent incubator when observed. A ruler that hanged in the middle of transparent was used as the passage in order to make the ant walk straightly. To stimulate the ant to walk, some nut was placed in the desired direction. The ants were filmed from three directions (above, front, and left) with MotinXtra HG-TH 100K digital high-speed camera (Redlake corporation, USA) which can get images as 1000 frame/s at the resolution of 752×564 (see Figure 1). To ensure the images' quality, the image acquisition frequency was set as 500 frame/s. The images were recorded as AVI format at 24fps and then the converted films were analyzed frame by frame using the software Image-Pro Plus6.0 (Media Cybernetics, USA).

The gait was identified by distinguishing the posterior extreme position (PEP, lift-off) and the anterior extreme position (AEP, touch-down). The software Image-Pro Plus6.0 was used to capture the foot locomotion trajectory of ant from the movies (Figure 2), and then the PEP and AEP could be found out easily. Frame-by-frame analysis accounts for an error of ±1frame (i.e., 2ms) when determining PEP and AEP. For convenience, continuous 250 frames were regarded as a section. Each section lasted for 0.5s and contained three to six steps according to the walking speed. The walking speed of each step can also be calculated.

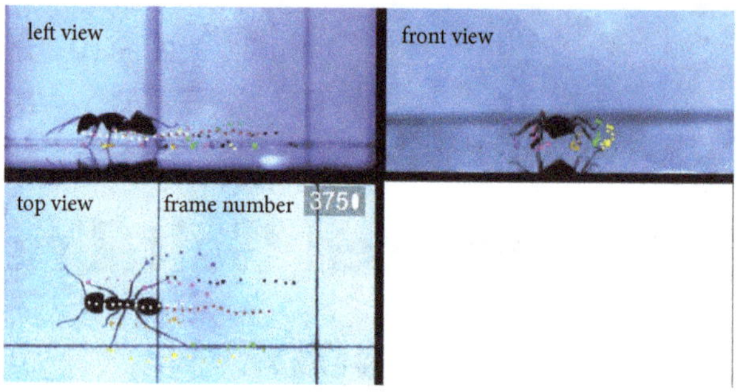

Figure 2: The foot trajectory of locomotion.

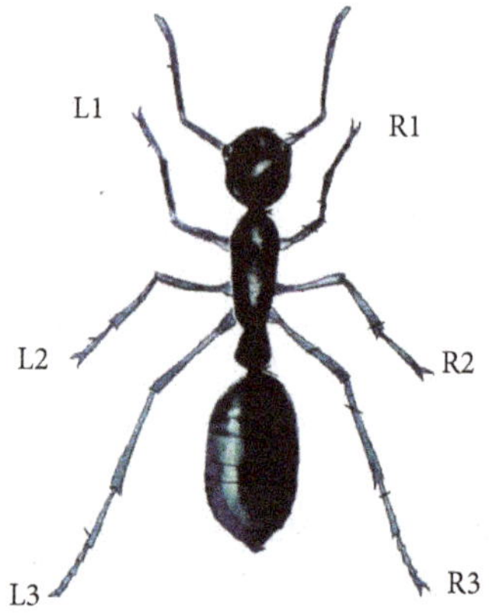

Figure 3: The serial number of ant's leg.

Foothold pattern diagram is a useful tool to depict the gait in which the white bars represent the swing phase of a leg and the black bars represent the stance phase. If the leg's number was defined as shown in Figure 3, then the foothold pattern diagram of regular tripod gait is shown in Figure 4.

2.2. Sliding Gait. The ruler remained horizontally when the four ants walked on it separately. A total of 20 effective sections were analyzed. Through observation, the exact regular gait as shown in Figure 4 rarely emerged. However, most steps had a trend to behave like the idealized gaits. Hence, a deviation from ideal phase relationship during the swing by ±0.15 was tolerated. The section's gait is assigned by the gait that appeared most in the section. According to the modifications above, there were 13 sections (65%) that performed the regular gait and 7 sections (35%) that performed the irregular gait. The distance between the start point and the end point was 50mm. Figure 5 shows the velocity where the sections began to be recorded.

It can be seen clearly in Figure 5 that the irregular gait mostly emerged at the beginning of the locomotion with a lower speed. With the increment of the speed, the irregular gait transited to the regular gait gradually and the speed of the regular gait tends to be stable after the transition. It can be concluded that this regular gait only appeared in low speed.

The regular gait of insect was considered to be composed of the swing phase and stance phase. However, what surprised us was that in irregular gait the hind legs of ant represented a

FIGURE 4: The foothold pattern diagram of regular tripod gait.

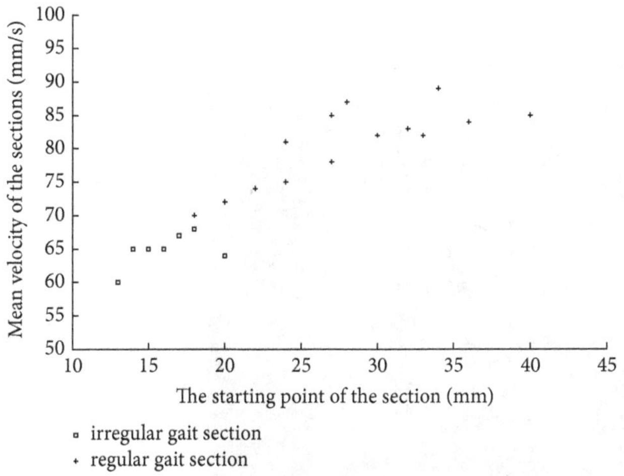

FIGURE 5: The mean velocity and the starting position of different gait sections.

sliding mode named as sliding phase in this paper. And this irregular gait was called sliding gait. For example, when the ant was in sliding gait as shown in Figure 6, the leg R3 slid on the surface of the ruler while the other five legs moved. And we could also find out that the abdomen of the ant kept very close to the ruler when R3 began sliding and then lifted up when R3 ended sliding.

To have better understanding of the sliding gait, the left view of typical sliding gait was analyzed using Image-Pro Plus 6.0. In order to highlight the sliding phase of the hind leg, the section that we selected was cut to 123 frames, i.e., 246ms. The trajectories of the foot end of L1, R3, and L3 and the tips of the abdomen were depicted in Figure 7. From the intuitive trajectories, we can see clearly that the hind legs L3 and R3 underwent one and two siding phases, respectively. At the same time, the other four legs walked normally, e.g., L1 in Figure 7, and the abdomen was close to the ruler as a result of lacking support from the hind leg. The abdomen lifted up gradually when the irregular gait almost came to an end.

The precise step phrase can be distinguished by analyzing kinematic parameters such as joint angle and displacement of leg [11]. To acquire an accurate foothold pattern diagram of this section, the vertical and horizontal coordinate of the trajectories in the image coordinate system was exported to MATLAB. After data processing, the foothold pattern diagram of an irregular gait section was shown in Figure 8. Indeed, there was no significant sign that six legs walked in synchrony in sliding gait. However, if only the front and middle legs were taken into consideration, coordination among the four legs could be found easily. To some extent, the front and middle legs performed trot gait like quadruped animal. The duty factors (i.e., the fraction of time a leg spends on the ground relative to the step period) of hind legs (L3 54.9%, R3 51.2%) were far less than that of front and middle legs (L1 73.5%, R1 63.4%, L2 67.5%, and R2 81.3%), and this was convincing evidence that the hind legs contributed less to the locomotion compared with the other legs. The hind legs were mainly used to support the abdomen of the ant, so the abdomen lifted up when the irregular gait ended (see Figure 8).

However, when the ruler was replaced by a rough board, the sliding gait disappeared. I supposed that the reason why ant performed the sliding gait is that the sliding gait could save the energy of locomotion since only two-thirds of the legs worked normally. But when higher speed was needed, the hind legs must join in to offer enough power. In addition, the sliding gait will not appear in the rough surface in case of hurt the hind legs while sliding.

2.3. Probing Gait. Apart from sliding gait, another kind of irregular gait was observed when ant crossed the obstacle (see Figure 9), i.e., the number of steps performed by front legs (NR1=13, NL1=12) in this process was drastically higher than that of middle legs or hind legs (NR2=NL3=7, NR3=NL2=8). In addition, the step length of front legs tends to be smaller. Surprisingly, the middle and hind legs coupled with each other just like what the front and middle legs did during the sliding phase. The same phenomenon has long been discovered in stick insect, another arthropod, which is one of the best subjects in studying insect locomotion. These studies suggest that the front legs not only function as motion organs but also serve an additional function, for example, to probe the environment. So this gait is named probing gait in this paper.

2.4. Discussion. All legs coordinate with a certain frequency performing regular gait. Thus, CPG neurons of all legs couple with each other and send the control signals with same frequency and amplitude to the motor neurons. However, this coupling relationship is broken partially in some special cases such as sliding gait and probing gait. The hind legs or front legs decouple with the main locomotion cycle in sliding phase and probing gait. The hexapod can remain stable locomotion if the middle legs are in the normal state even if the front and hind legs are amputated or act as sensors. The CPG neurons of other legs can couple with the main CPG in regular gait and decouple with the main CPG in irregular gait through the regulation of central neural system (CNS).

Some hypothesis can be drawn from the discussion above:

(1) The locomotion rhythm of some arthropod mainly depends on the main CPG that connect with middle legs. And without the participation of middle legs, none of the stable gaits, no matter regular or irregular, can be realized.

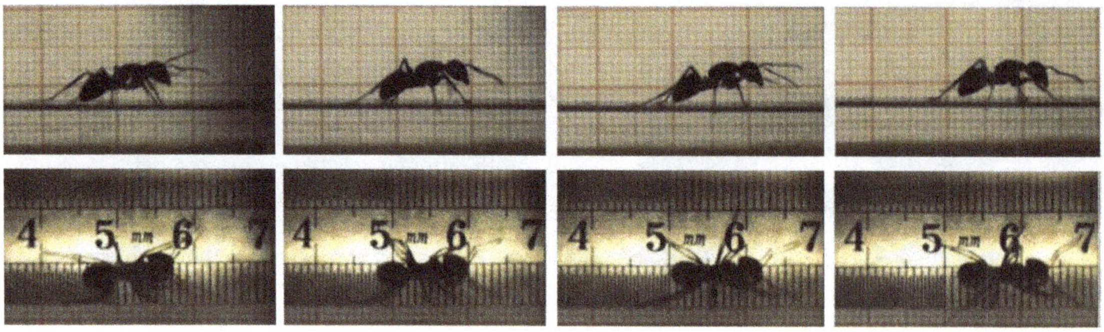

FIGURE 6: The left view and top view of sliding phase of leg R3.

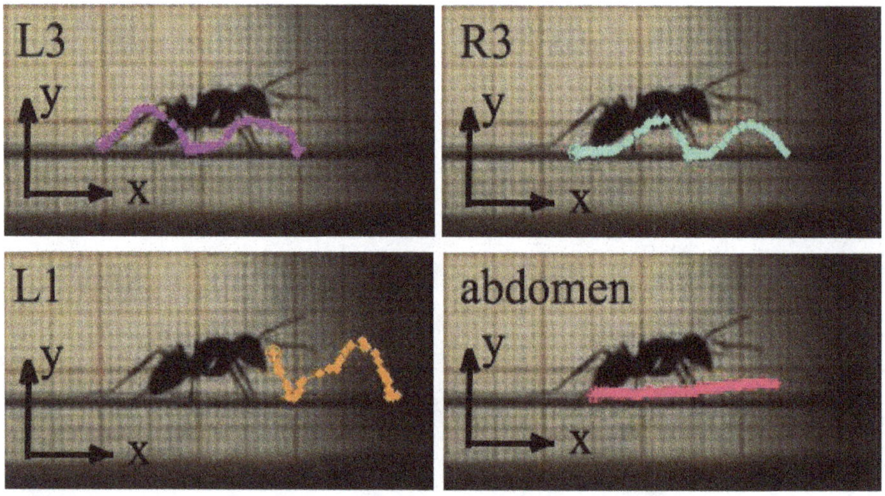

FIGURE 7: The trajectories of the foot end of L1, R3, and L3 and the tip of the abdomen.

(2) The CPG neurons of the front leg and hind leg are able to couple or decouple with main CPG through the regulation of CNS. The regulation is rather smooth and will not affect the stability of locomotion.

(3) The regulation implemented by CNS is a real-time feedback process; therefore, the locomotion state and environment such as velocity and ground roughness will affect the regulation.

(4) The real-time regulation contributes to a proper gait that adapts to the variable environment.

3. Bioinspired Gait Transition Model

Inspired by the sliding gait and probing gait of ant, a bionic gait transition model for a hexapod robot (see Figure 10) is put forward in this section. The bionic gait transition model which depends on the hypothesis proposed in section 3 can realize a smooth transition among regular gait, sliding gait, and probing gait.

3.1. Model Configuration. As shown in Figure 11, the gait transition model consists of CPGs, central neural system (CNS), and neural network (NN). As the core of the model, CNS is a charge of information acquisition and determines the basic locomotion parameters such as velocity and direction. The three parts mentioned above contribute to the desired gait transition that adapts to the variable environment. The NN is a three-layer perceptron neural network which can adjust the gait transition factors fp (probing) and fs (sliding) real-time based on the velocity (V) and surface roughness of ground (S) acquired from CNS. The CPG neurons of L2 and R2 couple with each other to generate the main CPG signal of locomotion. The frequency and amplitude of the CPG signal are regulated by CNS. The CPG neurons of other legs cannot only couple with L2 or R2 to obtain coordinate CPG signal for regular gait, but also decouple with L2 or R2 to realize irregular gait. Whether they couple with main CPG or not is decided by the discrimination results of NN.

3.2. Information Acquisition of CNS. The information used in our model includes the locomotion velocity and surface roughness of ground. Since middle legs participate in all gaits discussed in this paper, the velocity is calculated according to the step length and locomotion frequency of middle legs; for convenience, the velocity is normalized in

$$V = \frac{l \times f}{v_{\max}} \tag{1}$$

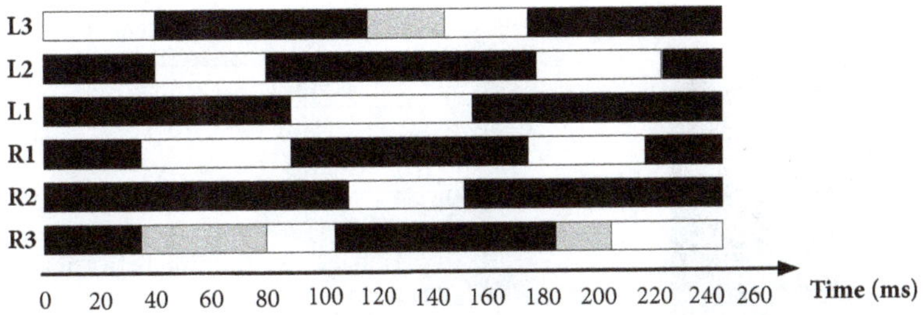

FIGURE 8: Foothold pattern diagram of the irregular gait.

FIGURE 9: The snapshot of obstacle crossing.

FIGURE 10: Prototype of the hexapod robot.

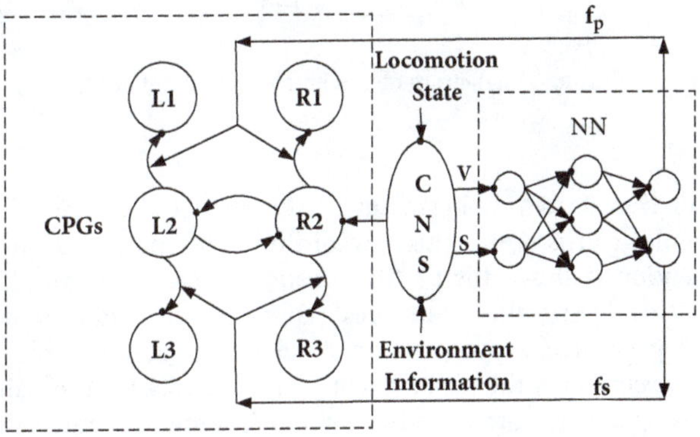

FIGURE 11: Schematic diagram of the model configuration.

where l is the step length and f is the locomotion frequency of middle legs. v_{max} represents the max velocity that the robot can obtain. V is the normalized velocity with a value between 0 and 1.

The surface roughness of ground is represented by the difference of contact force between the two front legs since the front legs are the first to contact with the unknown environment. After normalization, the surface roughness S is expressed in

$$S = \frac{|F_l - F_r|}{\max(F_l, F_r)} \quad (2)$$

where F_l and F_r are the contact force of legs L1 and R1, respectively, and $\max(F_l, F_r)$ is the max value between F_l and F_r. S represents the normalized surface roughness of ground with a value between 0 and 1.

The normalization of locomotion velocity and surface roughness of ground is the precondition of using a neural network to distinguish the desired gaits. Then the next step is to design a proper neural network to realize the classification function.

3.3. Neural Network (NN) Design. What can be ensured in our model is that the gaits are determined by V and S; however, this relationship is nonlinear. Therefore, the neural network is introduced in our model to achieve the nonlinear classification.

It can be concluded from the experiment above that the sliding gait appears in smooth ground with a low speed. And when the ground becomes uneven, the gait

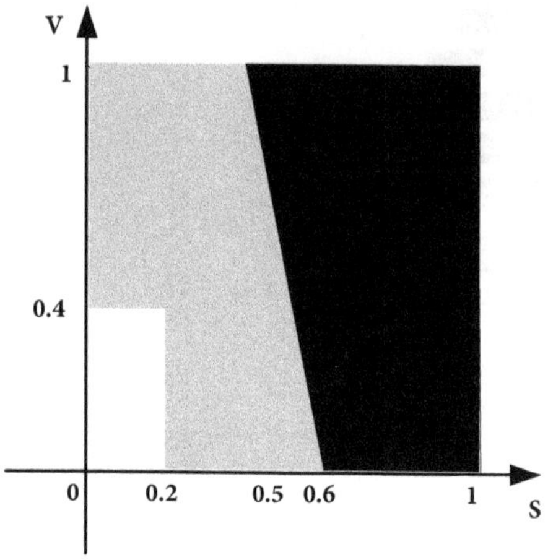

FIGURE 12: The figure of gait classification.

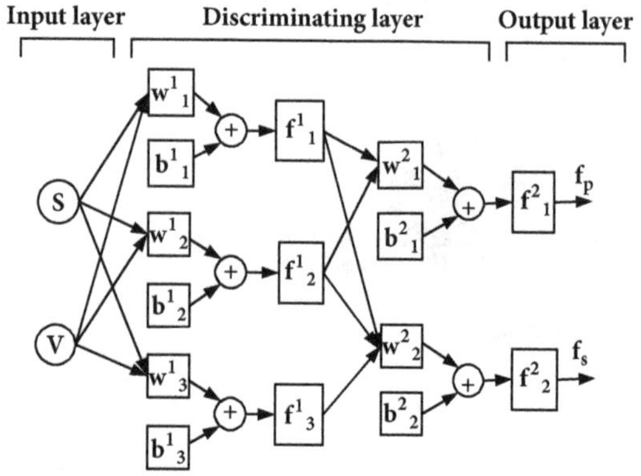

FIGURE 13: Diagram of perceptron neural network.

will transit to probing. And in the rest of the time, the regular gait is represented. According to the partition rules mentioned above, the partition result is shown in Figure 12. In this figure, the white area represents the sliding gait, the grey area represents the regular gait, and the black area represents the probing gait. Thus, the robot can obtain a proper gait corresponding to the locomotion state and environment.

To realize the gait classification shown in Figure 13, a three-layer perceptron neural network is inducted. The perceptron neural network includes the input layer, the discriminating layer, and output layer. Normalized velocity V and surface roughness S are introduced to the neural network through input layer and then discriminated by the five neurons of discriminating layer. The classification result is obtained through the output layer in the form of gait transition factors f_p and f_s.

The five neurons in the discriminating layer are divided into two rows. The neurons in the first row are used to partition the boundaries, and the neurons in second row implement logic operations of the partition result. w^i_j, b^i_j, and f^i_j represent the weight, bias, and transmission function of the jth neuron which locates in ith row. According to the boundary functions shown in (3), the parameters used in the

TABLE 1: Parameters of perceptron neural network.

	i=1,j=1	i=1,j=2	i=1,j=3	i=2,j=1	i=2,j=2
w^i_j	[1, 0]	[0, 1]	[10, 1]	[-1, -1]	[1, 1, 1]
b^i_j	-0.2	-0.4	-6	-1	-2

TABLE 2: Relationship between gait and gait transition factors.

	sliding	regular	probing
f_S	1	0	0
f_P	0	0	1

perceptron neural network can be determined in Table 1. And the transmission functions are expressed in (4)-(5).

$$S - 0.2 = 0$$
$$V - 0.4 = 0 \quad (3)$$
$$V + 10S - 6 = 0$$

$$f^1_j = \begin{cases} -1, & n < 0 \\ 1, & n \geq 0 \end{cases} \quad (4)$$

$$f^2_j = \begin{cases} 0, & n < 0 \\ 1, & n \geq 0 \end{cases} \quad (5)$$

After the classification of the perceptron neural network, whatever the state of the robot is, there will be a gait to correspond to. The gait is determined based on the gait transition factors f_p and f_s; see Table 2. When transmitted to CPGs, the gait transition will affect the CPGs to generate the proper gait.

3.4. Reconfigurable CPG Model. Though it has been widely accepted by the biologists that animals' rhythmic locomotion is generated by CPGs in the lower central nervous system, different researchers have different ideas on how the CPGs works. Most of the theories about CPGs argue that CPGs consist of doubles of CPG neurons, the CPG neurons which may array in the chain [14] or in reticulation [15] are exactly the same, and they couple with each other to generate a cycle signal. Unlike most of the CPG model that has been proposed, the CPG neurons in our model are not exactly the same. The CPG neurons of L2 and R2 act as the main CPG to generate main CPG signal, and the other CPG neurons can couple or decouple with main CPG according to the gait transition factors. In other words, the CPGs model is reconfigurable. To have a good knowledge of the numerical expressions of our model, the CPGs are numbered as shown in Figure 14.

The model proposed by Auke et al. [16] indicated the relationship between intrinsic amplitude, intrinsic frequency, and external stimulation; therefore variable and stable input can be obtained by adjusting the external stimulation.

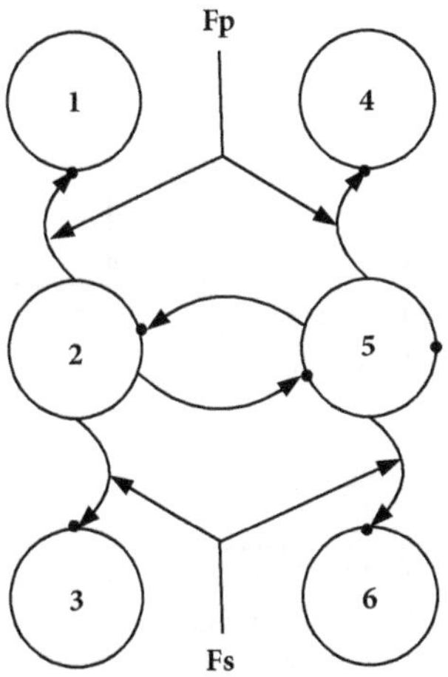

Figure 14: The serial number of CPG neurons.

Adapting from this model, our CPG model is shown as follows.

The CPG neurons of middle legs are implemented with the following differential equations:

$$\begin{aligned}
\dot{\theta}_2 &= 2\pi v_2 + \omega_{25} \sin(\theta_5 - \theta_2 - \phi_{25}) \\
\ddot{r}_2 &= a_2 \left(\frac{a_2}{4} (R_2 - r_2) - \dot{r}_2 \right) \\
x_2 &= r_2 (1 + \cos(\theta_2)) \\
\dot{\theta}_5 &= 2\pi v_5 + \omega_{52} \sin(\theta_2 - \theta_5 - \phi_{52}) \\
\ddot{r}_5 &= a_5 \left(\frac{a_5}{4} (R_5 - r_5) - \dot{r}_5 \right) \\
x_5 &= r_5 (1 + \cos(\theta_5))
\end{aligned} \tag{6}$$

The CPG neurons of the front legs are expressed as follows:

$$\begin{aligned}
\dot{\theta}_1 &= 2\pi \left(1 + f_p\right) v_1 + \omega_{12} \left(1 - f_p\right) \sin(\theta_2 - \theta_1 - \phi_{12}) \\
\ddot{r}_1 &= a_1 \left(\frac{a_1}{4} \left(\frac{R_1}{\left(1 + f_p\right)} - r_1 \right) - \dot{r}_1 \right) \\
x_1 &= r_1 (1 + \cos(\theta_1)) \\
\dot{\theta}_4 &= 2\pi \left(1 + f_p\right) v_4 + \omega_{45} \left(1 - f_p\right) \sin(\theta_5 - \theta_4 - \phi_{45}) \\
\ddot{r}_4 &= a_4 \left(\frac{a_4}{4} \left(\frac{R_4}{\left(1 + f_p\right)} - r_4 \right) - \dot{r}_4 \right) \\
x_4 &= r_4 (1 + \cos(\theta_4))
\end{aligned} \tag{7}$$

Finally, the CPG neurons of hind legs are shown in

$$\begin{aligned}
\dot{\theta}_3 &= 2\pi v_3 + \omega_{32} (1 - f_s) \sin(\theta_2 - \theta_3 - \phi_{23}) \\
\ddot{r}_3 &= a_3 \left(\frac{a_3}{4} (R_3 - r_3) - \dot{r}_3 \right) \\
x_3 &= r_3 (1 - f_s) (1 + \cos(\theta_3)) \\
\dot{\theta}_6 &= 2\pi v_6 + \omega_{65} (1 - f_s) \sin(\theta_5 - \theta_6 - \phi_{65}) \\
\ddot{r}_6 &= a_6 \left(\frac{a_6}{4} (R_6 - r_6) - \dot{r}_6 \right) \\
x_6 &= r_6 (1 - f_6) (1 + \cos(\theta_6))
\end{aligned} \tag{8}$$

where θ_i and r_i represent the phase and amplitude of the neurons and v_i and R_i are the intrinsic frequency and intrinsic amplitude of neurons. ai is a positive constant which determines the speed that r_i converges to R_i. Coupling weights ω_{ij} and phase biases Φ_{ij} affect the couplings between oscillators. And the phase lag is determined by phase bias. Periodic and positive signal X_i represents the output produced by the CPG neurons.

It can be seen clearly from the expressions of our model that the neurons of a middle legs couple with each other all the time, but whether the front or hind leg couple with the middle leg at the same side is decided by gait transition factor f_p and f_s.

4. Numerical Simulation and Experiment

4.1. Simulation Conditions Setting. To satisfy the simulation requirement mentioned above, the simulation will last for 30 s and the normalized velocity V and surface stiffness S used in the simulation are shown in Figure 15.

Then the parameters of the CPG model must be determined. From the expressions of CPG, we can see that the intrinsic frequency should be exactly the same in regular and sliding gait to generate the coordinate locomotion. Taking the electromechanical properties of the robot into consideration, the intrinsic frequency of all CPG neurons is set at 0.4 Hz in regular and sliding gait. In probing gait, the intrinsic frequency of front legs' CPG neurons increases to 0.8 Hz to acquire more environmental information.

Intrinsic amplitude hind legs' CPG neurons in sliding gait transits to zero, and except for the above case all the CPG neurons have the same intrinsic amplitude. Since the velocity is proportional to intrinsic frequency and intrinsic amplitude of middle legs' CPG neurons, the intrinsic frequency is constant, and then the intrinsic amplitude of middle legs' CPG neurons must change according to velocity, so the intrinsic amplitude of CPG neurons is expressed in

$$R_i = \begin{cases} 0.03t + 0.2 & 0 < t \leq 20 \\ 0.8 & 20 < t \end{cases} \tag{9}$$

When talking about the phase difference of the two coupled oscillators, it is useful to introduce the phase difference

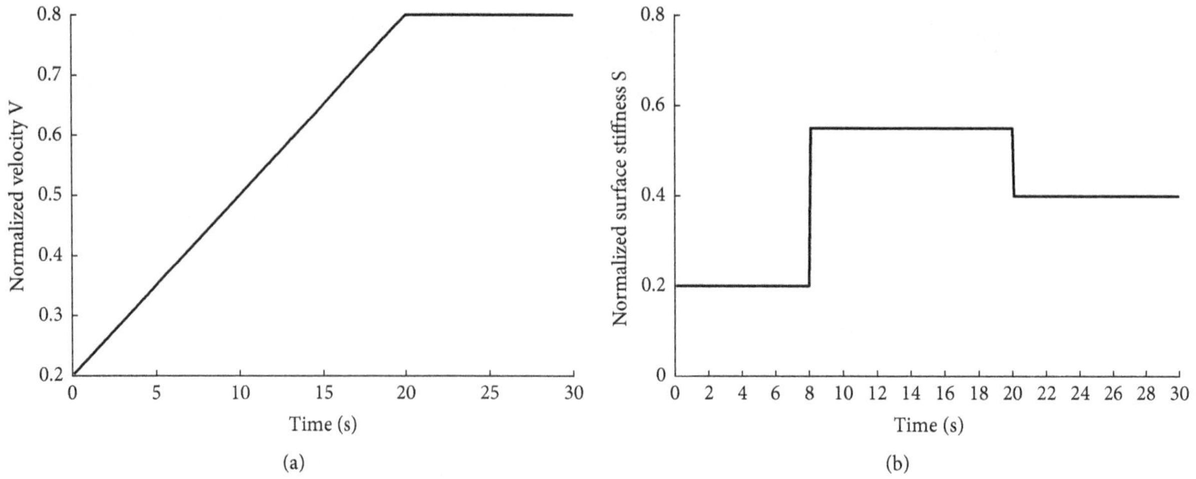

FIGURE 15: Normalized velocity surface stiffness used in the simulation.

$\varphi = \theta_i - \theta_j$. The time evolution of the phase difference is determined by

$$\dot{\varphi} = f(\varphi) = \dot{\theta}_i - \dot{\theta}_j = 2\pi(v_i - v_j) + 2r_i\omega_{ij}\sin(\varphi - \phi_{ij}) \quad (10)$$

When the oscillators synchronize, φ tends to be zero and $\varphi 0$ can be figured out by

$$\varphi_0 = \arcsin\left(\frac{\pi(v_i - v_j)}{R_i\omega_{ij}}\right) + \phi_{ij} \quad (11)$$

The function has no solution if $|\pi(v_i - v_j)/R_i\omega_{ij}| > 1$, which means that the oscillators will not oscillate. If $|\pi(v_i - v_j)/R_i\omega_{ij}| = 1$, the function has a single solution which is asymptotically stable and the oscillators will finally oscillate with that phase difference form any initial phase values. If $|\pi(v_i - v_j)/R_i\omega_{ij}| < 1$ the function has two solutions: one is stable and the other one is unstable, and we can judge the stability of the solutions by the value of

$$\frac{\partial f(\varphi_0)}{\partial \varphi} = R_i\omega_{ij}\cos(\varphi_0 - \phi_{ij}) \quad (12)$$

And if the value is positive, the solution is unstable, and if the value is negative, the solution is stable. The two oscillators will asymptotically synchronize to a phase difference corresponding to the sable solution. When the intrinsic frequency of both oscillators is the same, the phase difference is directly set by the phase bias Φ_{ij}. But in probing gait vi is not equal to vj, so wij between the front leg and middle leg should be large enough to ensure $|\pi(v_i - v_j)/R_i\omega_{ij}| < 1$. The selected values of ω_{ij} and Φ_{ij} are shown in Table 3. The last parameter ai is set as 10 to decrease the time that r_i converges to R_i.

4.2. Simulation Results. The simulation is performed using MATLAB software when the simulation condition has been set. The gait classification result of the neuron network is shown in Figure 16, and the final output of the gait transition model is expressed in Figure 17.

TABLE 3: Value of ω_{ij} and Φ_{ij}.

	i=1,j=2	i=2,j=3	i=2,j=5	i=4,j=5	i=5,j=2	i=6,j=5
ω_{ij}	30	10	20	30	20	10
Φ_{ij}	π	π	π	π	$-\pi$	π

4.3. Conversion of CPGs Signals to Control Signals. It can be seen from Figure 17 that the CPGs signals are just a group of rhythm signals and not be able to control the robot's motors directly. Therefore, some proper conversion must be implemented to obtain the control signals of the robot's motors.

See Figure 18; each leg of the hexapod robot has three joints driven by DC motors, namely J1, J2, and J3. And motors of the robot are controlled by angular displacement, so the CPGs signals must be converted to angle signals. Then we will take a leg R1 for example. The tree joints are all revolute joints. And J1, which links body and coxa, rotates in both swing phase and stance phase, while the other two joints rotate only in swing phase.

The conversion process expressed in the following:

$$Y_{i1} = K_1\int_0^t (x_i - 1)\,dt + c_1 \quad (13)$$

$$Y_{i2} = \begin{cases} K_2(x_i - 1) + c_2 & x_i \geq 1 \\ 0 & x_i < 1 \end{cases} \quad (14)$$

$$Y_{i3} = \begin{cases} K_3(x_i - 1) + c_3 & x_i \geq 1 \\ 0 & x_i < 1 \end{cases} \quad (15)$$

where Yij represents the angular control signal of Ji. xi represents the output produced by ith CPG neuron and K_i and c_i are the gain and the bias of joint Ji.

For the hexapod robot shown in Figure 12, the value of Ki and ci is represented in Table 4.

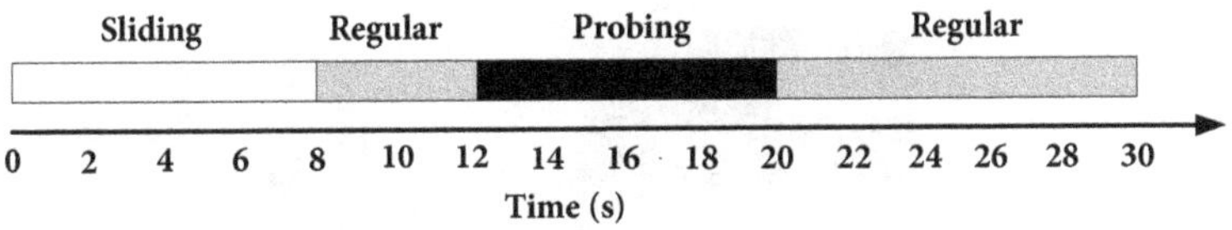

FIGURE 16: The gait classification result of the neuron network.

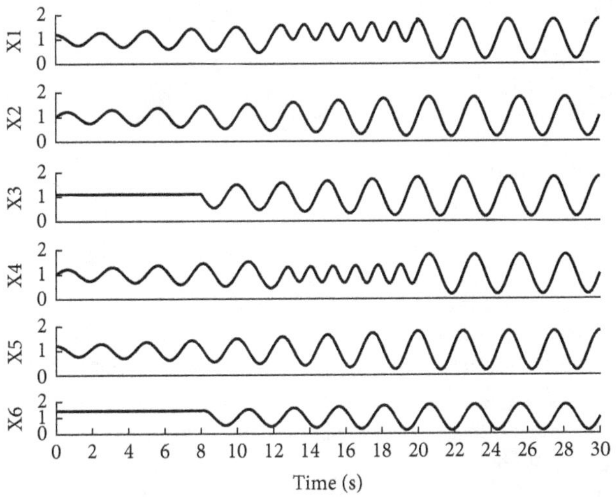

FIGURE 17: The output of CPG neurons.

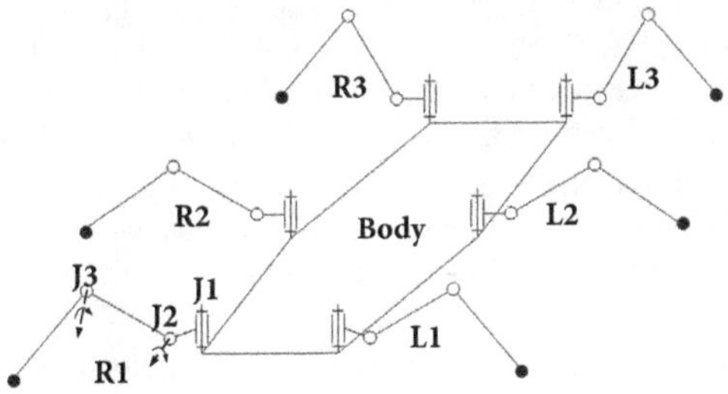

FIGURE 18: Structure diagram of a hexapod robot.

Then the CPG signal and corresponding angular control signals of leg R1 are shown in Figure 19. The angular control signals of other legs can be obtained using the same method. The control system of the hexapod robot sends the angular control signals to the eighteen DC motors, and variable gaits can be realized.

4.4. Experiment. To verify the model's practicality, an experiment is implemented on the hexapod robot. The biological gait transition model is embedded in the CPU based on PC/104-Plus compliant architecture; the DC motors and sensors communicate with CPU through CAN bus. The angular control signals sent by CPU are processed by the Elmo drive before arriving at DC motors. In the experiment the hexapod robot moved from the ground to a slope with the proposed model and the screen of experiment is shown

TABLE 4: the value of K_i and c_i.

	i=1	i=2	i=3
K_i	$\frac{\pi}{12}$	$\frac{\pi}{12}$	$\frac{\pi}{18}$
c_i	0	$\frac{\pi}{6}$	$\frac{\pi}{2}$

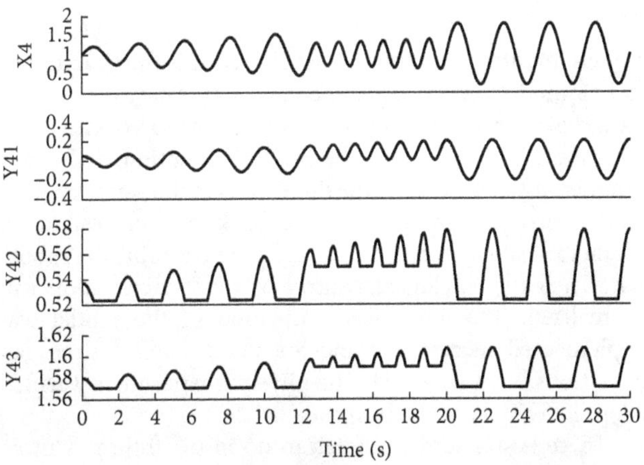

FIGURE 19: CPG and angular control signals of leg R1.

in Figure 20. The hexapod transited the gait automatically to realize the locomotion on the uneven ground, which verifies the model's practicability to some extent.

5. Conclusion

In this paper, inspired by the analysis of the ant locomotion observed by the high-speed camera, an ant-like gait transition model for the hexapod robot is proposed. The model which consists of central neural system (CNS), neural network (NN), and central pattern generators (CPGs) can adjust the gait automatically. The nonlinear gait classification is implemented by the three-layer article neural network which can realize arbitrary division of the objects. Obviously, this classification method can offer greater flexibility. Unlike of most CPG models, The CPGs in our model is reconfigurable; i.e., the couple relationship of the CPG neurons can adjust as required. Observations of some hexapod reveal that the CPGs structure in hexapod are not so immutable, and the CPG neurons have the ability to couple or decouple with certain neurons. Therefore, our CPG model satisfies the phenomenon well. And what is more, as a result of lacking definite meaning, the proper parameters of other CPG

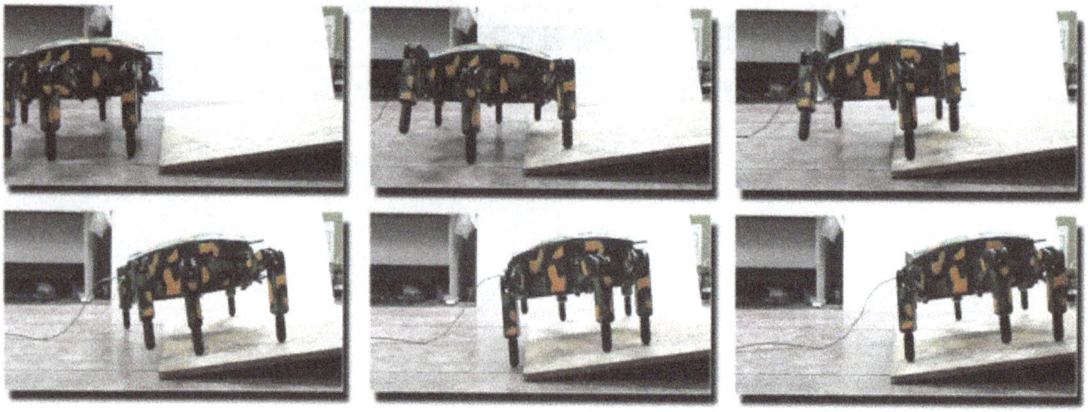

Figure 20: The screenshot of the experiment.

models are difficult to obtain, which reduces the practicality of their models. And the parameters in CPGs have the definite meaning and are more flexible to adjust compared with other CPG models. Fortunately, the problem mentioned above is conquered by our model. The three parts of the gait transition model can be regarded as a feedback control system, in which the external states are induced to determine the robot's gait. Then the intelligent control of robot locomotion can be realized. The numerical simulation of the model was implemented successfully, and the smooth and desired gait transitions were acquired. The physical experiment verified the practicability of the proposed model.

There is still something left to do in the future. A more detailed classification mechanism of gait is needed; the other factors which affect the gait transition of hexapod or other animals such as leg injuries are the key studies contents from now on. And how to extend our model to robots with four or eight legs is also what we are interested in.

Conflicts of Interest

The authors declare no conflicts of interest.

Authors' Contributions

Qing Chang analyzed the experimental data and designed the algorithm. Fanghua Mei designed and carried out the experiments and wrote the paper.

Acknowledgments

This study is supported by the National Natural Science Foundation of China (no. 61175108).

References

[1] A. Cully, J. Clune, D. Tarapore, and J.-B. Mouret, "Robots that can adapt like animals," *Nature*, vol. 521, no. 7553, pp. 503–507, 2015.

[2] Y. Zhu, T. Guo, Q. Liu, Q. Zhu, X. Zhao, and B. Jin, "Turning and radius deviation correction for a hexapod walking robot based on an ant-inspired sensory strategy," *Sensors*, vol. 17, no. 12, 2017.

[3] H. Shim, B.-H. Jun, and P.-M. Lee, "Mobility and agility analysis of a multi-legged subsea robot system," *Ocean Engineering*, vol. 61, pp. 88–96, 2013.

[4] F. Corucci, N. Cheney, S. Kriegman, J. Bongard, and C. Laschi, "Evolutionary Developmental Soft Robotics As a Framework to Study Intelligence and Adaptive Behavior in Animals and Plants," *Frontiers in Robotics and AI*, vol. 4, 2017.

[5] X. Chen, Z. Yu, W. Zhang, Y. Zheng, Q. Huang, and A. Ming, "Bio-inspired Control of Walking with Toe-off, Heel-strike and Disturbance Rejection for a Biped Robot," *IEEE Transactions on Industrial Electronics*, vol. 64, no. 10, pp. 7962–7971, 2017.

[6] N. T. Thinh, N. T. Tuyen, and D. T. Son, "Gait of Quadruped Robot and Interaction Based on Gesture Recognition," *Journal of Automation and Control Engineering*, vol. 3, no. 6, pp. 53–58, 2015.

[7] T. D. Barfoot, E. J. P. Earon, and G. M. T. D'Eleuterio, "Experiments in learning distributed control for a hexapod robot," *Robotics and Autonomous Systems*, vol. 54, no. 10, pp. 864–872, 2006.

[8] X. Tian, F. Gao, C. Qi, X. Chen, and D. Zhang, "External disturbance identification of a quadruped robot with parallel-serial leg structure," *International Journal of Mechanics and Materials in Design*, vol. 12, no. 1, pp. 109–120, 2016.

[9] F. Delcomyn, "Walking robots and the central and peripheral control of locomotion in insects," *Autonomous Robots*, vol. 7, no. 3, pp. 259–270, 1999.

[10] H. Rostro-Gonzalez, P. A. Cerna-Garcia, G. Trejo-Caballero et al., "A CPG system based on spiking neurons for hexapod robot locomotion," *Neurocomputing*, vol. 170, pp. 47–54, 2015.

[11] G. Zhong, L. Chen, Z. Jiao, J. Li, and H. Deng, "Locomotion Control and Gait Planning of a Novel Hexapod Robot Using Biomimetic Neurons," *IEEE Transactions on Control Systems Technology*, vol. 26, no. 2, pp. 624–636, 2018.

[12] P. Arena, L. Fortuna, and M. Frasca, "Multi-template approach to realize central pattern generators for artificial locomotion control," *International Journal of Circuit Theory and Applications*, vol. 30, no. 4, pp. 441–458, 2002.

[13] G. Ren, W. Chen, S. Dasgupta et al., "Multiple chaotic central pattern generators with learning for legged locomotion and malfunction compensation," *Information Sciences*, vol. 294, pp. 666–682, 2015.

[14] G. Wang, X. Chen, and S. Han, "Central pattern generator and feedforward neural network-based self-adaptive gait control for a crab-like robot locomoting on complex terrain under two reflex mechanisms," *International Journal of Advanced Robotic Systems*, vol. 14, no. 4, 2017.

[15] S. A. Haghpanah, F. Farahmand, and H. Zohoor, "Modular neuromuscular control of human locomotion by central pattern generator," *Journal of Biomechanics*, vol. 53, pp. 154–162, 2017.

[16] J. Auke, C. Alessandro, R. Dimitri et al., "From swimming to walking with a salamander robot driven by a spinal cord model," *Science*, vol. 315, no. 12, pp. 1416–1419, 2007.

9

Error Analysis and Adaptive-Robust Control of a 6-DoF Parallel Robot with Ball-Screw Drive Actuators

Navid Negahbani,[1] Hermes Giberti,[2] and Enrico Fiore[1]

[1]Dipartimento di Meccanica, Politecnico di Milano, Campus Bovisa Sud, Via La Masa 1, 20156 Milano, Italy
[2]Dipartimento di Ingegneria Industriale e dell'Informazione, Università degli Studi di Pavia, Via A. Ferrata 5, 27100 Pavia, Italy

Correspondence should be addressed to Hermes Giberti; hermes.giberti@polimi.it

Academic Editor: Andrew A. Goldenberg

Parallel kinematic machines (PKMs) are commonly used for tasks that require high precision and stiffness. In this sense, the rigidity of the drive system of the robot, which is composed of actuators and transmissions, plays a fundamental role. In this paper, ball-screw drive actuators are considered and a 6-degree of freedom (DoF) parallel robot with prismatic actuated joints is used as application case. A mathematical model of the ball-screw drive is proposed considering the most influencing sources of nonlinearity: sliding-dependent flexibility, backlash, and friction. Using this model, the most critical poses of the robot with respect to the kinematic mapping of the error from the joint- to the task-space are systematically investigated to obtain the workspace positional and rotational resolution, apart from control issues. Finally, a nonlinear adaptive-robust control algorithm for trajectory tracking, based on the minimization of the tracking error, is described and simulated.

1. Introduction

Different drive systems are used for moving PKMs. Electrical linear motors, ball-screw-driven units, and belt-driven units are the three most commonly used solutions to actuate prismatic joints. Linear motors directly provide the thrust force, while ball-screw- and belt-driven units convert a rotational motion into a linear one. Ball-screw-driven units have good stiffness and good precision for short and medium travel [1]. Belt-driven units can be used for high speed and long travel strokes [2, 3], but with a low accuracy because of the flexibility of the belt. Linear motors put together the advantageous features of ball-screw and belt: zero backlash, high stiffness, high velocities, and acceleration. Nevertheless, this is paid for in terms of higher costs, lower energy efficiency, smaller force-to-size ratio, and higher constructive complexity [4]. Moreover, linear motors are not available as ready-to-use units, especially when a large actuation force is required. In this case, custom-made solutions are necessary including an adequate liquid cooling system. Accordingly a ball-screw drive actuator often represents the best compromise solution.

Many researchers have modeled ball-screw-driven transmission by considering its structural flexibility. Flexible ball-screw models range from fairly simple ones consisting of lumped masses connected by springs, such as those in [5, 6], to more complicated ones with distributed parameter employing FEM, such as those in [7, 8]. Hybrid methods are also proposed, taking into account distributed stiffness and inertia of the screw while modeling other components as lumped masses connected by springs, as set out in [9, 10]. However, only lumped-masses equivalent models are used for control purposes, and, frequently, 2 DoFs are considered enough: a rotational DoF and a translational one [11–13]. In support of this approach, Frey et al. in [14] show that an accurate representation of the screw elastic behaviour can be obtained through a simplified lumped model. Complementary, other researchers such as Han and Lee in [15] model a rigid ball-screw-driven system by considering backlash and friction. In this work, flexibility, backlash, and friction are taken into account at the same time.

The control strategies for a PKM can be largely divided into two schemes: joint-space control and task-space control. The joint-space control scheme can be readily implemented as a collection of multiple, independent, single-input single-output (SISO) control systems using only information on each actuator displacement [16]. On the other hand, the task-space control implements direct control of the 6 DoFs by measuring or estimating them solving the forward kinematics. In addition, robot manipulator control strategies can be classified into model-based control and performance-based control [17]. Model-based control requires an accurate dynamic model of the robot while performance-based control adjusts the control parameters according to the tracking error, as has been done in this work with an estimated model.

Due to the closed-loop kinematic chains of PKMs, strong interactions between different actuators appear. This, combined with high speed motion, makes the dynamics of parallel manipulators highly nonlinear. In order to perform suitable compensation for these sources of nonlinearity, the parameters of the motion equations should be known exactly [17]. Since PKMs require generally a relatively small workspace compared to their overall size, even if this characteristic represents a drawback compared to serial manipulators, the configuration-dependent coefficient matrices of the dynamic equations may be approximated as constant ones without introducing large modeling errors. Based on these constant matrices, calculation of the approximated inverse dynamics becomes much simpler than solving the full inverse dynamics. However, to prevent deterioration of tracking performance that may cause this approximation with a controller based on the inverse dynamics [18], performance-based control methods can be used. The most important of these methods are (1) sliding mode control, (2) robust control, (3) adaptive control, and (4) adaptive-robust control.

In spite of the properties of robustness claimed, the real implementation of sliding mode control techniques presents a major drawback that is the so-called chattering effect with dangerous high-frequency vibrations of the controlled system. This phenomenon is due to the fact that, in real applications, it is not reasonable to assume that the control signal can switch at infinite frequency. Robust control means to design a controller such that some level of performance of the controlled system is guaranteed independently from changes in the plant dynamics. In the robust approach, the controller has a fixed structure that yields acceptable performance for a class of plants which include the one in question. Robust control is widely used for 6-DoF parallel robot, especially for Gough-Stewart platform [16, 18, 19]. A disadvantage of commonly used robust controllers is that they do not have the ability to learn. This inability implicates conservative design and the stability of the system is achieved at the cost of performance. Adaptive control measures performance indexes of the controlled system using the input, the state, the output, and the known disturbances [20]. From the comparison between the measured performance indexes and the required ones, the adaptation mechanism modifies the parameters of the adjustable controller and/or generates an auxiliary control input. According to [21], we can classify adaptive controllers in three schemes: adaptive control by a computed-torque approach, adaptive control by an inertia-related approach, and adaptive control based on passivity. Honegger et al. use adaptive feed-forward control method in [22, 23] because it is simpler to implement than other algorithms and it is robust to noise and leads to similar or better results [22]. Disadvantages of adaptive control are limited to nonlinear systems in which the uncertain parameters appear linearly, and adaptive controls often exhibit poor transients [24]. To simultaneously achieve the advantages of both adaptive and robust control, these methods can be used together, such as in [25].

In this paper, a systematic methodology to find the error of a 6-DoF robotic device with parallel kinematic and Hexaglide architecture [22] is presented. This robot was designed to equip the Politecnico di Milano wind tunnel with a motion simulator for Hardware-in-the-Loop (HIL) large-amplitude dynamic test. In particular, this works as an emulator to reproduce the hydrodynamic interaction between floating bodies and sea water, in the case of floating offshore wind turbines [26] and sailing boat scale models. In previous works [27], the drive system is mechanically sized without considering flexibility, backlash, and friction.

The experiments that will be conducted in the civil environmental chamber of the wind tunnel require turbine and boats scale models to be aeroelastic. In order to obtain consistent results, these models are designed to have specific modal shapes, by the way the coupling with the robot and the transmission units might affect the desired behaviour. For these reasons, the manipulator has to be as rigid as possible and in addition all the worsening effects coming from the transmission units (backlashes, flexibility, and friction) should be properly modeled and compensated by means of a suitable control algorithm; otherwise the experimental results would be invalidated.

A mathematical model of the ball-screw drive is proposed considering the most influential sources of nonlinearity: sliding-dependent flexibility, backlash, and friction. Using this model, the most critical poses of the robot with respect to the kinematic mapping of the error from the joint to the task-space are systematically investigated to obtain the workspace positional and rotational resolution, apart from control issues. Finally, a nonlinear adaptive-robust control algorithm for trajectory tracking, based on the minimization of the tracking error, is described and simulated. This device is developed for the experimental simulation of the dynamic working conditions of the above-mentioned hydroaeroelastic structures in our wind tunnel (specially for wind turbines).

This paper is organized as follows. Section 2 introduces the inverse kinematics and dynamics of the Hexaglide. Section 3 sets out the equations of motion of a ball-screw drive actuator. Section 4 describes a systematic error evaluation methodology. Section 5 presents adaptive-robust control method for trajectory tracking, whose performances are simulated and compared with other methods.

2. Hexaglide Robot

Among the PKMs that can provide 6 DoFs, attention has been focused on Hexaglide, which is also reported in the literature as 6-PUS kinematic architecture with parallel linear guideways. Its linkage consists of six closed-loop kinematic chains which connect the fixed base to the mobile platform with

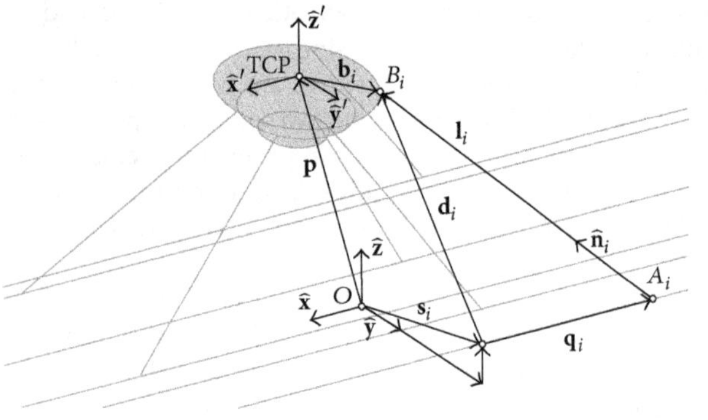

FIGURE 1: ith closed-loop kinematic chain of the Hexaglide.

the same sequence of joints: actuated prism (P), universal (U), and spherical (S). The links have a fixed length, while the actuation takes place through linear guideways, which are not necessarily on the same plane but lie on parallel planes.

2.1. Inverse Kinematics. To perform the inverse kinematics analysis of the Hexaglide means to find each slider position q_i from TCP position and mobile platform orientation; $\Theta = \{\alpha, \beta, \gamma\}$ [28], where α is the rotation around x-axis, β is the rotation around y-axis, and γ is the rotation around z-axis. With reference to the quantities shown in Figure 1, the inverse kinematics of the Hexaglide is solved as described in [26]. In particular, it is possible to write

$$\mathbf{l}_i = \mathbf{d}_i + q_i\hat{\mathbf{u}}_i \quad \text{with } \mathbf{d}_i = \mathbf{p} + \mathbf{R}\mathbf{b}_i' - \mathbf{s}_i, \tag{1}$$

where $\hat{\mathbf{u}}_i$ is a unitary vector aligned with the ith guide axis, $\mathbf{s}_i$ is the position of the same guide with respect to the fixed frame, $\mathbf{l}_i$ is the vector aligned with the ith link, $\mathbf{p}$ is the vector containing the position and orientation of the mobile platform, $\mathbf{R}$ is the rotation matrix, and $\mathbf{b}_i'$ expresses the position of the ith platform joint with respect to the TCP in the relative frame, while q_i represents the position of the ith slider along the corresponding guide axis. After some simple mathematical passages, the following expression is found:

$$q_i = d_{i,x} \pm \sqrt{l_i^2 - d_{i,y}^2 - d_{i,z}^2}. \tag{2}$$

The velocity and acceleration of the each slider can be derived from (2) as

$$\dot{q}_i = \dot{d}_{i,x} \mp \frac{\dot{d}_{i,y}d_{i,y} + \dot{d}_{i,z}d_{i,z}}{\sqrt{l_i^2 - d_{i,y}^2 - d_{i,z}^2}},$$
$$\ddot{q}_i = \ddot{d}_{i,x} \mp \frac{\left(\ddot{d}_{i,y}d_{i,y} + \dot{d}_{i,y}^2 + \ddot{d}_{i,z}d_{i,z} + \dot{d}_{i,z}^2\right)\Delta + \left(\dot{d}_{i,y}d_{i,y} + \dot{d}_{i,z}d_{i,z}\right)^2}{\Delta\sqrt{\Delta}}, \tag{3}$$

where $\Delta = l_i^2 - d_{i,y}^2 - d_{i,z}^2$, $\dot{\mathbf{d}}_i = \left[\dot{d}_{i,x}\ \dot{d}_{i,y}\ \dot{d}_{i,z}\right]^T = \mathbf{v}_C + \boldsymbol{\omega}\mathbf{b}_i$, and $\ddot{\mathbf{d}}_i = \left[\ddot{d}_{i,x}\ \ddot{d}_{i,y}\ \ddot{d}_{i,z}\right]^T = \mathbf{a}_C + \boldsymbol{\Xi}\mathbf{b}_i + \boldsymbol{\omega}(\boldsymbol{\omega}\mathbf{b}_i)$, whereas $\mathbf{v}_C$ and $\mathbf{a}_C$ are TCP velocity and acceleration, while $\boldsymbol{\omega}$ and $\boldsymbol{\Xi}$ are absolute angular velocity and acceleration of the platform. It must be noted that the components of the angular velocity and acceleration of the platform do not coincide with the time derivative of the angular coordinates:

$$\{\omega_x, \omega_y, \omega_z\} \neq \{\dot{\alpha}, \dot{\beta}, \dot{\gamma}\},$$
$$\{\alpha_x, \alpha_y, \alpha_z\} \neq \{\ddot{\alpha}, \ddot{\beta}, \ddot{\gamma}\}. \tag{4}$$

$\boldsymbol{\omega}$ and $\boldsymbol{\Xi}$ are defined as skew-symmetric matrices:

$$\boldsymbol{\omega} = \begin{bmatrix} 0 & -\omega_z & \omega_y \\ \omega_z & 0 & -\omega_x \\ -\omega_y & \omega_x & 0 \end{bmatrix},$$
$$\boldsymbol{\Xi} = \begin{bmatrix} 0 & -\alpha_z & \alpha_y \\ \alpha_z & 0 & -\alpha_x \\ -\alpha_y & \alpha_x & 0 \end{bmatrix}. \tag{5}$$

2.2. Inverse Dynamics. Inverse dynamics is the calculation of forces and torques on the robot actuated joints in order to produce the required motion of the mobile platform. A multibody model is implemented in Simulink using the SimMechanics library. Inverse dynamics is solved taking into consideration two payloads: a scale model of a sailing boat and a scale model of an offshore wind turbine. In order to perform the simulation, the required slider motions are calculated using (2) with first required platform motions as input parameters. Subsequently, required slider motions are applied to the SimMechanics robot model and the forces at the robot joints are computed. Figure 2 shows the block scheme of the inverse dynamics model in SimMechanics.

3. Dynamic Model of Ball-Screw Drive

Typical ball-screw drive consists of a motor and reducer, coupling, ball-screw, table, and end bearings. In this section, the dynamic model of ball-screw drive is studied taking into account flexibility (in gearbox, coupling, screw-ball, and bearing), backlash (in gearbox and between screw and nut), and friction. Figure 3 illustrates the lumped-mass-spring model of ball-screw drive.

The translational movement of the nut was modeled by taking into account two elements: the ball-screw transmission ratio R and the axial elastic deformation of the screw. In particular, the first contribution is easily described by the relation $R\theta$ where $R = h_p/2\pi$ is the ball-screw transmission ratio and θ is the rotational movement of the screw while the second one is represented by x_{bs} that is the translation movement due to axial elastic deformation of screw.

In addition, ball-screw has torsional flexibility (K_θ is ball-screw torsional stiffness) and axial flexibility (K_{eq} is ball-screw axial stiffness).

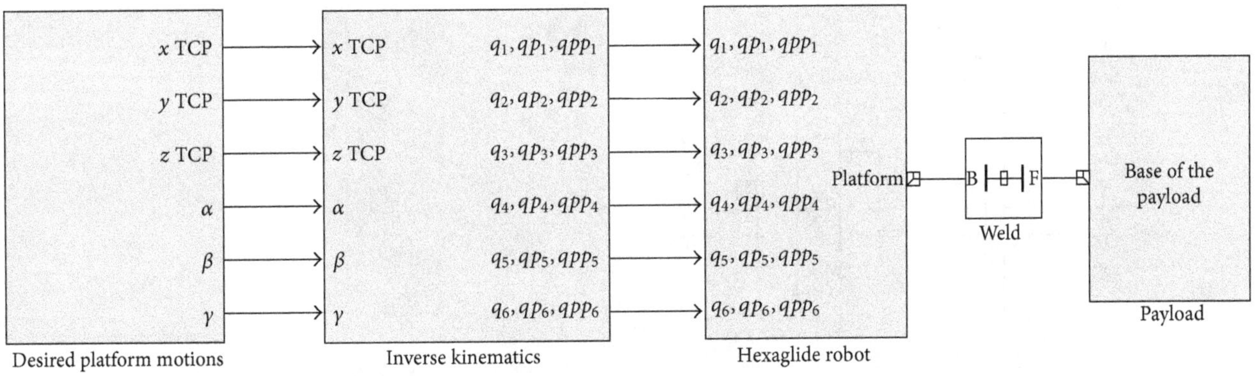

FIGURE 2: SimMechanics robot model for inverse dynamic solution.

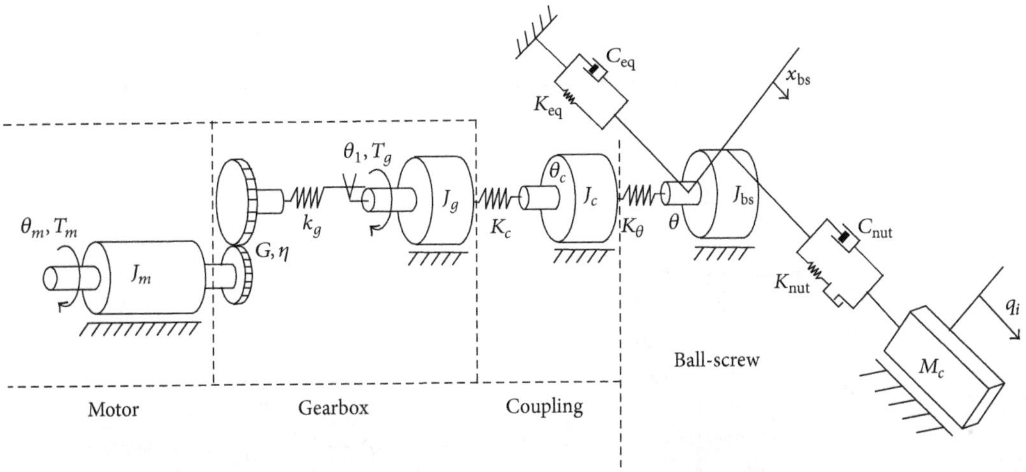

FIGURE 3: Lumped mass-spring model of ball-screw drive.

According to Figure 3, the kinetic energy (E_k), potential energy (V), dissipative function (D), and virtual work of the external forces (δW_{ext}) can be, respectively, expressed as

$$E_k = \frac{1}{2}\left[J_m\dot{\theta}_m^2 + J_g\dot{\theta}_1^2 + J_c\dot{\theta}_c^2 + J_{\text{bs}}\dot{\theta}^2 + M_c\dot{q}_i^2 + M_{\text{bs}}\dot{x}_{\text{bs}}^2\right], \quad (6)$$

$$V = \frac{1}{2}\left[k_c\left(\theta_c-\theta_1\right)^2 + k_\theta\left(q_i\right)\left(\theta-\theta_c\right)^2 + k_{\text{eq}}\left(q_i\right)x_{\text{bs}}^2\right], \quad (7)$$

$$D = \frac{1}{2}\left[C_{\text{nut}}\left(R\dot{\theta} + \dot{x}_{\text{bs}} - \dot{q}_i\right)^2 + C_{\text{eq}}\dot{x}_{\text{bs}}^2\right], \quad (8)$$

$$\begin{aligned}\delta W_{\text{ext}} &= T_m\delta\theta_m + T_g\left(\delta\theta_1 - \frac{1}{\eta G}\delta\theta_m\right) \\ &\quad + F_{\text{nut}}\left(\delta q_i - \delta x_{\text{bs}} - R\delta\theta\right) - F_f\delta q_i + F_{A_{i,x}}\delta q_i,\end{aligned} \quad (9)$$

where T_m, J_m, J_g, and M_c are the motor torque, motor inertia, gearbox inertia, and slider mass, respectively. Also F_f and $F_{A_{i,x}}$ are the slider friction force and reaction force between the ith slider and the ith robot link, respectively. The terms θ_m and θ_1 are the angular position of driver motor and the gearbox angle, whereas M_{bs} and J_{bs} are the mass and moment of inertia of screw ($J_{\text{bs}} = (M_{\text{bs}}d_{\text{screw}}^2)/8$, where d_{screw} is the diameter of screw shaft). The backlash torque T_g is calculated by means of (10), whereas η is gearbox mechanical efficiency and G is the gearbox ratio $\omega_{\text{in}}/\omega_{\text{out}}$. The angular position of coupling and the moment of inertial of coupling are θ_c and J_c, respectively. Finally F_{nut} is the backlash force in nut and K_c, C_{eq}, and C_{nut} are coupling stiffness, screw axial equivalent damping, and nut damping, respectively.

The following systems of equations allow evaluating T_g and F_{nut}:

$$T_g = k_g\begin{cases}\dfrac{\theta_m}{G}-\theta_1-\theta_b & \dfrac{\theta_m}{G}-\theta_1\geq\theta_b \\ 0 & -\theta_b\leq\dfrac{\theta_m}{G}-\theta_1\leq\theta_b \\ \dfrac{\theta_m}{G}-\theta_1+\theta_b & \dfrac{\theta_m}{G}-\theta_1\leq-\theta_b,\end{cases} \quad (10)$$

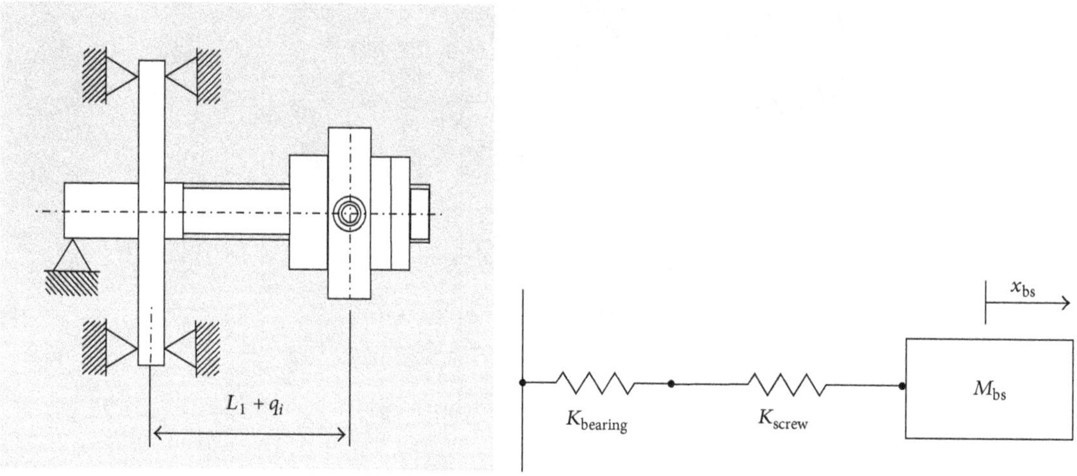

(a) (b)

FIGURE 4: (a) Ball-screw shaft free at one end and (b) mass-spring model of the axial deformation of the ball-screw.

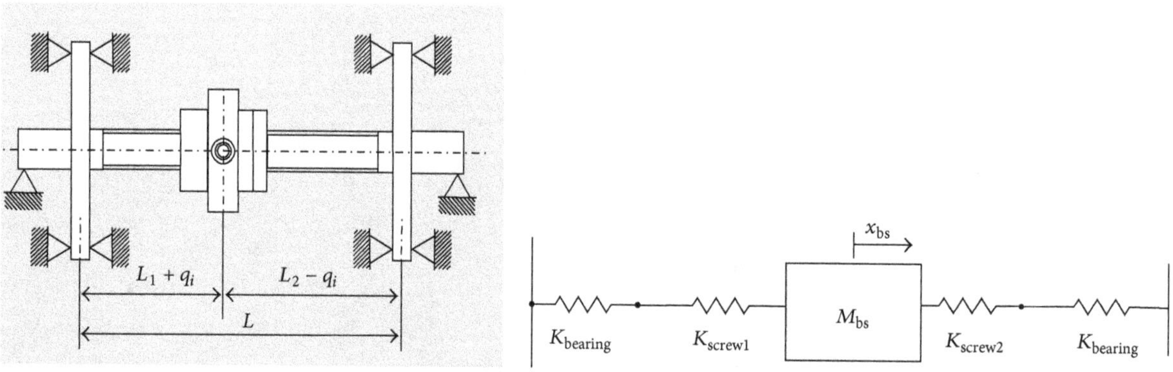

(a) (b)

FIGURE 5: (a) Ball-screw shaft fixed at both ends and (b) mass-spring model of the axial deformation of the ball-screw.

$$F_{\text{nut}} = K_{\text{nut}} \begin{cases} R\theta + x_{\text{bs}} - q_i - \Delta & R\theta + x_{\text{bs}} - q_i \geq \Delta \\ 0 & -\Delta \leq R\theta + x_{\text{bs}} - q_i \leq \Delta \\ R\theta + x_{\text{bs}} - q_i + \Delta & R\theta + x_{\text{bs}} - q_i \leq -\Delta, \end{cases} \quad (11)$$

where K_{nut} is the stiffness between nut and screw, θ_b is the backlash in the gearbox, and Δ is the backlash between nut and screw.

Torsional stiffness of the screw, K_θ, is calculated as explained in [29] and shown below:

$$K_\theta(q_i) = \frac{G_{\text{screw}} J_{\text{screw}}}{L_1 + q_i}, \quad (12)$$

where G_{screw} is the shear module of the screw, L_1 is the distance between the home position and the bearing near to the motor, and J_{screw} is the polar moment of the screw. Axial stiffness of the screw depends on type of bearing which is used in the ball-screw drive. As a result, axial stiffness of the screw and K_{eq} are found in the two conditions set out as follows.

Ball-Screw Shaft Fixed at One End. Figure 4 shows this case, where the screw shaft has axial force only between the axial-fixed bearing and the nut. Therefore, only this part of the screw shaft has axial deformation and can be modeled with a spring. Also the axial-fixed bearing can be modeled by a spring that is series with an equivalent spring model of the shaft screw. Equivalent stiffness is calculated by the following equation:

$$\frac{1}{K_{\text{eq}}} = \frac{1}{K_{\text{bearing}}} + \frac{1}{K_{\text{screw}_1}(q_i)}, \qquad K_{\text{screw}_1}(q_i) = \frac{EA}{L_1 + q_i}. \quad (13)$$

According to [9, 12], stiffness of the screw in this condition is function of the table position.

Ball-Screw Shaft Fixed at Both Ends. In this case there are two springs connected in series on each side of the shaft, as shown in Figure 5. Therefore, equivalent stiffness is calculated by using the formula of a parallel spring as shown in the following equation:

$$K_{\text{eq}} = \left(\frac{1}{K_{\text{bearing}}} + \frac{1}{K_{\text{screw}_1}(q_i)}\right)^{-1} + \left(\frac{1}{K_{\text{bearing}}} + \frac{1}{K_{\text{screw}_2}(q_i)}\right)^{-1},$$
$$K_{\text{screw}_1}(q_i) = \frac{EA}{L_1 + q_i},$$
$$K_{\text{screw}_2}(q_i) = \frac{EA}{L_2 - q_i}. \quad (14)$$

The friction force is estimated using this exponential form:

$$F_f = (M_c g - F_{i,z})\left[\mu_k + (\mu_s - \mu_k)\, e^{-(\dot{q}_i/V_s)^2}\right]\operatorname{sign}(\dot{q}_i) + C_v \dot{q}_i, \quad (15)$$

where μ_s is the static coefficient of friction, μ_k is the kinetic coefficient of friction, $F_{i,z}$ is the normal reaction force between ith slider and ith robot link, C_v is the viscous friction parameter, and V_s is the characteristic velocity of the Stribeck friction. Ball-screw drive equations of motion are resolved via Simulink and the results are used in Hexaglide model.

Considering (6), (7), (8), and (9) and applying Lagrange method, the following equations are obtained:

$$J_m \ddot{\theta}_m = T_m - \frac{T_g}{(\eta G)},$$
$$J_g \ddot{\theta}_1 + k_c(\theta_c - \theta_1) - T_g = 0,$$
$$J_c \ddot{\theta}_c + k_c(\theta_1 - \theta_c) + k_\theta(q_i)(\theta_c - \theta) = 0,$$
$$J_{\text{bs}} \ddot{\theta} + k_\theta(q_i)(\theta - \theta_c) + R F_{\text{nut}} + C_{\text{nut}} R\left(R\dot{\theta} + \dot{x}_{\text{bs}} - \dot{q}_i\right) = 0, \quad (16)$$
$$M_{\text{bs}} \ddot{x}_{\text{bs}} + F_{\text{nut}} + C_{\text{nut}}\left(R\dot{\theta} + \dot{x}_{\text{bs}} - \dot{q}_i\right) + k_{\text{eq}}(q_i)\, x_{\text{bs}} + C_{\text{eq}} \dot{x}_{\text{bs}} = 0,$$
$$M_c \ddot{q}_i - F_{\text{nut}} - C_{\text{nut}}\left(R\dot{\theta} + \dot{x}_{\text{bs}} - \dot{q}_i\right) + F_f = F_{A_{i,x}}.$$

4. Error Evaluation

To evaluate the positioning error of the TCP in the task-space, a method based on a kinematics analysis has been performed. First of all, the critical poses of the Hexaglide workspace are found via kinematic mapping of the error taking into account as a source of error only the slider position. Subsequently the critical poses found are used as the initial positions and orientations of the TCP from which to begin the dynamic simulation of the behaviour of the robot. The aim of these steps is to detect the point which has the worst condition in terms of the maximum error. This method is explained in more depth hereafter.

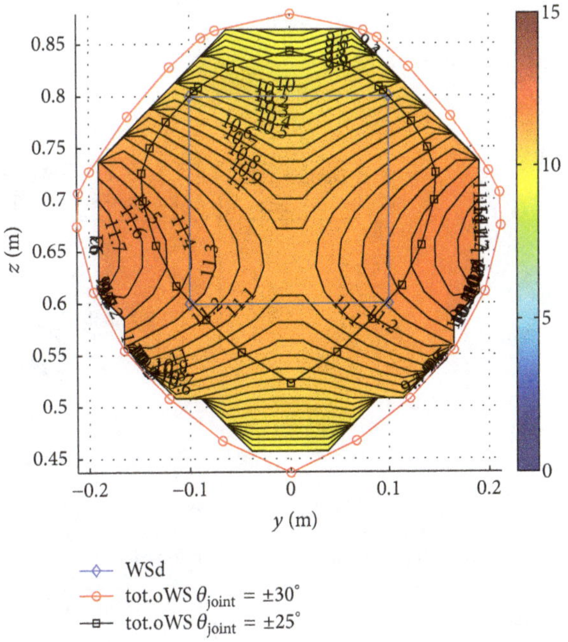

FIGURE 6: Error of TCP position in y direction.

From kinematic analysis of the Hexaglide, we know that $\mathbf{W} = [J]\dot{\mathbf{q}}$, where $[J]$ is a Jacobian matrix, $\dot{\mathbf{q}} = [\dot{q}_1, \dot{q}_2, \ldots, \dot{q}_6]^T$ is the sliders velocity vector, and $\mathbf{W} = [\dot{x}, \dot{y}, \dot{z}, \omega_x, \omega_y, \omega_z]^T$ is the velocity vector of the platform. For small variations, it is correct to write

$$\Delta\mathbf{X} = [J]\,\Delta\mathbf{q}, \quad (17)$$

where $\mathbf{X} = [x, y, z, \alpha, \beta, \gamma]^T$ is the robot pose. If all the actuators and transmissions are equal and the robot is considered rigid with ideal joints, the drive systems are the only source of errors. It is reasonable to assume that the errors in the sliders positions are limited by a maximum value $\Delta q_{\max}$ (infinity norm is the best suited norm when it comes to representing this situation: $\|\Delta q\|_\infty \le \Delta q_{\max}$). Considering that all the sliders have the same errors $\Delta q_{\max}$, the maximum errors of TCP position and platform orientations are defined by

$$\Delta\mathbf{X}_{\max} = [J]\begin{Bmatrix} 1 \\ 1 \\ \vdots \\ 1 \end{Bmatrix}_{6\times 1} \Delta q_{\max}. \quad (18)$$

Figure 6 shows an example of the density distribution of error contours in the workspace of the Hexaglide in y direction. Critical TCP positions are summarised in Table 1; furthermore critical orientations are ±10° for sailboats and are ±7.5° for wind turbines.

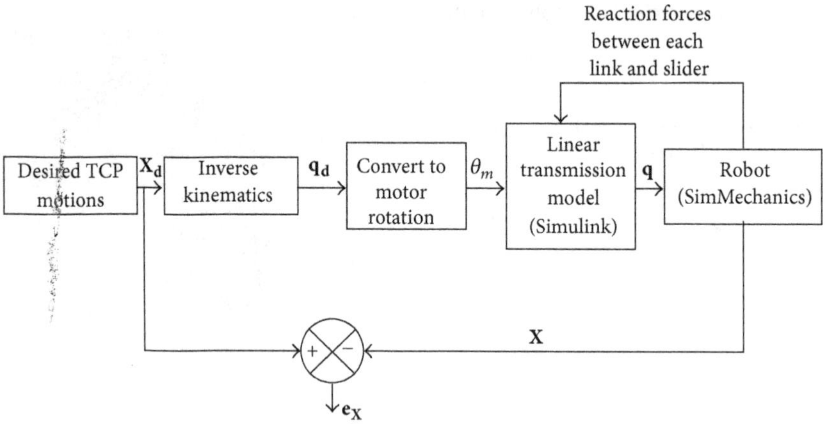

FIGURE 7: Block scheme to evaluate the error pose.

TABLE 1: Critical points of initial position for each movement.

Movement	TCP position [m]		
	x_0	y_0	z_0
x mov.	0	0.1	0.8
y mov.	0	0	0.7
z mov.	0	0.1	0.7
α mov.	0	0.1	0.6
β mov.	0	0.1	0.6
γ mov.	0	0.1	0.6

TABLE 2: Maximum amplitudes and frequencies of the desired movements.

Movement	Amplitude		Frequency (Hz)	
	Wind turbine	Sailboat	Wind turbine	Sailboat
x mov.	0.25 m	—	0.7	1.2
y mov.	0.15 m	—	0.7	1.2
z mov.	0.15 m	—	0.7	1.2
α mov.	7.5°	10°	0.7	1.2
β mov.	7.5°	10°	0.7	1.2
γ mov.	7.5°	10°	0.7	1.2

Error of TCP position and platform orientation is calculated by means of

$$\mathbf{e_X} = \mathbf{X_d} - \mathbf{X}, \tag{19}$$

where $\mathbf{X_d}$ and $\mathbf{e_X}$ are the required pose of the robot and the pose error, respectively. Then, the slider position setpoints are calculated from the required TCP position using the inverse kinematic equation (2). This analysis has been performed taking into consideration each degree of freedom of the TCP (x, y, z, α (rotation around x-axis), β (rotation around y-axis), and γ (rotation around z-axis)) separately and the motion law used for each DoF is sinusoidal. In order to set the correct initial conditions of the simulation and to prevent impulse forces in the robot joints, a five-order polynomial function has been used to fade in the sinusoidal function. In this way, the simulation begins with zero velocity and acceleration.

Thus, the required platform motions are defined by the following equation:

$$X_i = X_{0_i} + \begin{cases} a_5 t^5 + a_4 t^4 + a_3 t^3 & t \leq t_c \\ A_i \sin\left(2\pi f \left(t - t_d\right)\right) & t > t_c, \end{cases} \tag{20}$$

where $\mathbf{X}_o$ is the initial platform pose (Table 1) whereas A_i and f are amplitude and frequency of movement, respectively (Table 2).

For the ideal situation (linear transmission rigid and gearbox without backlash), the angular motor positions are calculated from the slider positions by $\theta_{m,\text{ref}} = q_{i,\text{ref}} G/R$ and the dynamic behaviour of the sliders is evaluated by the equations shown in (16). Figure 7 summarises the main steps necessary to obtain the TCP position and the platform orientation for the nonideal system using the SimMechanics model, earlier described. As can be see, the error between ideal system and the system which includes backlash, flexibility, and friction is calculated. The worst operating condition, in other words the maximum error, is found by comparing the results achieved by each DoF under different motion laws.

The previous analysis has been performed using data of a Rexroth Bosch CKK 25-200 ball-screw drive. A ball-screw drive with shaft free at one end has been investigated. In fact this condition has a lower stiffness with respect to the situation in which the shaft is fixed at both ends, so it represents the worst situation. On the other hand, the ball-screw configuration with both ends fixed in general is not used because any gradient of temperature on the screw can generate high stress, reducing the screw and bearings life. Instead, backlash is a real problem in precise control; therefore, screw with preload must be chosen to reduce it. Below parameters are used in the simulations:

$$K_{\text{nut}} = 137e6\,\text{N/m},$$

$$K_{\text{bearing}} = 113e6\,\text{N/m},$$

$$\Delta = 0.05\,\text{mm},$$

TABLE 3: Values of the maximum error and RMS of error in the desired robot workspace.

Position or orientation		Maximum	
		Sailboat	Wind turbine
Maximum error	x	0.68 mm	0.70 mm
	y	0.68 mm	0.70 mm
	z	0.4 mm	0.36 mm
	α	0.17°	0.16°
	β	0.10°	0.10°
	γ	0.05°	0.05°
RMS error	x	0.35 mm	0.37 mm
	y	0.36 mm	0.38 mm
	z	0.28 mm	0.26 mm
	α	0.09°	0.08°
	β	0.06°	0.06°
	γ	0.03°	0.03°

$$\begin{aligned} \theta_b &= 4\,\text{arcmin}, \\ h_p &= 32\,\text{mm}, \\ d_{\text{screw}} &= 32\,\text{mm}, \\ K_c &= 39543\,\text{N}\cdot\text{m/rad}, \\ k_g &= 7.5287e4\,\text{N}\cdot\text{m/rad}, \\ G &= 2, \\ L &= 1.6\,\text{m}. \end{aligned} \tag{21}$$

Note that the ball-screw damping (C_{eq} and C_{nut}) has to be identified by means of experimental tests on the real machine, but it does not have too much significance in predicting the elastic deformation [12]. Consequently, values of C_{eq} and C_{nut} are chosen from the literature [12, 30]. The same approach is used to choose the values of the friction parameters in (15) in particular by the cited reference Okwudire [31].

Figure 8 shows maximum and RMS error for TCP position and platform orientation in each movement when the requirements are defined for sailboats simulations. Table 3 summarises maximum value of TCP positioning and orientation errors while Table 4 shows in which poses these maximum error conditions have been achieved.

According to Tables 3 and 4 and Figure 8, the critical condition is found when α is moving and $\beta = -10°$ and $\gamma = -10°$ are the initial platform orientation for the sailboat simulations. When α is moving and $\beta = -7.5°$ and $\gamma = -7.5°$ are the initial platform orientation, we have the maximum error in the case of wind turbine simulation.

5. Control of the Hexaglide

The design of a control system of a six-degree of freedom parallel kinematic machine is a very difficult task. Usually it is very expensive to measure the end effector position of a 6-DoF robot. Instead of the pose of the robot platform, the position of each slider or angular position of each motor is measured and the pose of the end effector is estimated using the direct kinematics. By the way this method can only be applied if the machine is properly calibrated and if all the machine components are realized respecting very strict tolerances; otherwise the estimation of the pose of the mobile platform would be erroneous. If these requirements cannot be met, it is necessary to use specific measurements devices like CMM or 3D positioning systems as the ones described in [32, 33]. Furthermore 6-DoF PKM robot has complex dynamics without an analytical solution made more complex by the nonlinearity of the actuation systems. Therefore, to setup a control strategy it is necessary to pass through a linear form of dynamic equations as described in [22, 23].

A promising approach for developing a control algorithm in these conditions is the adaptive and robust nonlinear control as presented in [34, 35]. In this paper, starting from a literature review, a PID adaptive-robust control for the Hexaglide is developed. Figure 9 shows the block diagram of the control proposed.

We can write the dynamic equation of the robot, including the actuator dynamics, as follows:

$$\mathbf{T_m} = \mathbf{M}(\mathbf{q})\,\ddot{\mathbf{q}} + \mathbf{f}(\mathbf{q}, \dot{\mathbf{q}}), \tag{22}$$

where $\mathbf{T_m}$ is the vector of motor torques and $\mathbf{q}$ is the vector of the joint position. The manipulator mass matrix, $\mathbf{M}(\mathbf{q})$, is symmetric and positive definite. The vector $\mathbf{f}(\mathbf{q}, \dot{\mathbf{q}})$ represents torque or force arising from centrifugal, Coriolis, gravity, and friction forces.

The control action can be obtained by means of a suitable input motor torque, defined in this way:

$$\begin{aligned} \mathbf{T_m} &= \mathbf{\Psi}(\mathbf{q}, \dot{\mathbf{q}}, \dot{\mathbf{q}}_\mathbf{r}, \ddot{\mathbf{q}}_\mathbf{r})\,\mathbf{p} + \mathbf{K_D}\mathbf{s} + \mathbf{K_I}\int \mathbf{s}\,dt \\ &\quad + \boldsymbol{\eta} sat\left(\mathbf{\Phi}^{-1}\mathbf{s}\right). \end{aligned} \tag{23}$$

In this expression, $\mathbf{\Psi}$ is a matrix containing nonlinear equation and $\mathbf{p}$ is a vector containing dynamical parameters, whereas $\mathbf{K_D}$, $\mathbf{K_I}$, $\boldsymbol{\eta}$, and $\mathbf{\Phi}$ are positive diagonal matrices. Position error vector of the sliders is defined as

$$\mathbf{e} = \mathbf{q_d} - \mathbf{q}, \tag{24}$$

where $\mathbf{q_d}$ is the desired position of slider found from the desired platform pose via the inverse kinematics. The vector $\mathbf{s}$ represents the combined error and it is defined in a similar way to the sliding control approach:

$$\mathbf{s} = \dot{\mathbf{e}} + \mathbf{\Lambda}\mathbf{e}, \tag{25}$$

where $\mathbf{\Lambda}$ is a positive and diagonal matrix. The vector $\mathbf{s}$ can be also defined as

$$\mathbf{s} = \dot{\mathbf{q}}_\mathbf{r} - \dot{\mathbf{q}}, \tag{26}$$

where $\dot{\mathbf{q}}_\mathbf{r}$ is called reference value of $\dot{\mathbf{q}}$ and it is obtained by modifying $\dot{\mathbf{q}}_\mathbf{d}$ according to the tracking error. $\dot{\mathbf{q}}_\mathbf{r}$ is defined as

$$\dot{\mathbf{q}}_\mathbf{r} = \dot{\mathbf{q}}_\mathbf{d} + \mathbf{\Lambda}\mathbf{e}. \tag{27}$$

TABLE 4: Conditions of maximum error and maximum RMS of the error.

Position or orientation		Condition	
		Sailboat	Wind turbine
Maximum error	x	β mov. $\alpha = 10^\circ, \gamma = -10^\circ$	β mov. $\alpha = -7.5^\circ, \gamma = 7.5^\circ$
	y	α mov. $\beta = -10^\circ, \gamma = 10^\circ$	β mov. $\alpha = -7.5^\circ, \gamma = 7.5^\circ$
	z	α mov. $\beta = -10^\circ, \gamma = -10^\circ$	γ mov. $\alpha = 7.5^\circ, \beta = -7.5^\circ$
	α	α mov. $\beta = -10^\circ, \gamma = -10^\circ$	α mov. $\beta = -7.5^\circ, \gamma = -7.5^\circ$
	β	β mov. $\alpha = 10^\circ, \gamma = -10^\circ$	β mov. $\alpha = -7.5^\circ, \gamma = 7.5^\circ$
	γ	α mov. $\beta = -10^\circ, \gamma = -10^\circ$	α mov. $\beta = -7.5^\circ, \gamma = -7.5^\circ$
RMS error	x	β mov. $\alpha = -10^\circ, \gamma = 10^\circ$	x-mov. $\alpha, \beta = 7.5^\circ, \gamma = -7.5^\circ$
	y	γ mov. $\alpha = -10^\circ, \beta = -10^\circ$	γ mov. $\alpha = -7.5^\circ, \beta = -7.5^\circ$
	z	γ mov. $\alpha = 10^\circ, \beta = -10^\circ$	γ mov. $\alpha = 7.5^\circ, \beta = -7.5^\circ$
	α	α mov. $\beta = -10^\circ, \gamma = -10^\circ$	α mov. $\beta = -7.5^\circ, \gamma = -7.5^\circ$
	β	β mov. $\alpha = -10^\circ, \gamma = 10^\circ$	β mov. $\alpha = -7.5^\circ, \gamma = 7.5^\circ$
	γ	α mov. $\beta = -10^\circ, \gamma = 10^\circ$	α mov. $\beta = -7.5^\circ, \gamma = -7.5^\circ$

In order to simplify the computational aspects of this control structure, it is possible to omit the dynamic behaviour of the six links. In this way, the model of the robot is made up of only seven bodies: six linear transmission drive servomechanisms and one mobile platform. The function $\mathbf{\Psi}(\mathbf{q}, \dot{\mathbf{q}}, \dot{\mathbf{q}}_r, \ddot{\mathbf{q}}_r)\mathbf{p}$ can be modified in the following way:

$$\mathbf{T}_{\mathbf{robot}} = \begin{bmatrix} \mathbf{\Psi}_{1\cdots 6} & \mathbf{\Psi}_{\mathbf{b}} & \mathbf{\Psi}_7 \end{bmatrix} \begin{bmatrix} \mathbf{p}_{1\cdots 6} \\ \mathbf{p}_{\mathbf{b}} \\ \mathbf{p}_7 \end{bmatrix}, \quad (28)$$

where $\mathbf{\Psi}_{1\cdots 6}$ and $\mathbf{p}_{1\cdots 6}$ describe the dynamics of the six linear transmission drive servomechanisms, $\mathbf{\Psi}_{\mathbf{b}}$ and $\mathbf{p}_{\mathbf{b}}$ describe Coulomb friction within the linear transmissions, and $\mathbf{\Psi}_7$ and $\mathbf{p}_7$ describe the dynamics of the platform. The terms $\mathbf{\Psi}_{1\cdots 6}$ and $\mathbf{p}_{1\cdots 6}$ are defined by

$$\mathbf{\Psi}_{1\cdots 6} = \begin{bmatrix} \ddot{q}_{r_1} & \dot{q}_{r_1} & q_1 & & 0 \\ & \ddots & \ddots & \ddots & \\ 0 & & \ddot{q}_{r_6} & \dot{q}_{r_6} & q_6 \end{bmatrix},$$
$$\mathbf{p}_{1\cdots 6} = \begin{bmatrix} m_1 & c_1 & k_1 & \cdots & m_6 & c_6 & k_6 \end{bmatrix}^T, \quad (29)$$

where $m_1, c_1, k_1, \ldots, m_6, c_6, k_6$ are estimated mass, damper, and spring coefficient, respectively, for each linear transmission, whereas $\mathbf{\Psi}_{\mathbf{b}}$ and $\mathbf{p}_{\mathbf{b}}$ are defined by

$$\mathbf{\Psi}_{\mathbf{b}} = \begin{bmatrix} \operatorname{sign}(\dot{q}_{r_1}) & & 0 \\ & \ddots & \\ 0 & & \operatorname{sign}(\dot{q}_{r_6}) \end{bmatrix},$$
$$\mathbf{p}_{\mathbf{b}} = \begin{bmatrix} b_1 \\ \vdots \\ b_6 \end{bmatrix}, \quad (30)$$

where $b_1, \ldots, b_6$ are Coulomb friction coefficients in each linear transmission. The definition of $\mathbf{\Psi}_7$ and $\mathbf{p}_7$ is more complicated. The vector of dynamical parameters is set out by

$$\mathbf{p}_7 = \begin{bmatrix} m_7 & m_7 r_x & m_7 r_y & m_7 r_z & I_{xx} & I_{yy} & I_{zz} \end{bmatrix}^T. \quad (31)$$

It is made up of the mass of the platform, the payload, and the inertia moments I_{xx}, I_{yy}, and I_{zz}. The frame connected to the TCP is supposed to be oriented in order not to consider the inertia moments I_{xy}, I_{xz}, and I_{yz}. The matrix $\mathbf{\Psi}_7$ is given by

$$\mathbf{\Psi}_7 = \mathbf{J} \begin{bmatrix} \mathbf{a}_7 & \mathbf{R}\mathbf{\Omega}_7 & \mathbf{0}_{3\times 3} \\ \mathbf{0}_{1\times 3} & -\hat{\mathbf{a}}_7 \mathbf{R} & \mathbf{R}\psi_{\omega_7} \end{bmatrix}, \quad (32)$$

where $\mathbf{J}$ is the Jacobian matrix and $\mathbf{R}$ is the rotation matrix. The acceleration $\mathbf{a}_7$ and the skew-symmetric matrix $\hat{\mathbf{a}}_7$ corresponding to the cross product are defined as follows:

$$\mathbf{a}_7 = \begin{bmatrix} \ddot{x}_d \\ \ddot{y}_d \\ \ddot{z}_d + g \end{bmatrix},$$
$$\hat{\mathbf{a}}_7 = \begin{bmatrix} 0 & -\ddot{z}_d - g & \ddot{y}_d \\ \ddot{z}_d + g & 0 & -\ddot{x}_d \\ -\ddot{y}_d & \ddot{x}_d & 0 \end{bmatrix}. \quad (33)$$

The matrices $\mathbf{\Omega}_7$ and ψ_{ω_7} are defined as

$$\mathbf{\Omega}_7 = \begin{bmatrix} -\omega_y^2 - \omega_z^2 & -\dot{\omega}_z + \omega_x\omega_y & \dot{\omega}_y + \omega_x\omega_z \\ \dot{\omega}_z + \omega_x\omega_y & -\omega_x^2 - \omega_z^2 & -\dot{\omega}_x + \omega_z\omega_y \\ -\dot{\omega}_y + \omega_z\omega_x & \dot{\omega}_x + \omega_z\omega_y & -\omega_x^2 - \omega_y^2 \end{bmatrix},$$
$$\psi_{\omega_7} = \begin{bmatrix} \dot{\omega}_x & -\omega_z\omega_y & \omega_y\omega_z \\ \omega_x\omega_z & \dot{\omega}_y & -\omega_z\omega_x \\ -\omega_y\omega_x & \omega_x\omega_y & \dot{\omega}_z \end{bmatrix}, \quad (34)$$

where $\boldsymbol{\omega}$ is the absolute angular velocity of the platform.

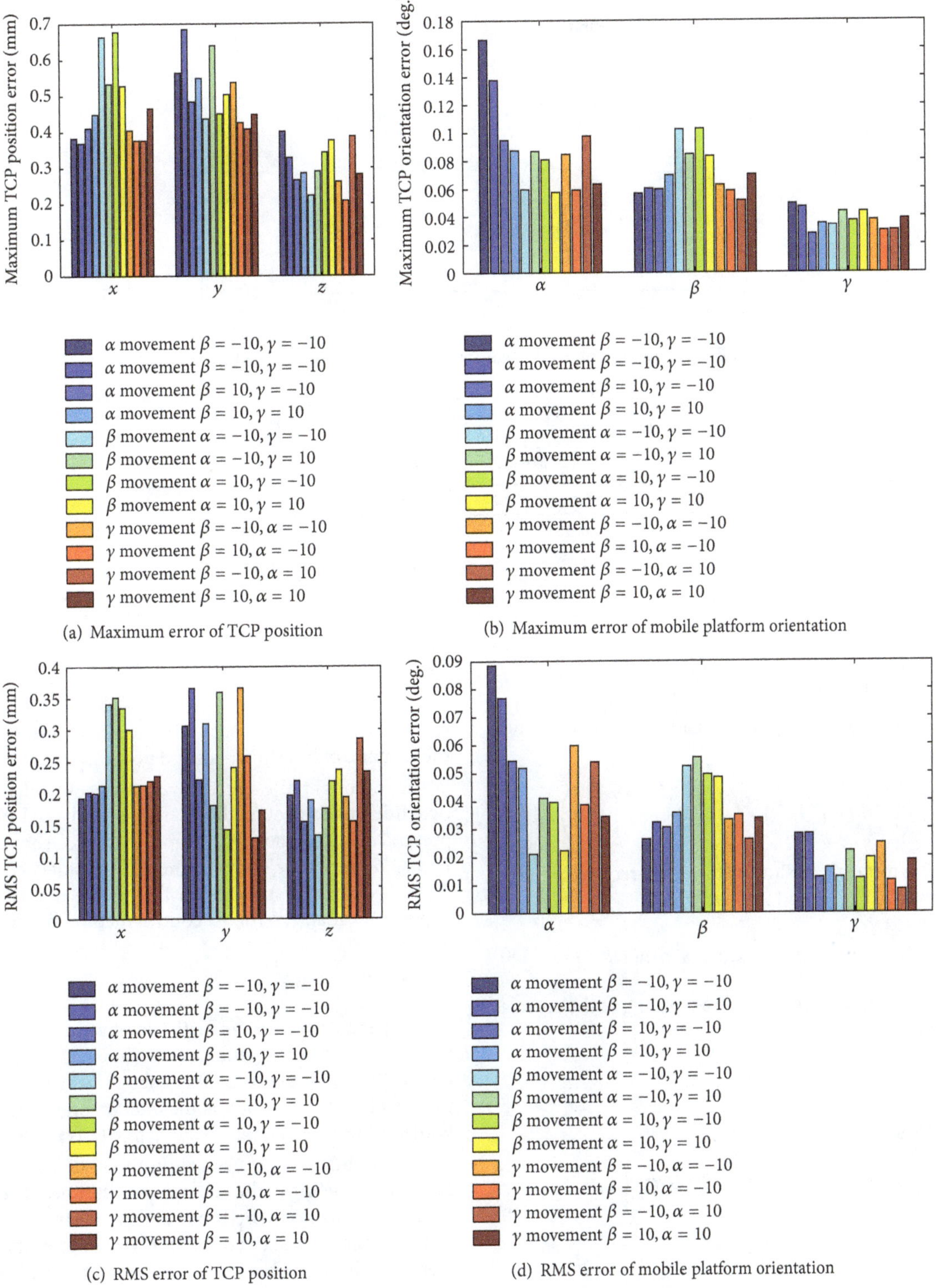

(a) Maximum error of TCP position

(b) Maximum error of mobile platform orientation

(c) RMS error of TCP position

(d) RMS error of mobile platform orientation

FIGURE 8: Maximum and RMS of pose error when sailboat is installed on platform.

Given the desired trajectory $\mathbf{q_d}$ (we shall assume that the desired position, velocity, and acceleration are all bounded), and with some or all manipulator parameters unknown, the adaptive controller design problem is to obtain a control law for the actuator torques and an estimation of the unknown parameters, in such a way that the manipulator follows the required trajectory in the best way possible.

To do that we define a function to estimate the parameter error $\widetilde{\mathbf{p}} = \widehat{\mathbf{p}} - \mathbf{p}$ as a difference between a vector of unknown parameters describing the manipulator's mass properties and

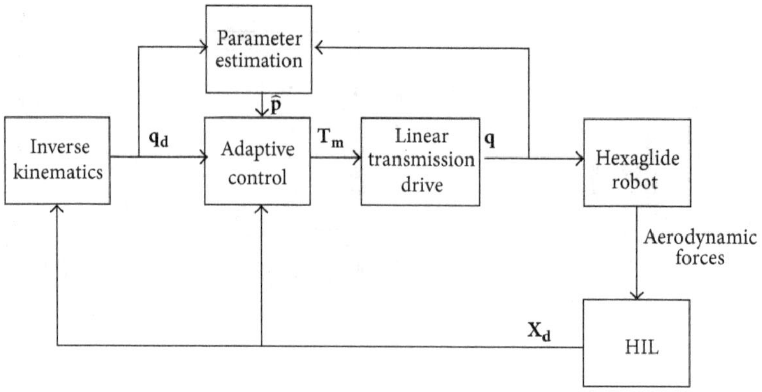

FIGURE 9: Block diagram of the adaptive-robust control for controlling the Hexaglide.

TABLE 5: Position errors of the sliders.

	Slider 1	Slider 2	Slider 3	Slider 4	Slider 5	Slider 6
Maximum error [mm]	0.57	0.40	0.93	0.30	0.30	0.38
Error percentage	0.23	1.00	0.26	0.30	0.73	0.22
RMS error [mm]	0.23	0.18	0.38	0.10	0.12	0.13

its estimate. By considering as a candidate the Lyapunov function,

$$V(t) = \frac{1}{2}\left(\mathbf{s}^T\mathbf{M}\mathbf{s} + \tilde{\mathbf{p}}^T\mathbf{\Gamma}^{-1}\tilde{\mathbf{p}} + \left(\int \mathbf{s}\,dt\right)^T \mathbf{K}_\mathbf{I} \int \mathbf{s}\,dt\right), \quad (35)$$

where $\mathbf{\Gamma}$ is a symmetric positive definite matrix. Differentiating and using (22), (23), and (26) yield

$$\dot{V}(t) = \mathbf{s}^T\mathbf{\Psi}\tilde{\mathbf{p}} + \dot{\hat{\mathbf{p}}}^T\mathbf{\Gamma}^{-1}\tilde{\mathbf{p}} - \mathbf{s}^T\mathbf{K}_\mathbf{D}\mathbf{s} - \mathbf{s}^T\eta sat\left(\mathbf{\Phi}^{-1}\mathbf{s}\right). \quad (36)$$

Updating the parameter estimates $\hat{\mathbf{p}}$ according to the correlation integrals [35],

$$\dot{\hat{\mathbf{p}}} = \mathbf{\Gamma}\mathbf{\Psi}^T\mathbf{s} \quad (37)$$

and (23) changes to

$$\mathbf{T_m} = \mathbf{\Psi}\left(\mathbf{q}, \dot{\mathbf{q}}, \dot{\mathbf{q}}_\mathbf{r}, \ddot{\mathbf{q}}_\mathbf{r}\right)\hat{\mathbf{p}} + \mathbf{K_D}\mathbf{s} + \mathbf{K_I}\int \mathbf{s}\,dt + \eta sat\left(\mathbf{\Phi}^{-1}\mathbf{s}\right). \quad (38)$$

By choosing $\mathbf{K_D} = 100\mathbf{I}$, $\mathbf{K_I} = 100\mathbf{I}$, and $\mathbf{\Lambda} = 200\mathbf{I}$, robot is controlled for tracking six-degree of freedom sinusoidal movements ((20) and Table 2) in its task-space, simultaneously. The position error of each slider is shown in Figure 10 whereas Table 5 presents the maximum and RMS of the position error of sliders.

Figure 11 shows the tracking errors of the platform. Table 6 presents the maximum tracking error and the RMS tracking error. According to this table, maximum percentage of the tracking error occurs in γ orientation with 1.29% of the γ movement amplitude. According to these results, adaptive-robust control has shown a good performance.

To highlight the efficiency of the controller designed, two different control methods are analysed: dual PID control presented in [36] and PD adaptive control shown in [23]. Figure 12 shows the comparison of the results achieved with these two control strategies with the one described in this paper: PID adaptive-robust control method has minor error.

6. Conclusions

In this paper, a systematic methodology to find the error of a 6-DoF robotic device with parallel kinematic and Hexaglide architecture is presented. This robot works as an emulator to reproduce the hydrodynamic interaction between floating bodies and sea water, for aerodynamic tests in wind tunnel.

A systematic error evaluation methodology is based on accurate modeling of the behaviour of the linear transmission actuators that move the robot and by means of a mapping of the robot working volume in order to identify the worst work conditions. The critical poses of the end effector in the workspace for each desired movement have been found through a kinematic analysis whereas the dynamic analysis of Hexaglide actuated by ball-screw linear transmissions has been performed, in these critical poses, for obtaining the worst cases. The robot has been simulated into Simulink-SimMechanics environment and an adaptive-robust control strategy has been designed to control the end effector position in order to track spatial complex trajectory. Finally the control strategy performances have been compared with other control methods.

Table 6: Poses errors of the platform when adaptive-robust control is used.

	x [mm]	y [mm]	z [mm]	α [degree]	β [degree]	γ [degree]
Maximum error	0.64	0.68	0.31	0.06	0.09	0.10
Error percentage	0.43	0.68	0.31	0.81	1.20	1.29
RMS error	0.18	0.16	0.11	0.02	0.02	0.02

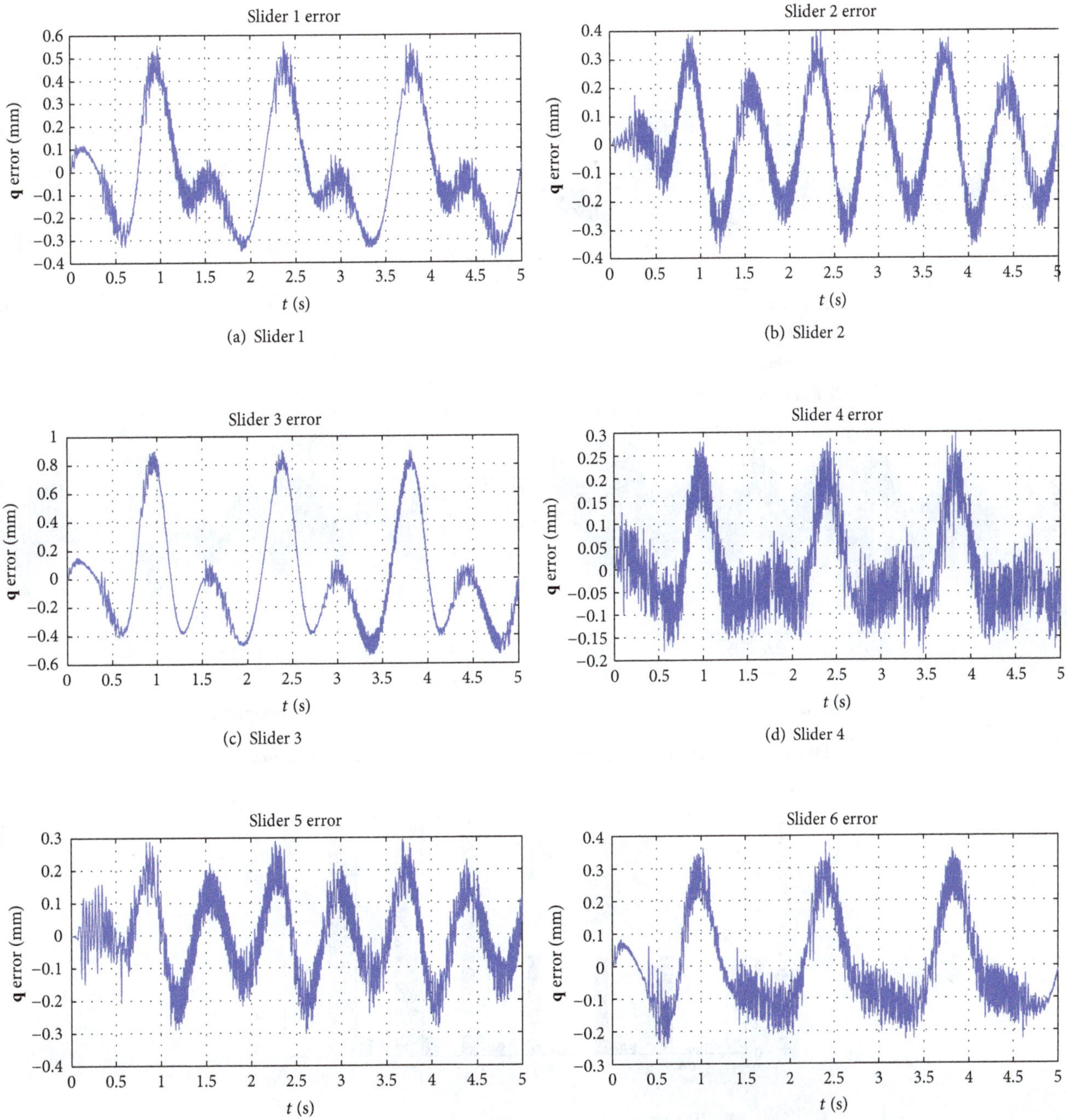

(a) Slider 1

(b) Slider 2

(c) Slider 3

(d) Slider 4

(e) Slider 5

(f) Slider 6

Figure 10: Position errors of the sliders when adaptive-robust control is used.

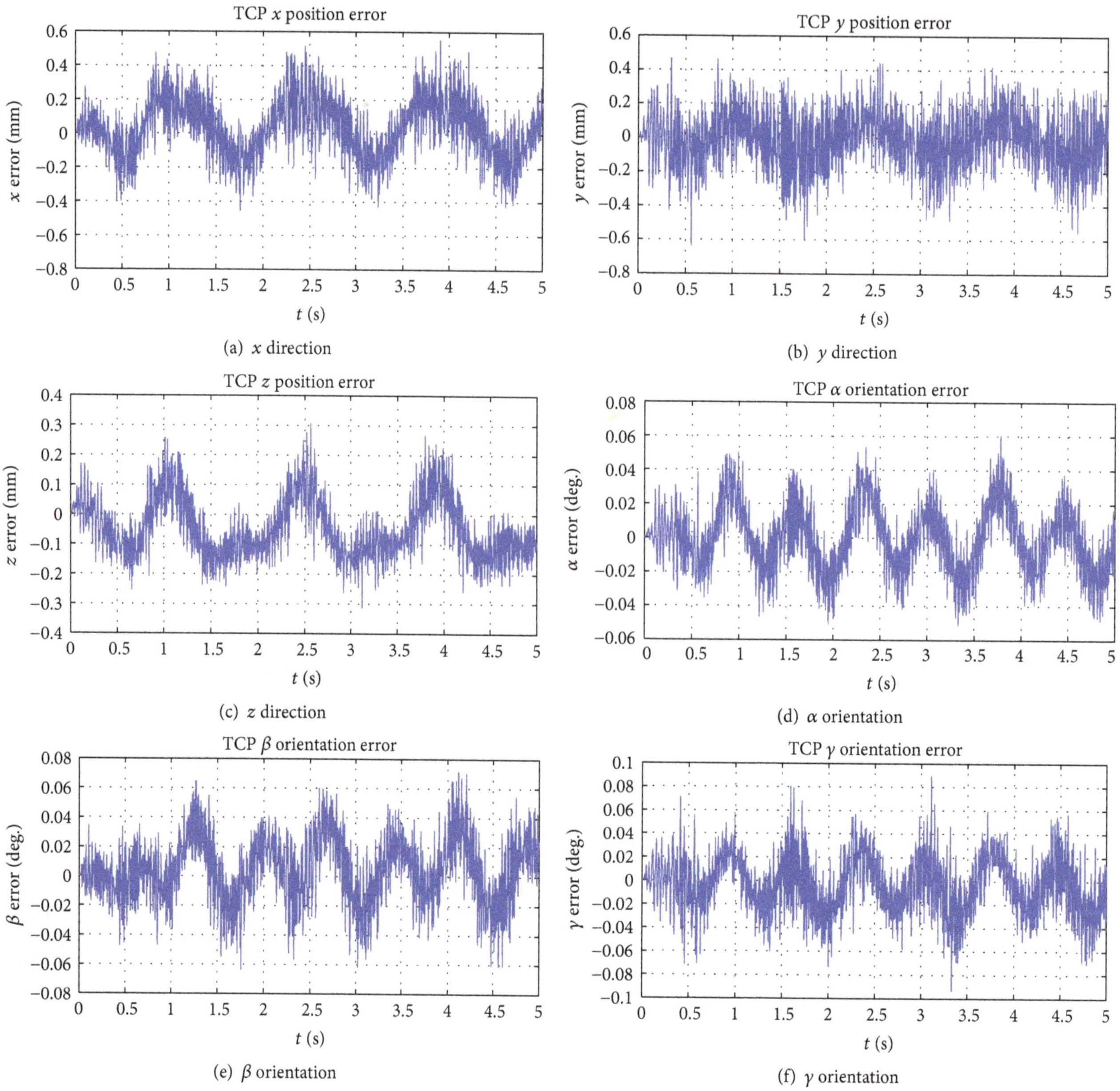

FIGURE 11: Pose error of the platform when adaptive-robust control is used.

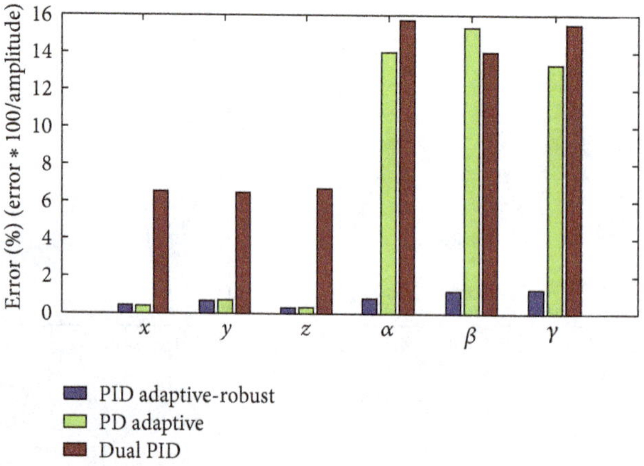

FIGURE 12: Pose error percentage in three types of the control method.

The results of the work demonstrate that the ball-screw linear actuator used to move the Hexaglide architecture developed and the PID adaptive-robust control allows one to achieve accuracy of approximately 0.7 mm in TCP position and of 0.17 degrees in platform orientation. These results are in line with our required performance and consolidate the design choices with respect to the actuation system and the algorithm control strategy.

Competing Interests

The authors declare that they have no competing interests.

References

[1] V. Scheinman and J. M. McCarthy, "Mechanisms and actuation," in *Springer Handbook of Robotics*, B. Siciliano and O. Khatib, Eds., chapter 3, pp. 67–86, 2008.

[2] A. S. Kulkarni and M. A. El-Sharkawi, "Intelligent precision position control of elastic drive systems," *IEEE Transactions on Energy Conversion*, vol. 16, no. 1, pp. 26–31, 2001.

[3] A. Hace, K. Jezernik, and A. Šabanović, "SMC with disturbance observer for a linear belt drive," *IEEE Transactions on Industrial Electronics*, vol. 54, no. 6, pp. 3402–3412, 2007.

[4] D. Tosi, G. Legnani, N. Pedrocchi, P. Righettini, and H. Giberti, "Cheope: a new reconfigurable redundant manipulator," *Mechanism and Machine Theory*, vol. 45, no. 4, pp. 611–626, 2010.

[5] J.-S. Chen, Y.-K. Huang, and C.-C. Cheng, "Mechanical model and contouring analysis of high-speed ball-screw drive systems with compliance effect," *International Journal of Advanced Manufacturing Technology*, vol. 24, no. 3-4, pp. 241–250, 2004.

[6] L. Liu, Z. Wu, and H. Liu, "Modeling and analysis of the crossfeed servo system of a heavy-duty lathe with friction," *Mechanics Based Design of Structures and Machines*, vol. 41, no. 1, pp. 1–20, 2013.

[7] E. Schafers, J. Denk, and J. Hamann, "Mechatronic modeling and analysis of machine tools," in *Proceedings of the 2nd International Conference on High Performance Cutting (CIRP-HPC '06)*, Vancouver, Canada, June 2006.

[8] S. J. Ma, G. Liu, G. Qiao, and X. J. Fu, "Thermo-mechanical model and thermal analysis of hollow cylinder planetary roller screw mechanism," *Mechanics Based Design of Structures and Machines*, vol. 43, no. 3, pp. 359–381, 2015.

[9] K. K. Varanasi and S. A. Nayfeh, "The dynamics of lead-screw drives: low-order modeling and experiments," *Journal of Dynamic Systems, Measurement and Control*, vol. 126, no. 2, pp. 388–396, 2004.

[10] D. A. Vicente, R. L. Hecker, F. J. Villegas, and G. M. Flores, "Modeling and vibration mode analysis of a ball screw drive," *International Journal of Advanced Manufacturing Technology*, vol. 58, no. 1–4, pp. 257–265, 2012.

[11] C. Okwudire and Y. Altintas, "Minimum tracking error control of flexible ball screw drives using a discrete-time sliding mode controller," *Journal of Dynamic Systems, Measurement and Control*, vol. 131, no. 5, pp. 1–12, 2009.

[12] A. Kamalzadeh, D. J. Gordon, and K. Erkorkmaz, "Robust compensation of elastic deformations in ball screw drives," *International Journal of Machine Tools and Manufacture*, vol. 50, no. 6, pp. 559–574, 2010.

[13] L. Dong and W. C. Tang, "Adaptive backstepping sliding mode control of flexible ball screw drives with time-varying parametric uncertainties and disturbances," *ISA Transactions*, vol. 53, no. 1, pp. 125–133, 2014.

[14] S. Frey, A. Dadalau, and A. Verl, "Expedient modeling of ball screw feed drives," *Production Engineering*, vol. 6, no. 2, pp. 205–211, 2012.

[15] S. I. Han and J. M. Lee, "Adaptive dynamic surface control with sliding mode control and RWNN for robust positioning of a linear motion stage," *Mechatronics*, vol. 22, no. 2, pp. 222–238, 2012.

[16] S. K. Hag, M. C. Young, and L. Kyo, "Robust nonlinear task space control for 6 {DOF} parallel manipulator," *Automatica*, vol. 41, no. 9, pp. 1591–1600, 2005.

[17] J. F. He, H. Z. Jiang, D. C. Cong, Z. M. Ye, and J. W. Han, "A survey on control of parallel manipulator," *Key Engineering Materials*, vol. 339, pp. 307–313, 2007.

[18] S.-H. Lee, J.-B. Song, W.-C. Choi, and D. Hong, "Position control of a Stewart platform using inverse dynamics control with approximate dynamics," *Mechatronics*, vol. 13, no. 6, pp. 605–619, 2003.

[19] H. Abdellatif and B. Heimann, "Advanced model-based control of a 6-DOF hexapod robot: a case study," *IEEE/ASME Transactions on Mechatronics*, vol. 15, no. 2, pp. 269–279, 2010.

[20] Z. Ma, Y. Hu, J. Huang et al., "A novel design of in pipe robot for inner surface inspection of large size pipes," *Mechanics Based Design of Structures and Machines*, vol. 35, no. 4, pp. 447–465, 2007.

[21] F. L. Lewis, D. M. Dawson, and T. A. Chaouki, *Robot Manipulator Control: Theory and Practice*, Marcel Dekker, New York, NY, USA, 2nd edition, 2004.

[22] M. Honegger, A. Codourey, and E. Burdet, "Adaptive control of the Hexaglide, a 6 dof parallel manipulator," in *Proceedings of the IEEE International Conference on Robotics and Automation (ICRA '97)*, vol. 1, pp. 543–548, Albuquerque, NM, USA, April 1997.

[23] M. Honegger, R. Brega, and G. Schweitzer, "Application of a nonlinear adaptive controller to a 6 dof parallel manipulator," in *Proceedings of the IEEE International Conference on Robotics and Automation (ICRA '00)*, pp. 1930–1935, San Francisco, Calif, USA, April 2000.

[24] G. Song, R. W. Longman, R. Mukherjee, and J. Zhang, "Integrated sliding-mode adaptive-robust control," in *Proceedings of the IEEE International Conference on Control Applications*, pp. 656–661, Dearborn, Mich, USA, September 1996.

[25] X. Zhu, G. Tao, B. Yao, and J. Cao, "Adaptive robust posture control of a parallel manipulator driven by pneumatic muscles," *Automatica*, vol. 44, no. 9, pp. 2248–2257, 2008.

[26] I. Bayati, M. Belloli, D. Ferrari, F. Fossati, and H. Giberti, "Design of a 6-dof robotic platform for wind tunnel tests of floating wind turbines," *Energy Procedia Journal*, vol. 53, pp. 313–323, 2014.

[27] H. Giberti and D. Ferrari, "Drive system sizing of a 6-Dof parallel robotic platform," in *Proceedings of ASME 12th Biennial Conference on Engineering Systems Design and Analysis (ESDA '14)*, pp. 25–27, Copenhagen, Denmark, June 2014.

[28] M. Vallés, M. Díaz-Rodríguez, Á. Valera, V. Mata, and A. Page, "Mechatronic development and dynamic control of a 3-DOF parallel manipulator," *Mechanics Based Design of Structures and Machines*, vol. 40, no. 4, pp. 434–452, 2012.

[29] K. K. Varanasi and S. Nayfeh, "The dynamics of lead-screw drives: low-order modeling and experiments," *Journal of Dynamic Systems, Measurement and Control*, vol. 126, no. 2, pp. 388–396, 2004.

[30] A. Kamalzadeh and K. Erkorkmaz, "Compensation of axial vibrations in ball screw drives," *CIRP Annals—Manufacturing Technology*, vol. 56, no. 1, pp. 373–378, 2007.

[31] C. Okwudire, *Finite element modeling of ballscrew feed drive systems for control purposes [Master of Applied Science thesis]*, The University of British Columbia, 2005.

[32] M. Verner, F. Xi, and C. Mechefske, "Optimal calibration of parallel kinematic machines," *Journal of Mechanical Design*, vol. 127, no. 1, pp. 62–69, 2005.

[33] Y. M. Zhao, Y. Lin, F. Xi, and S. Guo, "Calibration-based iterative learning control for path tracking of industrial robots," *IEEE Transactions on Industrial Electronics*, vol. 62, no. 5, pp. 2921–2929, 2015.

[34] J. Craig, P. Hsu, and S. S. Sastry, "Adaptive control of mechanical manipulators," in *Proceedings of the IEEE International Conference on Robotics and Automation*, Slotine, April 1986.

[35] J.-J. Slotine and W. Li, *Applied Nonlinear Control*, Prentice-Hall International, New York, NY, USA, 1991.

[36] B. Ding, R. M. Stanley, B. S. Cazzolato, and J. J. Costi, "Real-time FPGA control of a hexapod robot for 6-DOF biomechanical testing," in *Proceedings of the 37th Annual Conference of the IEEE Industrial Electronics Society (IECON '11)*, pp. 252–257, Melbourne, Australia, November 2011.

10

On the Direct Kinematics Problem of Parallel Mechanisms

Arthur Seibel, Stefan Schulz, and Josef Schlattmann

Workgroup on System Technologies and Engineering Design Methodology, Hamburg University of Technology, 21073 Hamburg, Germany

Correspondence should be addressed to Arthur Seibel; arthur.seibel@tuhh.de

Academic Editor: Gordon R. Pennock

The direct kinematics problem of parallel mechanisms, that is, determining the pose of the manipulator platform from the linear actuators' lengths, is, in general, uniquely not solvable. For this reason, instead of measuring the lengths of the linear actuators, we propose measuring their orientations and, in most cases, also the orientation of the manipulator platform. This allows the design of a low-cost sensor system for parallel mechanisms that completely renounces length measurements and provides a unique solution of their direct kinematics.

1. Introduction

A typical six-degrees-of-freedom parallel mechanism consists of a (fixed) base platform and a (movable) manipulator platform. The position and orientation (also known as pose) of the manipulator platform are commanded by fixing the distances between n points on the base platform and m points on the manipulator platform, where $n, m \in \{3, \ldots, 6\}$. There may be different ways for realizing such a mechanism. The most common one is to use six linear actuators for connecting the platforms together.

Determining the pose of the manipulator platform from the linear actuators' lengths (also known as direct kinematics problem) generally leads to a system of algebraic equations that has at most 40 different solutions [1–8]. This number of solutions can be further reduced by introducing additional constraints, for example, combinatorial or planarity constraints [9]. Nonetheless, a closed-form solution cannot be realized by only measuring the lengths of the linear actuators.

Current sensor concepts for solving the direct kinematics problem can be basically classified into two groups [10]. The first group consists of using the minimal number of sensors, in our case, six length sensors, and then including additional numerical procedures to uniquely identify the parallel mechanism's pose [11–20]. These procedures, however, are generally not real-time capable, require an initial estimate of the solution, and may exhibit convergence problems or even converge to a wrong solution. The requirement of an initial solution estimate is especially then problematic when starting the mechanism at an arbitrary pose.

In contrast, the second approach consists of adding extra sensors for obtaining additional information about the parallel mechanism's state [21–28]. These can be, for example, angular sensors that are placed on the base or the manipulator platform joints or linear and/or angular sensors that are placed on supplementary passive legs. Here, the number and location of the sensors must be carefully chosen because, otherwise, this may cause specific problems such as workspace limitations due to the passive leg or joint arrangement. Furthermore, using different sensor types leads to a higher complexity and may even negatively affect the performance due to possible time delays and/or conflicting measurement values.

For this reason, in order to provide a unique solution of the direct kinematics problem without using additional numerical procedures or sensors, instead of measuring the lengths of the linear actuators, we propose measuring their orientations and, if necessary, also the orientation of the manipulator platform. The orientations of the linear actuators and the roll-pitch orientation of the manipulator platform can be measured, for example, by acceleration sensors with three axes, and the measurement of the yaw orientation of the manipulator platform can be realized, for example, by using a magnetic sensor [29].

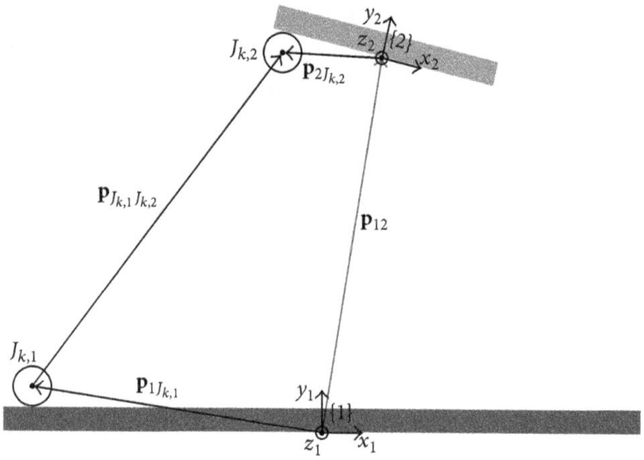

FIGURE 1: Nomenclature for the description of parallel mechanisms.

(a) 3-3 (⋀⋀)   (b) 6-3 (⋀ ⋀ |)   (c) 6-6 (|⁶)

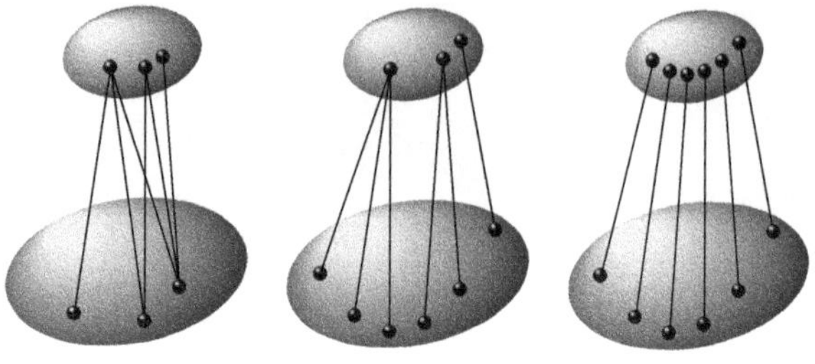

FIGURE 2: Examples of combinatorial classes of n-m mechanisms.

The remainder of this paper is organized as follows. In Section 2, a classification of six-degrees-of-freedom parallel mechanisms based on the number of base and manipulator platform joints as well as combinatorial classes is introduced. In Section 3, we investigate if a closed-form solution for the direct kinematics problem of the mechanism types presented in Section 2 is possible by only measuring the orientations of the linear actuators. In Section 4, for the mechanism types where a closed-form solution of the direct kinematics problem is not possible by only measuring the linear actuators' orientations, we also include the information about the roll-pitch orientation of the manipulator platform. In Section 5, we discuss the last remaining case where also the information about the manipulator platform's yaw orientation is included. In order to complete our systematic investigation, in Section 6, we extend our results to three-degrees-of-freedom planar mechanisms. Section 7 discusses some practical considerations regarding the sensor selection and implementation of the proposed algorithms in a real-time control. Finally, in Section 8, our results are summarized and discussed.

Throughout the paper, we use the following notation, referring to Figure 1. The body-fixed frame of the base platform is denoted as {1} and the body-fixed frame of the manipulator platform as {2}. The position vector of the kth joint $J_{k,i}$ of platform {i} is denoted as $\mathbf{p}_{iJ_{k,i}}$ and the connection vector between the joints $J_{k,1}$ and $J_{k,2}$ of platforms {1} and {2} as $\mathbf{p}_{J_{k,1}J_{k,2}}$ with $k \in \{1,\ldots,6\}$. Using inverse kinematics, this vector can be determined from

$$^1\mathbf{p}_{J_{k,1}J_{k,2}} = {}^1\mathbf{p}_{12} + {}^1\mathbf{R}_2 \cdot {}^2\mathbf{p}_{2J_{k,2}} - {}^1\mathbf{p}_{1J_{k,1}} \tag{1}$$

with respect to platform {1}. Here, $^1\mathbf{R}_2$ denotes the rotation matrix from frame {2} into frame {1}, and $\mathbf{p}_{12}$ is the vector connecting the origins of platforms {1} and {2}. The roll, pitch, and yaw angles of the manipulator platform shall be denoted as α, β, and γ, and the direction, or orientation, of $\mathbf{p}_{J_{k,1}J_{k,2}}$ is referred to as $\mathbf{r}_{J_{k,1}J_{k,2}}$, which has unit length.

2. Classification of Six-Degrees-of-Freedom Parallel Mechanisms

Typically, parallel mechanisms are classified by the number of joints on the base and the manipulator platform. This type of classification, however, is not sufficient for catching all descriptive parameters of a parallel mechanism. For this reason, Faugère and Lazard [9] introduced the notion of a combinatorial class, which is represented by a graph where the edges are the linear actuators and the vertices are the joints (see Figure 2). Here, we use both approaches together to classify parallel mechanisms.

2.1. n-3 Mechanisms. This group of mechanisms contains n base platform joints, with $n \in \{3,\ldots,6\}$, and three manipulator platform joints. Each manipulator platform joint is connected to one, two, or three linear actuators. We can classify this group into two types. In the first type, which shall be referred to as n-3-I mechanisms, all three manipulator platform joints are connected to exactly two linear actuators. According to [9], this type of mechanisms corresponds to seven combinatorial classes:

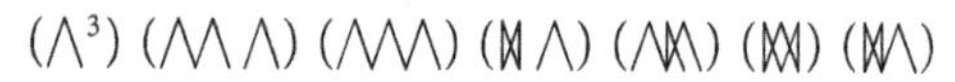

$$(\wedge^3)\ (\wedge\wedge\,\wedge)\ (\wedge\wedge\wedge)\ (\mathrm{N}\wedge)\ (\wedge\mathrm{N})\ (\mathrm{N}\mathrm{N})\ (\mathrm{N}\wedge) \tag{2}$$

In the second type, which shall be referred to as n-3-II mechanisms, the first manipulator platform joint is connected to three linear actuators, the second manipulator platform joint to two linear actuators, and the third manipulator platform joint to one linear actuator. This type of mechanisms corresponds to ten combinatorial classes [9]:

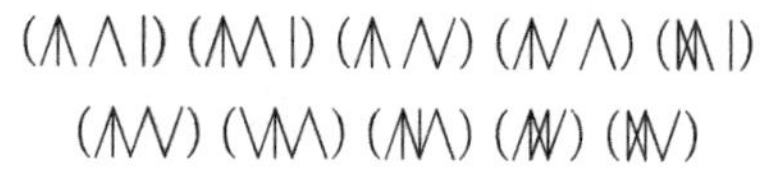

$$(\wedge\wedge|)\ (\mathrm{M}|)\ (\wedge\wedge\mathrm{N})\ (\mathrm{N}\wedge)\ (\mathrm{M}|)$$
$$(\wedge\mathrm{N})\ (\mathrm{V}\mathrm{M})\ (\mathrm{N}\wedge)\ (\mathrm{N})\ (\mathrm{N}\mathrm{N}) \tag{3}$$

2.2. n-4 Mechanisms. Similar to n-3 mechanisms, this group of mechanisms can be also classified into two types. The first type, which shall be referred to as n-4-I mechanisms, is characterized by two manipulator platform joints with each of them connected to two linear actuators and two further manipulator platform joints with each of them connected to one linear actuator. According to [9], this type of mechanisms corresponds to sixteen combinatorial classes:

$$
\begin{gathered}
(\wedge^2 |^2)\ (\wedge\wedge |^2)\ (\wedge\!\vee \wedge |)\ (\wedge^2 \vee)\ (\wedge\!\vee\!\vee |)\ (\wedge\!\vee^2) \\
(\bowtie |^2)\ (\wedge\wedge \vee)\ (\vee\!\!\wedge\!\wedge\ \wedge)\ (\wedge\!\vee\!\!\wedge |)\ (\vee\vee \wedge)\ (\wedge\!\vee\!\vee) \\
(\vee\!\wedge\!\wedge)\ (\bowtie\!\vee |)\ (\bowtie \vee)\ (\vee\vee\vee)
\end{gathered} \tag{4}
$$

The second type, which shall be referred to as n-4-II mechanisms, is characterized by one manipulator platform joint connected to three linear actuators and three further manipulator platform joints with each of them connected to one linear actuator. This type of n-4 mechanisms corresponds to nine combinatorial classes [9]:

$$
\begin{gathered}
(\pitchfork |^3)\ (\pitchfork\!\vee |^2)\ (\pitchfork \vee |)\ (\pitchfork\!\vee\!\vee |)\ (\vee\!\pitchfork\!\vee |) \\
(\pitchfork\!\vee \vee)\ (\vee\!\pitchfork |)\ (\vee\!\pitchfork\!\vee)\ (\vee\!\bowtie)
\end{gathered} \tag{5}
$$

2.3. n-5 Mechanisms. This group of mechanisms is described by twelve combinatorial classes [9]:

$$
\begin{gathered}
(\wedge |^4)\ (\wedge\!\vee |^3)\ (\wedge \vee |^2)\ (\Downarrow\!\vee \wedge |)\ (\vee\!\wedge\!\wedge |^2)\ (\vee^2 \wedge) \\
(\vee\!\wedge \vee |)\ (\vee\vee |^2)\ (\vee\!\wedge\!\wedge \vee)\ (\Downarrow\!\vee\ \vee\!\wedge)\ (\vee\!\vee\!\vee |)\ (\vee\vee \vee)
\end{gathered} \tag{6}
$$

2.4. n-6 Mechanisms. This group of mechanisms is associated with six combinatorial classes [9]:

$$
(|^6)\ (\vee |^4)\ (\Downarrow\!\vee |^3)\ (\vee^2 |^2)\ (\Downarrow\!\vee \vee |)\ (\vee^3) \tag{7}
$$

3. Closed-Form Solution by Only Measuring the Linear Actuators' Orientations

In this section, we will investigate if a closed-form solution for the direct kinematics problem of the mechanisms introduced in the previous section is possible by only measuring the orientations of the linear actuators.

3.1. *n-3 Mechanisms*

3.1.1. Type I. Consider an n-3-I mechanism where the linear actuators $k = 1$ and $k = 2$ are connected to the manipulator platform joint $J_{1,2}$, the linear actuators $k = 3$ and $k = 4$ to the manipulator platform joint $J_{2,2}$, and the linear actuators $k = 5$ and $k = 6$ to the manipulator platform joint $J_{3,2}$. The positions of these joints, ${}^1\mathbf{p}_{1J_{1,2}}$, ${}^1\mathbf{p}_{1J_{2,2}}$, and ${}^1\mathbf{p}_{1J_{3,2}}$, are defined by the intersection points between the straight lines g_k through the linear actuators $k = 1, \ldots, 6$ with

$$
g_k : {}^1\mathbf{p}_k = {}^1\mathbf{p}_{1J_{k,1}} + \lambda_k \cdot {}^1\mathbf{r}_{J_{k,1}J_{k,2}}, \quad \lambda_k \in \mathbb{R}, \tag{8}
$$

where ${}^1\mathbf{p}_k$ denote the coordinates of g_k and ${}^1\mathbf{r}_{J_{k,1}J_{k,2}}$ the measured orientations of the linear actuators. In particular, the intersection point between g_1 and g_2 defines ${}^1\mathbf{p}_{1J_{1,2}}$, the intersection point between g_3 and g_4 defines ${}^1\mathbf{p}_{1J_{2,2}}$, and the intersection point between g_5 and g_6 defines ${}^1\mathbf{p}_{1J_{3,2}}$. These three joint positions, on the other hand, define a plane with the normal vector

$$
{}^1\mathbf{n} = {}^1\mathbf{p}_{J_{1,2}J_{2,2}} \times {}^1\mathbf{p}_{J_{1,2}J_{3,2}}, \tag{9}
$$

for example, where

$$
\begin{aligned}
{}^1\mathbf{p}_{J_{1,2}J_{2,2}} &= {}^1\mathbf{p}_{1J_{2,2}} - {}^1\mathbf{p}_{1J_{1,2}}, \\
{}^1\mathbf{p}_{J_{1,2}J_{3,2}} &= {}^1\mathbf{p}_{1J_{3,2}} - {}^1\mathbf{p}_{1J_{1,2}}.
\end{aligned} \tag{10}
$$

The orientation angles α, β, and γ of the manipulator platform can be then determined from

$$
\begin{aligned}
\alpha &= \cos^{-1}\left(\frac{{}^1\mathbf{e}_{z_1} \cdot {}^1\mathbf{n}^{y_1 z_1}}{\left|{}^1\mathbf{n}^{y_1 z_1}\right|}\right), \\
\beta &= \cos^{-1}\left(\frac{{}^1\mathbf{e}_{z_1} \cdot {}^1\mathbf{n}^{x_1 z_1}}{\left|{}^1\mathbf{n}^{x_1 z_1}\right|}\right), \\
\gamma &= \cos^{-1}\left(\frac{{}^1\mathbf{e}_{x_1} \cdot {}^1\mathbf{p}^{x_1 y_1}_{J_{1,2}J_{2,2}}}{\left|{}^1\mathbf{p}^{x_1 y_1}_{J_{1,2}J_{2,2}}\right|}\right) \\
&= \cos^{-1}\left(\frac{{}^1\mathbf{e}_{y_1} \cdot {}^1\mathbf{p}^{x_1 y_1}_{J_{1,2}J_{2,2}}}{\left|{}^1\mathbf{p}^{x_1 y_1}_{J_{1,2}J_{2,2}}\right|}\right),
\end{aligned} \tag{11}
$$

where ${}^1\mathbf{e}_{x_1}$, ${}^1\mathbf{e}_{y_1}$, and ${}^1\mathbf{e}_{z_1}$ are the unit vectors of the base platform in x_1, y_1, and z_1 direction, ${}^1\mathbf{n}^{y_1 z_1}$ is the projection of ${}^1\mathbf{n}$ on the y_1-z_1 plane, ${}^1\mathbf{n}^{x_1 z_1}$ is the projection of ${}^1\mathbf{n}$ on the x_1-z_1 plane, and ${}^1\mathbf{p}^{x_1 y_1}_{J_{1,2}J_{2,2}}$ is the projection of ${}^1\mathbf{p}_{J_{1,2}J_{2,2}}$ on the x_1-y_1 plane with

$$
\begin{aligned}
{}^1\mathbf{n}^{y_1 z_1} &= \left({}^1\mathbf{n} \cdot {}^1\mathbf{e}_{y_1}\right) \cdot {}^1\mathbf{e}_{y_1} + \left({}^1\mathbf{n} \cdot {}^1\mathbf{e}_{z_1}\right) \cdot {}^1\mathbf{e}_{z_1}, \\
{}^1\mathbf{n}^{x_1 z_1} &= \left({}^1\mathbf{n} \cdot {}^1\mathbf{e}_{x_1}\right) \cdot {}^1\mathbf{e}_{x_1} + \left({}^1\mathbf{n} \cdot {}^1\mathbf{e}_{z_1}\right) \cdot {}^1\mathbf{e}_{z_1}, \\
{}^1\mathbf{p}^{x_1 y_1}_{J_{1,2}J_{2,2}} &= \left({}^1\mathbf{p}_{J_{1,2}J_{2,2}} \cdot {}^1\mathbf{e}_{x_1}\right) \cdot {}^1\mathbf{e}_{x_1} + \left({}^1\mathbf{p}_{J_{1,2}J_{2,2}} \cdot {}^1\mathbf{e}_{y_1}\right) \\
&\quad \cdot {}^1\mathbf{e}_{y_1}.
\end{aligned} \tag{12}
$$

The manipulator platform's position ${}^1\mathbf{p}_{12}$, on the other hand, can be obtained, for example, from

$$
{}^1\mathbf{p}_{12} = {}^1\mathbf{p}_{1J_{1,2}} - {}^1\mathbf{R}_2 \cdot {}^2\mathbf{p}_{2J_{1,2}}, \tag{13}
$$

where

$$
{}^1\mathbf{R}_2 = {}^1\mathbf{R}_{2,\alpha} \cdot {}^1\mathbf{R}_{2,\beta} \cdot {}^1\mathbf{R}_{2,\gamma} \tag{14}
$$

with

$$
{}^1\mathbf{R}_{2,\alpha} = \begin{bmatrix} 1 & 0 & 0 \\ 0 & \cos\alpha & \sin\alpha \\ 0 & -\sin\alpha & \cos\alpha \end{bmatrix},
$$

$$
{}^{1}\mathbf{R}_{2,\beta} = \begin{bmatrix} \cos\beta & 0 & -\sin\beta \\ 0 & 1 & 0 \\ \sin\beta & 0 & \cos\beta \end{bmatrix},
$$

$$
{}^{1}\mathbf{R}_{2,\gamma} = \begin{bmatrix} \cos\gamma & \sin\gamma & 0 \\ -\sin\gamma & \cos\gamma & 0 \\ 0 & 0 & 1 \end{bmatrix}. \tag{15}
$$

We can see that, by only measuring the orientations of the linear actuators, a unique solution for the direct kinematics problem of n-3-I mechanisms can be found.

3.1.2. Type II. Now, consider an n-3-II mechanism where the linear actuators $k = 1$, $k = 2$, and $k = 3$ are connected to the manipulator platform joint $J_{1,2}$, the linear actuators $k = 4$ and $k = 5$ are connected to the manipulator platform joint $J_{2,2}$, and the linear actuator $k = 6$ is connected to the manipulator platform joint $J_{3,2}$. The position ${}^{1}\mathbf{p}_{1J_{1,2}}$ of the joint $J_{1,2}$ is defined by the intersection point between two of the straight lines g_1, g_2, and g_3 through the linear actuators $k = 1$, $k = 2$, and $k = 3$. Hence, only two orientations of these linear actuators are necessary for defining this position. The position ${}^{1}\mathbf{p}_{1J_{2,2}}$ of the joint $J_{2,2}$, on the other hand, is defined by the intersection point between the straight lines g_4 and g_5 through the linear actuators $k = 4$ and $k = 5$. Both positions define a straight line around which the manipulator platform can virtually rotate. We can now define a sphere, for example, with the center ${}^{1}\mathbf{p}_{1J_{2,2}}$ and the radius $|{}^{2}\mathbf{p}_{J_{2,2}J_{3,2}}|$. The intersection points between the sphere and the straight line g_6 through the linear actuator $k = 6$ define the two possible positions ${}^{1}\mathbf{p}_{1J_{3,2}}$ of the manipulator platform joint $J_{3,2}$. So, by only measuring the orientations of the linear actuators, it is not possible to find a unique solution for the direct kinematics problem of n-3-II mechanisms by only measuring the linear actuators' orientations.

3.2. n-4 Mechanisms

3.2.1. Type I. Consider an n-4-I mechanism where the linear actuators $k = 1$ and $k = 2$ are connected to the manipulator platform joint $J_{1,2}$, the linear actuators $k = 3$ and $k = 4$ are connected to the manipulator platform joint $J_{2,2}$, the linear actuator $k = 5$ is connected to the manipulator platform joint $J_{3,2}$, and the linear actuator $k = 6$ is connected to the manipulator platform joint $J_{4,2}$. The position ${}^{1}\mathbf{p}_{1J_{1,2}}$ of the manipulator platform joint $J_{1,2}$ is defined by the intersection point between the straight lines g_1 and g_2 through the linear actuators $k = 1$ and $k = 2$, and the position ${}^{1}\mathbf{p}_{1J_{2,2}}$ of the manipulator platform joint $J_{2,2}$ is defined by the intersection point between the straight lines g_3 and g_4 through the linear actuators $k = 3$ and $k = 4$. Both positions define a straight line around which the manipulator platform can virtually rotate. We can now define a sphere, for example, with the center ${}^{1}\mathbf{p}_{1J_{2,2}}$ and the radius $|{}^{2}\mathbf{p}_{J_{2,2}J_{3,2}}|$. The intersection points between the sphere and the straight line g_5 through the linear actuator $k = 5$ define the two possible positions ${}^{1}\mathbf{p}_{1J_{3,2}}$ of the manipulator platform joint $J_{3,2}$. So, it is not possible to find a unique solution for the direct kinematics problem of n-4-I mechanisms by only measuring the orientations of the linear actuators.

3.2.2. Type II. Now, consider an n-4-II mechanism where the linear actuators $k = 1$, $k = 2$, and $k = 3$ are connected to the manipulator platform joint $J_{1,2}$, the linear actuator $k = 4$ is connected to the manipulator platform joint $J_{2,2}$, the linear actuator $k = 5$ is connected to the manipulator platform joint $J_{3,2}$, and the linear actuator $k = 6$ is connected to the manipulator platform joint $J_{4,2}$. The position ${}^{1}\mathbf{p}_{1J_{1,2}}$ of the joint $J_{1,2}$ is defined by the intersection point between two of the straight lines g_1, g_2, and g_3 through the linear actuators $k = 1$, $k = 2$, and $k = 3$. Hence, only two orientations of these linear actuators are necessary for defining this position. We can now define two spheres, for example, the first sphere with the center ${}^{1}\mathbf{p}_{1J_{1,2}}$ and the radius $|{}^{2}\mathbf{p}_{J_{1,2}J_{2,2}}|$ and the second sphere with the same center but with the radius $|{}^{2}\mathbf{p}_{J_{1,2}J_{3,2}}|$. The intersection points between the first sphere and the straight line g_4 through the linear actuator $k = 4$ define the two possible positions ${}^{1}\mathbf{p}_{1J_{2,2}}$ of the manipulator platform joint $J_{2,2}$, and the intersection points between the second sphere and the straight line g_5 through the linear actuator $k = 5$ define the two possible positions ${}^{1}\mathbf{p}_{1J_{3,2}}$ of the manipulator platform joint $J_{3,2}$. So, in total, we obtain four different possible orientations of the manipulator platform, and it is hence not possible to find a unique solution for the direct kinematics problem of n-4-II mechanisms by only measuring the linear actuators' orientations.

3.3. n-5 Mechanisms. Consider an n-5 mechanism where the linear actuators $k = 1$ and $k = 2$ are connected to the manipulator platform joint $J_{1,2}$, the linear actuator $k = 3$ is connected to the manipulator platform joint $J_{2,2}$, the linear actuator $k = 4$ is connected to the manipulator platform joint $J_{3,2}$, the linear actuator $k = 5$ is connected to the manipulator platform joint $J_{4,2}$, and the linear actuator $k = 6$ is connected to the manipulator platform joint $J_{5,2}$. The position ${}^{1}\mathbf{p}_{1J_{1,2}}$ of the manipulator platform joint $J_{1,2}$ is defined by the intersection point between the straight lines g_1 and g_2 through the linear actuators $k = 1$ and $k = 2$. We can now define two spheres, for example, the first sphere with the center ${}^{1}\mathbf{p}_{1J_{1,2}}$ and the radius $|{}^{2}\mathbf{p}_{J_{1,2}J_{2,2}}|$ and the second sphere with the same center but with the radius $|{}^{2}\mathbf{p}_{J_{1,2}J_{3,2}}|$. The intersection points between the first sphere and the straight line g_3 through the linear actuator $k = 3$ define the two possible positions ${}^{1}\mathbf{p}_{1J_{2,2}}$ of the manipulator platform joint $J_{2,2}$, and the intersection points between the second sphere and the straight line g_4 through the linear actuator $k = 4$ define the two possible positions ${}^{1}\mathbf{p}_{1J_{3,2}}$ of the manipulator platform joint $J_{3,2}$. So, in total, we obtain four different possible orientations of the manipulator platform, and it is hence not possible to find a unique solution for the direct kinematics problem of n-5 mechanisms by only measuring the orientations of the linear actuators.

3.4. *n-6 Mechanisms.* Consider an n-6 mechanism where the linear actuators $k \in \{1, \ldots, 6\}$ are connected to the manipulator platform joints $J_{k,2}$. We can now choose three arbitrary linear actuators l, p, and q with $l, p, q \in \{1, \ldots, 6\}$ and $l \neq p \neq q$ and define three straight lines through these linear actuators:

$$g_l : {}^1\mathbf{p}_l = {}^1\mathbf{p}_{1J_{l,1}} + \lambda_l \cdot {}^1\mathbf{r}_{J_{l,1}J_{l,2}}, \quad \lambda_l \in \mathbb{R},$$
$$g_p : {}^1\mathbf{p}_p = {}^1\mathbf{p}_{1J_{p,1}} + \lambda_p \cdot {}^1\mathbf{r}_{J_{p,1}J_{p,2}}, \quad \lambda_p \in \mathbb{R}, \qquad (16)$$
$$g_q : {}^1\mathbf{p}_q = {}^1\mathbf{p}_{1J_{q,1}} + \lambda_q \cdot {}^1\mathbf{r}_{J_{q,1}J_{q,2}}, \quad \lambda_q \in \mathbb{R}.$$

Here, ${}^1\mathbf{p}_l$, ${}^1\mathbf{p}_p$, and ${}^1\mathbf{p}_q$ denote the coordinates of g_l, g_p, and g_q, and ${}^1\mathbf{r}_{J_{l,1}J_{l,2}}$, ${}^1\mathbf{r}_{J_{p,1}J_{p,2}}$, and ${}^1\mathbf{r}_{J_{q,1}J_{q,2}}$ denote the measured orientations of the linear actuators $k = l$, $k = p$, and $k = q$. Next, we can define two spheres, the first sphere with the center ${}^1\mathbf{p}_l$ and the radius $|{}^2\mathbf{p}_{J_{l,2}J_{p,2}}|$ and the second sphere with the same center but with the radius $|{}^2\mathbf{p}_{J_{l,2}J_{q,2}}|$. In this context, the first sphere has to intersect g_p and the second sphere g_q, so that we can write the following two equations:

$$\left({}^1\mathbf{p}_p - {}^1\mathbf{p}_l\right)^2 = \left|{}^2\mathbf{p}_{J_{l,2}J_{p,2}}\right|^2, \qquad (17)$$

$$\left({}^1\mathbf{p}_q - {}^1\mathbf{p}_l\right)^2 = \left|{}^2\mathbf{p}_{J_{l,2}J_{q,2}}\right|^2. \qquad (18)$$

Since we know the angle between ${}^2\mathbf{p}_{J_{l,2}J_{p,2}}$ and ${}^2\mathbf{p}_{J_{l,2}J_{q,2}}$, we can also write

$$\left({}^1\mathbf{p}_p - {}^1\mathbf{p}_l\right) \cdot \left({}^1\mathbf{p}_q - {}^1\mathbf{p}_l\right) = {}^2\mathbf{p}_{J_{l,2}J_{p,2}} \cdot {}^2\mathbf{p}_{J_{l,2}J_{q,2}}. \qquad (19)$$

We now have a system of three nonlinear equations, (17), (18), and (19), in three variables (λ_l, λ_p, and λ_q), which, in general, is uniquely not solvable. So, it is not possible to find a unique solution for the direct kinematics problem of n-6 mechanisms by only measuring the linear actuators' orientations.

4. Closed-Form Solution by Measuring the Linear Actuators' Orientations and the Roll-Pitch Orientation of the Manipulator Platform

In Section 3, we have shown that, by only measuring the orientations of the linear actuators, it is only possible to find a unique solution for n-3-I mechanisms. In this section, we will investigate if a closed-form solution for the direct kinematics problem is possible by also including the information about the roll-pitch orientation of the manipulator platform.

Consider an n-m mechanism with $n \in \{3, \ldots, 6\}$ and $m \in \{3, 4, 5\}$. Each of these mechanisms contains at least two linear actuators that are connected to one manipulator platform joint $J_{l,2}$. We assume that the measured roll-pitch orientation of the manipulator platform is given by the unit normal vector ${}^1\overline{\mathbf{n}}$. The position ${}^1\mathbf{p}_{1J_{l,2}}$ of the manipulator platform joint $J_{l,2}$ and the unit normal vector ${}^1\overline{\mathbf{n}}$ define a plane, and the desired positions ${}^1\mathbf{p}_{1J_{p,2}}$ and ${}^1\mathbf{p}_{1J_{q,2}}$ of two other

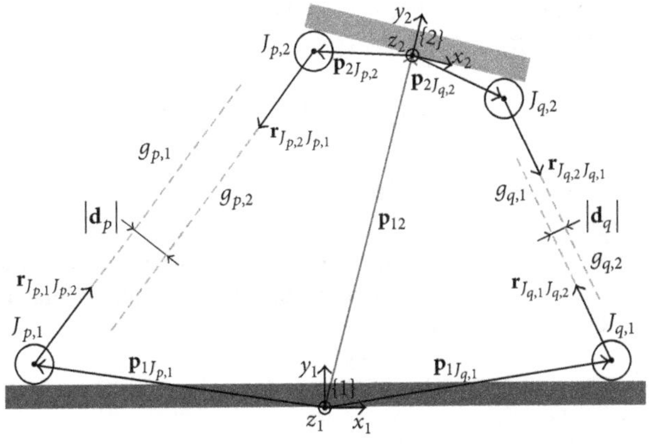

FIGURE 3: Schematic diagram for the algorithm from Section 5.

linear actuators $k = p$ and $k = q$, where $p \neq q$, are defined by the intersection points between these two linear actuators and the plane. So, by measuring the orientations of four linear actuators as well as the roll-pitch orientation of the manipulator platform, it is possible to obtain a closed-form solution for the direct kinematics problem of n-m mechanisms with $m \neq 6$. For $m = 6$, however, our solution strategy fails since n-6 mechanisms do not have a common connection of at least two linear actuators at the manipulator platform. In this case, the information about the manipulator platform's orientation leads to the following plane equation:

$$\left({}^1\mathbf{p}_p - {}^1\mathbf{p}_l\right) \cdot {}^1\overline{\mathbf{n}} = 0. \qquad (20)$$

We now have a system of two nonlinear equations, (17) and (20), in two variables (λ_l and λ_p), which, in general, is uniquely not solvable. So, by including the information about the roll-pitch orientation of the manipulator platform, it is not possible to find a unique solution for the direct kinematics problem of n-6 mechanisms.

5. Closed-Form Solution by Measuring the Linear Actuators' Orientations and the Roll-Pitch-Yaw Orientation of the Manipulator Platform

We have seen in Section 4 that the information about the orientations of the linear actuators and the roll-pitch orientation of the manipulator platform is not enough to obtain a closed-form solution for the direct kinematics problem of n-6 mechanisms. However, we have shown in [29] that, by also including the information about the yaw orientation of the manipulator platform, the direct kinematics problem of n-6 mechanisms can be uniquely solved. In the following, we will review our solution concept from [29] in the context of a general n-m mechanism with $n, m \in \{3, \ldots, 6\}$.

Consider the orientations ${}^1\mathbf{r}_{J_{p,1}J_{p,2}}$ and ${}^1\mathbf{r}_{J_{q,1}J_{q,2}}$ of two arbitrarily chosen linear actuators $k = p$ and $k = q$, where $p \neq q$. These orientations define two pairs of straight lines, $g_{p,1}$ and $g_{p,2}$ as well as $g_{q,1}$ and $g_{q,2}$, with the base vectors ${}^1\mathbf{p}_{1J_{p,1}}$ and ${}^1\mathbf{p}_{2J_{p,2}}$ as well as ${}^1\mathbf{p}_{1J_{q,1}}$ and ${}^1\mathbf{p}_{2J_{q,2}}$ (see Figure 3). We can now define two distance vectors between these two pairs of straight lines:

$$\mathbf{d}_p = {}^1\mathbf{r}_{J_{p,1}J_{p,2}} \times \left({}^1\mathbf{p}_{12} + {}^1\mathbf{R}_2 \cdot {}^2\mathbf{p}_{2J_{p,2}} - {}^1\mathbf{p}_{1J_{p,1}}\right),$$
$$\mathbf{d}_q = {}^1\mathbf{r}_{J_{q,1}J_{q,2}} \times \left({}^1\mathbf{p}_{12} + {}^1\mathbf{R}_2 \cdot {}^2\mathbf{p}_{2J_{q,2}} - {}^1\mathbf{p}_{1J_{q,1}}\right), \tag{21}$$

where the measured roll, pitch, and yaw orientation of the manipulator platform is summarized in the rotation matrix ${}^1\mathbf{R}_2$. Using the identity

$$\mathbf{a} \times \mathbf{b} \equiv \tilde{\mathbf{a}} \cdot \mathbf{b} \tag{22}$$

where

$$\tilde{\mathbf{a}} = \begin{bmatrix} 0 & -a_z & a_y \\ a_z & 0 & -a_x \\ -a_y & a_x & 0 \end{bmatrix}, \tag{23}$$

we can rewrite (21) as

$$\begin{aligned}
\mathbf{d}_p &= {}^1\tilde{\mathbf{r}}_{J_{p,1}J_{p,2}} \cdot \left({}^1\mathbf{p}_{12} + {}^1\mathbf{R}_2 \cdot {}^2\mathbf{p}_{2J_{p,2}} - {}^1\mathbf{p}_{1J_{p,1}}\right) \\
&= \underbrace{{}^1\tilde{\mathbf{r}}_{J_{p,1}J_{p,2}}}_{=:\mathbf{A}_p} \cdot \underbrace{{}^1\mathbf{p}_{12}}_{=:\mathbf{x}} \\
&\quad + \underbrace{{}^1\tilde{\mathbf{r}}_{J_{p,1}J_{p,2}} \cdot \left({}^1\mathbf{R}_2 \cdot {}^2\mathbf{p}_{2J_{p,2}} - {}^1\mathbf{p}_{1J_{p,1}}\right)}_{=:-\mathbf{c}_p}, \\
\mathbf{d}_q &= {}^1\tilde{\mathbf{r}}_{J_{q,1}J_{q,2}} \cdot \left({}^1\mathbf{p}_{12} + {}^1\mathbf{R}_2 \cdot {}^2\mathbf{p}_{2J_{q,2}} - {}^1\mathbf{p}_{1J_{q,1}}\right) \\
&= \underbrace{{}^1\tilde{\mathbf{r}}_{J_{q,1}J_{q,2}}}_{=:\mathbf{A}_q} \cdot \underbrace{{}^1\mathbf{p}_{12}}_{=:\mathbf{x}} \\
&\quad + \underbrace{{}^1\tilde{\mathbf{r}}_{J_{q,1}J_{q,2}} \cdot \left({}^1\mathbf{R}_2 \cdot {}^2\mathbf{p}_{2J_{q,2}} - {}^1\mathbf{p}_{1J_{q,1}}\right)}_{=:-\mathbf{c}_q}.
\end{aligned} \tag{24}$$

Now, in order to find the unknown position ${}^1\mathbf{p}_{12} =: \mathbf{x}$, we have to solve the linear least-squares problem

$$\|\mathbf{A}\mathbf{x} - \mathbf{c}\|_2 = \min!, \tag{25}$$

where

$$\mathbf{A} = \begin{bmatrix} \mathbf{A}_p \\ \mathbf{A}_q \end{bmatrix} \in \mathbb{R}^{6\times 3}, \qquad \mathbf{c} = \begin{bmatrix} \mathbf{c}_p \\ \mathbf{c}_q \end{bmatrix} \in \mathbb{R}^{6}. \tag{26}$$

This linear least-squares problem can be reduced to the set of linear equations

$$\Big(\underbrace{\mathbf{A}^\top\mathbf{A}}_{=:\overline{\mathbf{A}}}\Big)\mathbf{x} = \underbrace{\mathbf{A}^\top\mathbf{c}}_{=:\overline{\mathbf{c}}} \tag{27}$$

with the unique solution

$$\mathbf{x} = \overline{\mathbf{A}}^{-1}\overline{\mathbf{c}} = \left(\mathbf{A}^\top\mathbf{A}\right)^{-1}\mathbf{A}^\top\mathbf{c}. \tag{28}$$

The measurement of the orientations of the other four linear actuators is not necessary here since the positions of the corresponding manipulator platform joints are defined by the manipulator platform's geometry. Note that a robust way of computing $\mathbf{x}$ in this case is by using, for example, the QR decomposition [30].

6. Extension to Planar Mechanisms

In this section, we extend our previous findings for spatial mechanisms to three-degrees-of-freedom planar mechanisms. Note that, here, only the roll orientation of the manipulator platform is measured due to the planarity constraint.

6.1. n-2 Mechanisms. This group of mechanisms contains n base platform joints, with $n \in \{2, 3\}$, and two manipulator platform joints. It can be described by two combinatorial classes:

$$(\wedge|)\ (\wedge\!/) \tag{29}$$

Now, assume that the linear actuators $k = 1$ and $k = 2$ are connected to the manipulator platform joint $J_{1,2}$, and the linear actuator $k = 3$ is connected to the manipulator platform joint $J_{2,2}$. The position ${}^1\mathbf{p}_{1J_{1,2}}$ of the manipulator platform joint $J_{1,2}$ is defined by the intersection point between the straight lines g_1 and g_2 through the linear actuators $k = 1$ and $k = 2$. We can now define a circle with the center ${}^1\mathbf{p}_{1J_{1,2}}$ and the radius $|{}^2\mathbf{p}_{J_{1,2}J_{2,2}}|$. The intersection points between the circle and the straight line g_3 through the linear actuator $k = 3$ define the two possible positions ${}^1\mathbf{p}_{1J_{2,2}}$ of the manipulator platform joint $J_{2,2}$. So, by only measuring the orientations of the linear actuators, it is not possible to find a unique solution for the direct kinematics problem of n-2 mechanisms.

In the next step, we assume that the roll orientation of the manipulator platform is measured in terms of the unit normal vector ${}^1\overline{\mathbf{n}}$. Then, the angle γ between the manipulator and the base platform can be determined as follows:

$$\gamma = \cos^{-1}\left({}^1\mathbf{e}_{y_1} \cdot {}^1\overline{\mathbf{n}}\right), \tag{30}$$

where ${}^1\mathbf{e}_{y_1}$ denotes the unit vector of the base platform in y_1 direction. The manipulator platform's position ${}^1\mathbf{p}_{12}$, on the other hand, can be obtained, for example, from

$${}^1\mathbf{p}_{12} = {}^1\mathbf{p}_{1J_{1,2}} - {}^1\mathbf{R}_2 \cdot {}^2\mathbf{p}_{2J_{1,2}}, \tag{31}$$

where

$${}^1\mathbf{R}_2 = \begin{bmatrix} \cos\gamma & \sin\gamma \\ -\sin\gamma & \cos\gamma \end{bmatrix}. \tag{32}$$

The measurement of the orientation of the third linear actuator is not necessary here, since the position of the corresponding manipulator platform joint is defined by the manipulator platform's geometry.

6.2. n-3 Mechanisms. In contrast to n-2 mechanisms, this group of mechanisms contains three manipulator platform joints. It can be also described by two combinatorial classes:

$$(|^3)\ (\vee\,|) \tag{33}$$

The measured orientations ${}^1\mathbf{r}_{J_{k,1}J_{k,2}}$ define the straight lines g_k through the linear actuators $k = 1, \ldots, 3$ with

$$g_k : {}^1\mathbf{p}_k = {}^1\mathbf{p}_{1J_{k,1}} + \lambda_k \cdot {}^1\mathbf{r}_{J_{k,1}J_{k,2}}, \quad \lambda_k \in \mathbb{R}, \tag{34}$$

where ${}^1\mathbf{p}_k$ denote the coordinates of g_k. We can now define two circles, for example, the first circle with the center ${}^1\mathbf{p}_1$ and the radius $|{}^2\mathbf{p}_{J_{1,2}J_{2,2}}|$ and the second circle with the same center but with the radius $|{}^2\mathbf{p}_{J_{1,2}J_{3,2}}|$. In this context, the first circle has to intersect g_2 and the second circle g_3, so that we can write the following two equations:

$$\begin{aligned} \left({}^1\mathbf{p}_2 - {}^1\mathbf{p}_1\right)^2 &= \left|{}^2\mathbf{p}_{J_{1,2}J_{2,2}}\right|^2, \\ \left({}^1\mathbf{p}_3 - {}^1\mathbf{p}_1\right)^2 &= \left|{}^2\mathbf{p}_{J_{1,2}J_{3,2}}\right|^2. \end{aligned} \tag{35}$$

Since ${}^2\mathbf{p}_{J_{1,2}J_{2,2}}$ and ${}^2\mathbf{p}_{J_{1,2}J_{3,2}}$ are linearly dependent, we can also write

$$\left({}^1\mathbf{p}_2 - {}^1\mathbf{p}_1\right) \cdot \left({}^1\mathbf{p}_3 - {}^1\mathbf{p}_1\right) = 1. \tag{36}$$

We now have a system of three nonlinear equations, (35) and (36), in three variables (λ_1, λ_2, and λ_3), which, in general, is uniquely not solvable. So, it is not possible to find a unique solution for the direct kinematics problem of n-3 mechanisms by only measuring the orientations of the linear actuators. However, by also including the roll orientation in terms of the rotation matrix (32), we can always apply our general algorithm from Section 5, which always provides a closed-form solution of the direct kinematics problem. Note that, in this case, only the measurement of the orientations of two linear actuators is necessary, since the position of the third manipulator platform joint is defined by the manipulator platform's geometry.

7. Practical Considerations

Currently, there are many possible sensors available on the market, spreading from very expensive, precalibrated high-precision sensors to uncalibrated low-cost sensors. In [29], we used the InvenSense MPU-9150 inertial measurement units (IMUs) consisting of an acceleration sensor with three axes, a gyroscope, and a magnetic sensor for determining the closed-form solution for the direct kinematics problem of a general n-m mechanism. We obtained the correct solution for selected static poses, but the results showed relatively high mean errors and standard deviations, especially for the yaw orientation of the manipulator platform. This was primarily caused by the noisy and uncalibrated IMUs. The calibration problem, however, can be solved without any additional external equipment by using the approach from [31]. For example, the acceleration sensor has to be calibrated/corrected in terms of sensor bias, scaling error, and nonorthogonality. In this context, the calibrated measurement data $\mathbf{a}_{\text{cal}}$ can be obtained by the following transformation:

$$\mathbf{a}_{\text{cal}} = \mathbf{TS}\left(\mathbf{a}_{\text{u}} + \mathbf{b}\right), \tag{37}$$

where $\mathbf{a}_{\text{u}}$ is the uncalibrated acceleration vector and $\mathbf{b}$ a constant bias term. The matrix $\mathbf{S}$ is a diagonal matrix comprising of scaling factors in each axis, and $\mathbf{T}$ is an upper triangular matrix to correct nonorthogonality.

Another problem related to calibration is sensor placement. In [29], we mounted the IMUs on the gearboxes of the linear actuators and on top of the manipulator platform. In this context, the alignment of the IMUs regarding the base platform's coordinate system has to be determined very carefully. One possibility is to use very precise measurement equipment, for example, by using optical or angular sensors to obtain the location of the IMUs on the linear actuators. An alternative way is to determine the sensor alignment by comparing the target orientations with the measured orientations for several predefined poses.

In order to achieve a closed-form solution of the direct kinematics problem in hard real-time, the introduced algorithms where the linear actuators' orientations and, if necessary, the roll-pitch orientation of the manipulator platform are measured are preferable due to the precision of the acceleration sensors. However, the algorithm where also the yaw orientation of the manipulator platform is needed requires the usage of a magnetic sensor, which, in general, is very imprecise. Several information filters, such as Kalman filter or complementary filters, were proposed to obtain the optimal measurement data fusion (see, e.g., [32, 33]). These filters can perform very quickly (between 1.3 and 7 μs [32]), but they do not calculate the correct yaw orientation with the first measurement value. Instead, the calculated yaw orientation only converges towards the correct value within several measurements.

As already mentioned above, we tested our approach for the general n-m mechanism on several static poses [29]. In the static case, the acceleration sensors measure the constant gravity vector of the earth without any disturbances. Under dynamic conditions, however, in addition to the earth's gravity field, the acceleration sensors also measure the acceleration of the mechanism itself. Since we can also measure the angular velocities by the available gyroscopes, we are able to compensate these erroneous measurements. In particular, by implementing an information filter for fusing the measurement data of the acceleration sensors, the gyroscopes, and the magnetic sensors of the IMUs, the orientations of the linear actuators can be robustly obtained (see, e.g., [34]).

Figure 4 shows the concept for the pose control of parallel mechanisms associated with the introduced algorithms for solving the direct kinematics problem. Here, a target pose $\mathbf{p}_{\text{target}}$ is defined and compared with the actual pose $\mathbf{p}_{\text{is}}$ leading to the pose deviation $\Delta\mathbf{p}$. By using inverse kinematics, we can convert $\Delta\mathbf{p}$ into the required length deviation $\Delta\mathbf{l}$ of

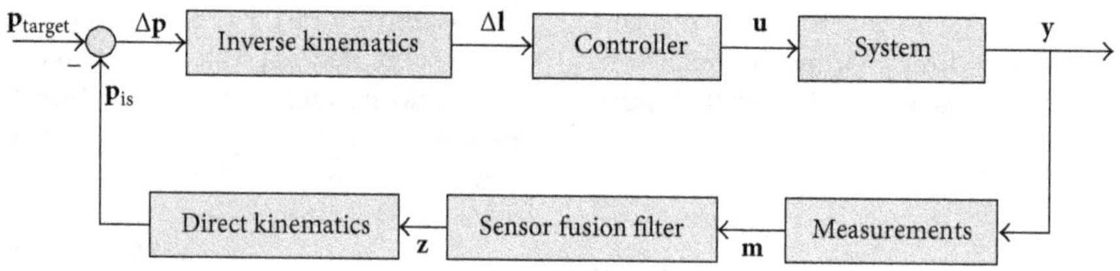

FIGURE 4: Pose control concept for parallel mechanisms using the algorithms from Sections 3–6.

the linear actuators and give it to the controller, for example, a PID controller, as input. The controller then generates the control input $\mathbf{u}$ for the system that, in turn, produces the system output $\mathbf{y}$. The measurement vector $\mathbf{m}$ that includes the raw data of the acceleration sensors, the gyroscopes, and the magnetic sensors is sent to the sensor fusion filter, for example, a Kalman filter. Here, the orientations of the linear actuators and, if necessary, also the orientation of the manipulator platform are calculated. The filter output $\mathbf{z}$ is then used to calculate the actual pose of the manipulator platform $\mathbf{p}_{is}$ by using the algorithms introduced in Sections 3–6.

In conclusion, the pose of an n-m parallel mechanism can be determined by only measuring the linear actuators' orientations and, if necessary, the orientation of the manipulator platform. The accuracy mainly depends on three things: (1) the precision of the used sensors, (2) their calibration and accurate alignment on the linear actuators and the manipulator platform, and (3) whether we have to measure the yaw orientation or not. For a dynamic pose determination, we have to estimate the linear actuators' orientations by a suitable sensor fusion.

8. Conclusions

We showed that, for n-3-I mechanisms, it is possible to find a unique solution for the direct kinematics problem by only measuring the orientations of the linear actuators. By also including the information about the roll-pitch orientation of the manipulator platform, it is also possible to uniquely solve the direct kinematics problem for n-3-II, n-4, and n-5 mechanisms. Finally, we demonstrated that the most general case of n-6 mechanisms also requires the information about the yaw orientation of the manipulator platform.

We then extended our approach to planar mechanisms and showed that the direct kinematics problem can be uniquely solved by measuring the linear actuators' orientations and the roll orientation of the manipulator platform.

The results suggest that, in most cases, it is not even necessary to measure the orientations of all six linear actuators. In particular, for n-3-II, n-4, and n-5 mechanisms, additionally to the roll-pitch orientation of the manipulator platform, only the orientations of four linear actuators are needed. By also measuring the yaw orientation of the manipulator platform, the number of required linear actuators' orientations can be even reduced to two.

The case where only the linear actuators' orientations are measured is advantageous because, then, the sensors can be placed close to the base platform, thus reducing the wiring effort. Furthermore, only measuring the roll-pitch orientation of the manipulator platform provides better results compared to an additional measurement of the yaw orientation [29]. This is especially advantageous, for example, for milling machines, where the yaw degree-of-freedom is not used.

Our results enable the design of a low-cost sensor system for parallel mechanisms that provides a unique solution of their direct kinematics problem. This concept is particularly important if no information about the previous states of the parallel mechanism is available, for example, if it is switched on in a certain pose. Furthermore, acceleration or magnetic sensors are significantly smaller than the usual sensors for measuring the linear actuators' lengths, thus allowing for a reduction of moving equipment as well as extending the workspace.

The real-time performance of the proposed sensor concept and the associated closed-form solutions for the direct kinematics problem of parallel mechanisms can be improved by sensor fusion including the information of additional linear actuators' orientations or sensors.

Conflicts of Interest

The authors declare that there are no conflicts of interest regarding the publication of this paper.

Acknowledgments

This work was supported by the German Research Foundation (DFG) (Grant SCHL 275/15-1). The publication of this work was also supported by the DFG and Hamburg University of Technology (TUHH) in the funding programme "Open Access Publishing."

References

[1] F. Ronga and T. Vust, "Stewart platforms without computer?" in *Proceedings of the International Conference on Real Analytic and Algebraic Geometry*, pp. 197–212, Trento, Italy, 1992.

[2] K. H. Hunt and E. J. F. Primrose, "Assembly configurations of some in-parallel-actuated manipulators," *Mechanism and Machine Theory*, vol. 28, no. 1, pp. 31–42, 1993.

[3] D. Lazard, "On the representation of rigid-body motions and its application to generalized platform manipulators," in *Computational Kinematics*, G. M. L. Gladwell, J. Angeles, G. Hommel, and P. Kovács, Eds., vol. 28 of *Solid Mechanics and Its Applications*, pp. 175–181, Springer, Dordrecht, The Netherlands, 1993.

[4] B. Mourrain, "The 40 "generic" positions of a parallel robot," in *Proceedings of the International Symposium on Symbolic and Algebraic Computation*, pp. 173–182, Kiev, Ukraine, July 1993.

[5] M. Raghavan, "Stewart platform of general geometry has 40 configurations," *Journal of Mechanical Design*, vol. 115, no. 2, pp. 277–280, 1993.

[6] M. Raghavan and B. Roth, "Solving polynomial systems for the kinematic analysis and synthesis of mechanisms and robot manipulators," *Journal of Mechanical Design*, vol. 117, pp. 71–79, 1995.

[7] M. L. Husty, "An algorithm for solving the direct kinematics of general Stewart-Gough platforms," *Mechanism and Machine Theory*, vol. 31, no. 4, pp. 365–379, 1996.

[8] P. Dietmaier, "The Stewart-Gough platform of general geometry can have 40 real postures," in *Proceedings of the International Symposium on Advances in Robot Kinematics*, pp. 7–16, Strobl, Austria, 1998.

[9] J. C. Faugère and D. Lazard, "Combinatorial classes of parallel manipulators," *Mechanism and Machine Theory*, vol. 30, no. 6, pp. 765–776, 1995.

[10] J.-P. Merlet, *Parallel Robots*, Springer, Dordrecht, The Netherlands, 2nd edition, 2006.

[11] Z. Geng and L. Haynes, "Neural network solution for the forward kinematics problem of a Stewart platform," in *Proceedings of the IEEE International Conference on Robotics and Automation*, pp. 2650–2655, Sacramento, CA, USA, April 1991.

[12] N. Mimura and Y. Funahashi, "New analytical system applying 6 DOF parallel link manipulator for evaluating motion sensation," in *Proceedings of the IEEE International Conference on Robotics and Automation*, pp. 227–233, Nagoya, Japan, May 1995.

[13] R. Boudreau and N. Turkkan, "Solving the forward kinematics of parallel manipulators with a genetic algorithm," *Journal of Robotic Systems*, vol. 13, no. 2, pp. 111–125, 1996.

[14] B. Dasgupta and T. S. Mruthyunjaya, "A constructive predictor-corrector algorithm for the direct position kinematics problem for a general 6-6 Stewart platform," *Mechanism and Machine Theory*, vol. 31, no. 6, pp. 799–811, 1996.

[15] C. M. Gosselin, "Parallel computational algorithms for the kinematics and dynamics of planar and spatial parallel manipulators," *Journal of Dynamic Systems, Measurement, and Control*, vol. 118, no. 1, pp. 22–28, 1996.

[16] P. R. McAree and R. W. Daniel, "A fast, robust solution to the Stewart platform forward kinematics," *Journal of Robotic Systems*, vol. 13, no. 7, pp. 407–427, 1996.

[17] K. Der-Ming, "Direct displacement analysis of a Stewart platform mechanism," *Mechanism and Machine Theory*, vol. 34, no. 3, pp. 453–465, 1999.

[18] A. K. Dhingra, A. N. Almadi, and D. Kohli, "A Gröbner-Sylvester hybrid method for closed-form displacement analysis of mechanisms," *Journal of Mechanical Design*, vol. 122, no. 4, pp. 431–438, 2000.

[19] K. H. Hunt and P. R. McAree, "The octahedral manipulator: geometry and mobility," *The International Journal of Robotics Research*, vol. 17, no. 8, pp. 868–885, 1998.

[20] Z. Šika, V. Kocandrle, and V. Stejskal, "An investigation of properties of the forward displacement analysis of the generalized Stewart platform by means of general optimization methods," *Mechanism and Machine Theory*, vol. 33, no. 3, pp. 245–253, 1998.

[21] R. Stoughton and T. Arai, "Optimal sensor placement for forward kinematics evaluation of a 6-DOF parallel link manipulator," in *Proceedings of the IEEE/RSJ International Workshop on Intelligent Robots and Systems*, pp. 785–790, Osaka, Japan, 1991.

[22] K. C. Cheok, J. L. Overholt, and R. R. Beck, "Exact methods for determining the kinematics of a Stewart platform using additional displacement sensors," *Journal of Robotic Systems*, vol. 10, no. 5, pp. 689–707, 1993.

[23] J.-P. Merlet, "Closed-form resolution of the direct kinematics of parallel manipulators using extra sensors data," in *Proceedings of the IEEE International Conference on Robotics and Automation*, pp. 200–204, Singapore, Singapore, May 1993.

[24] R. Nair and J. H. Maddocks, "On the forward kinematics of parallel manipulators," *The International Journal of Robotics Research*, vol. 13, no. 2, pp. 171–188, 1994.

[25] L. Baron and J. Angeles, "The direct kinematics of parallel manipulators under joint-sensor redundancy," *IEEE Transactions on Robotics and Automation*, vol. 16, no. 1, pp. 12–19, 2000.

[26] Y.-J. Chiu and M.-H. Perng, "Forward kinematics of a general fully parallel manipulator with auxiliary sensors," *The International Journal of Robotics Research*, vol. 20, no. 5, pp. 401–414, 2001.

[27] J. Hesselbach, C. Bier, I. Pietsch et al., "Passive joint-sensor applications for parallel robots," in *Proceedings of the IEEE/RSJ International Conference on Intelligent Robots and Systems*, pp. 3507–3512, Sendai, Japan, October 2004.

[28] R. Vertechy and V. Parenti Castelli, "Accurate and fast body pose estimation by three point position data," *Mechanism and Machine Theory*, vol. 42, no. 9, pp. 1170–1183, 2007.

[29] S. Schulz, A. Seibel, D. Schreiber, and J. Schlattmann, "Sensor concept for solving the direct kinematics problem of the Stewart-Gough platform," in *Proceedings of the IEEE/RSJ International Conference on Intelligent Robots and Systems*, pp. 1959–1964, Vancouver, Canada, September 2017.

[30] G. H. Golub and C. F. Van Loan, *Matrix Computations*, vol. 3 of *Johns Hopkins Series in the Mathematical Sciences*, Johns Hopkins, Baltimore, MD, USA, 1983.

[31] U. Qureshi and F. Golnaraghi, "An algorithm for the in-field calibration of a MEMS IMU," *IEEE Sensors Journal*, 2017.

[32] R. G. Valenti, I. Dryanovski, and J. Xiao, "Keeping a good attitude: a quaternion-based orientation filter for IMUs and MARGs," *Sensors*, vol. 15, no. 8, pp. 19302–19330, 2015.

[33] S. O. H. Madgwick, A. J. L. Harrison, and R. Vaidyanathan, "Estimation of IMU and MARG orientation using a gradient descent algorithm," in *Proceedings of the IEEE International Conference on Rehabilitation Robotics*, pp. 179–185, Zurich, Switzerland, July 2011.

[34] X. Yun, M. Lizarraga, E. R. Bachmann, and R. B. McGhee, "An improved quaternion-based Kalman filter for real-time tracking of rigid body orientation," in *Proceedings of the IEEE/RSJ International Conference on Intelligent Robots and Systems*, pp. 1074–1079, Las Vegas, NV, USA, October 2003.

11

Allocating Multiple Types of Tasks to Heterogeneous Agents based on the Theory of Comparative Advantage

Toma Morisawa, Kotaro Hayashi, and Ikuo Mizuuchi

Department of Mechanical Systems Engineering, Tokyo University of Agriculture & Technology, 2-24-16 Nakacho, Koganei, Tokyo 184-8588, Japan

Correspondence should be addressed to Kotaro Hayashi; hayashik@cc.tuat.ac.jp

Academic Editor: Yunyi Jia

We present a method to allocate multiple tasks with uncertainty to heterogeneous robots using the theory of comparative advantage: an economic theory that maximizes the benefit of specialization. In real applications, robots often must execute various tasks with uncertainty and future multirobot system will have to work effectively with people as a team. As an example, it may be necessary to explore an unknown environment while executing a main task with people, such as carrying, rescue, military, or construction. The proposed task allocation method is expected to reduce the total makespan (total length of task-execution time) compared with conventional methods in robotic exploration missions. We expect that our method is also effective in terms of calculation time compared with the time-extended allocation method (based on the solution of job-shop scheduling problems). We simulated carrying tasks and exploratory tasks, which include uncertainty conditions such as unknown work environments (2 tasks and 2 robots, multiple tasks and 2 robots, 2 robots and multiple tasks, and multiple tasks and multiple robots). In addition, we compared our method with full searching and methods that maximize the sum of efficiency in these simulations by several conditions: first, 2 tasks (carrying and exploring) in the four uncertain conditions (later time, new objects appearing, disobedient robots, and shorter carrying time) and second, many types of tasks to many types of robots in the three uncertain conditions (unknown carrying time, new objects appearing, and some reasonable agents). The proposed method is also effective in three terms: the task-execution time with an increasing number of objects, uncertain increase in the number of tasks during task execution, and uncertainty agents who are disobedient to allocation orders compared to full searching and methods that maximize the sum of efficiency. Additionally, we performed two real-world experiments with uncertainty.

1. Introduction

Research on a multirobot task allocation system and coordinating heterogeneous robots have become a hot topic of research in the field robotics. We propose a method of allocating multiple types of tasks to heterogeneous robots, based on the theory of comparative advantage [1], in order to minimize the makespan (the total length of task-execution time). The calculation time in the proposed method is negligibly short compared with the task-execution time, and it may also be effective for dynamic allocation (e.g., in an unknown real environment).

A key driving force of multirobot systems is their force of numbers. Multirobot systems are expected to improve the effectiveness of robotic systems in terms of the temporal efficiency [2]. Similar to humans, multiple robots work more effectively than a single robot, even those that do not have complex mechanisms or advanced intelligent control systems. Recent studies on decision-making of multirobot systems focus on two categories: allocation (centralized method for coordinating) and cooperation (decentralized method for coordinating). Yan et al. [3] valued the following two categories: the allocation that the central controller has a global view of the world, whereby the globally optimal plans can be produced, but it is not robust in relation to dynamic environments or failures in communications and other uncertainties and the cooperation can better respond to unknown or uncertain environments, but the solutions they reach are often suboptimal. In the real world, multirobot systems will be offered various tasks of the unknown and uncertain environments. With the advancement of information and communication technology, it will require that multirobot systems for heterogeneous robots have various perception and capability by many vendors. Moreover, effective teamwork between groups of humans and robots is one of the goals in the fields of robotics and AI [4]. Multirobot systems must have a view of the environment including the humans. For example, in a rescue or industrial mission, humans and robots are required to work together. Our ultimate goal is allocating tasks to heterogeneous robots and humans, or the cooperating of heterogeneous robots with humans with minimum makespan and the real world. First of all, this study is focused on the allocation of tasks to heterogeneous robots in unknown environments by using the theory of comparative advantage: an economic theory that maximizes the benefit of specialization.

Conventional task allocation methods are categorized as static or dynamic allocation [5]. Static allocation gives the task assignments at the start until all the tasks are completed. Static allocation calculates the assignments to minimize the required time. This type of problem, in which the makespan is minimized, is called a job-shop scheduling problem [6]. The makespan refers to the total length of the schedule. However, the calculation time of static allocation increases exponentially as the number of tasks increases. In static allocation, the execution time of each task should be calculated and is perturbed by uncertainty. Dynamic allocation frequently conducts reallocation of static allocation [7] and requires more calculation time.

Dispatching methods to reduce the calculation time have been studied. In multirobot task allocation, the methods are instances of optimal assignment problems [8]. In the auction based method proposed by Gerkey and Matarić [9], the goal of the problem is to maximize the efficiency per task of allocation, where efficiency refers to the value per unit work of task-execution. However, maximizing the efficiency does not minimize the makespan in the case of allocating tasks to heterogeneous robots. For example, if a robot executes a task that another robot should do, the latter robot must execute yet another task that it is either not proficient at or is hard to execute. This type of loss is called opportunity cost [10]. We propose an allocation method based on the theory of comparative advantage, which reduces the opportunity cost and minimizes the makespan.

In this study, the allocation based on our method reduces the makespan compared with methods that maximize the sum of efficiency. We perform simulation experiments in environments with uncertainty and find that allocation based on the proposed method reduces the makespan under any uncertainty, compared with methods that maximize the sum of efficiency. In uncertain environments, the makespan with the proposed method was almost the same compared with frequent reallocation of static allocation. We also conduct a heterogeneous robots task-execution experiment and a human-robot collaborative experiment in the real environment. We propose task allocation methods for various tasks execution by heterogeneous multirobots. The proposed method is expected to be effective in real world experiments, because uncertainty can actually appear.

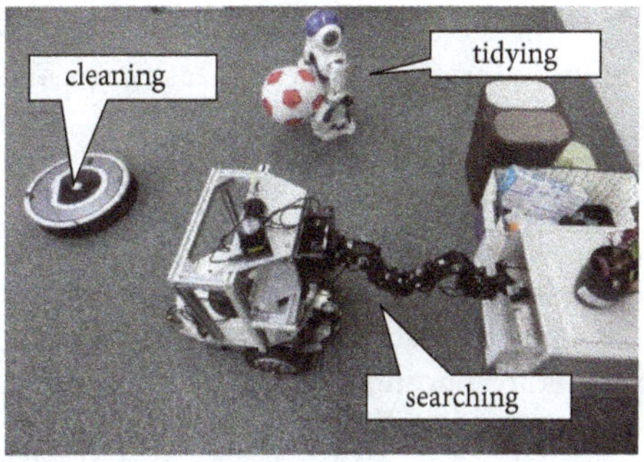

FIGURE 1: Task allocation to heterogeneous robots.

Currently, the area of the human-robot symbiosis is growing rapidly. As stated in the call of this special issue, assembly, warehouses, and home services need human-robot collaboration as soon as possible. Uncertainness of human is not only each individual difference but also effort input in human-robot collaboration. The definition of the theory of comparative advantage had to be extended to adapt to increases in the number of robots and tasks in the real world including humans.

2. Materials and Methods

2.1. Related Work. Because of the difficulties of constructing a single robot that has the required capability in real world, in recent literature, there are works on heterogeneous multirobot systems (e.g., in assembly of large-scale structures [2]). To allocate tasks to heterogeneous robots, each robot accomplishes its advantageous task, which increases the time performance [2, 11, 12]. How to optimally assign a set of robots to a set of tasks is well known as multirobot task allocation (MRTA) problem [13]. Liemhetcharat and Veloso [14] suggested that "the MRTA problem is categorized along three axes: single-task robots (ST) versus multitask robots (MT), single-robot tasks (SR) versus multirobot tasks (MR), and instantaneous assignment (IA) versus time-extended assignment (TA)." In the example shown in Figure 1, each robot accomplishes its advantageous task. Scheutz et al. [15] showed that swarms of high- and low-speed unmanned aerial vehicles exhibit higher performance than homogeneous swarms in locating and tracking chemical clouds. King et al. [16] performed experiments with a team consisting of rovers and blimps cooperatively executing two types of tasks. Zhang et al. [17] developed a fuzzy collaborative intelligence based algorithm to the collaboration of heterogeneous robots. In these studies, the types of robots are determined in advance, and each robot only executes its specific type of task. Ideally, task allocation methods should not depend on the types of robots, tasks, and environments. Many types of robots are expected to accomplish tasks flexibly as a team operating in any type of environment.

Under other circumstances, the problems of heterogeneous multirobot task allocation problems have been understood as instances of operations research, economics, scheduling, network flows, and combinatorial optimization

[5]. The allocating methods of these problems are categorized as either static allocation, in which tasks are allocated at the start until all the tasks are completed, or dynamic allocation, in which the next task is allocated to any robot as required. Gerkey and Matarić [5] have categorized task allocation problems in multirobot systems as instantaneous or time-extended assignment. Instantaneous assignment is similar to dynamic allocation, and time-extended assignment is similar to static allocation. Time-extended methods have been based on the scheduling algorithm [18]. However, it takes exponential time or longer for these methods to calculate the exact solution. In addition, static allocation is only initially efficient and will degrade over time [19]. We suppose that there are several types of uncertainty related to work in real environments. Especially in the human/multirobot collaboration, it is uncertain to predict a human's behavior such as in the following three cases:

(i) Uncertainty in task-execution time

(ii) Uncertainty increasing the number of tasks during task execution

(iii) Uncertainty agent who is disobedient to allocation orders

Dynamic allocation is a method that repeatedly allocates tasks dynamically during execution. There are many studies on rescheduling in job-shop scheduling problems in factories. In these works, jobs are added during execution [20], the processing speed is uncertain [7], efficiency under an uncertain environment based on the Intuitionistic Fuzzy Sets [21], slack-based techniques at low levels of uncertainty [22], and so forth. These methods are divided into two types: (1) all task assignments are given repeatedly until all the tasks are completed and (2) no planning for future allocations is performed. The latter is called a dispatching method (e.g., first in first out (FIFO) and shortest processing time (SPT) [18]). Allocations based on rescheduling all the tasks are thought to reduce the makespan, and allocations based on dispatching are thought to reduce the calculation time. The dispatching method is also used in CPU scheduling [23], where tasks (processes) are added successively. Compared with production scheduling in a factory [24], the makespan of multirobot task allocation is thought to be smaller. Computer performance may also be restricted in some multirobot task allocation problems. Therefore, dispatching methods have often been studied in multirobot task allocation (e.g., methods that maximize the sum of efficiency [9, 25–33]). These methods entail searching for the best combination of robots and tasks that maximize the sum of efficiency. In this case, efficiency is defined per robot per task, e.g., the inverse of the task-execution time for a robot to perform a task. These methods are based on the solution to the optimal assignment problem [8], the goal of which is to allocate tasks to robots to maximize the overall expected performance. However, allocation based on this method does not necessarily minimize the makespan. In allocation to heterogeneous agents, the opportunity cost can be added to the makespan [10]. Opportunity cost is a term used in economics, representing the value of an alternative forgone (unselected combination) to pursue a certain action. For example, consider that a robot executes a task that another robot should do, to maximize the sum of efficiency, and another robot executes a task for which it cannot minimize the makespan or that it cannot execute. If the former robot executes the task, the difference in task-execution times between the robots is added to the makespan as the opportunity cost. We focus on the makespan at the uncertainty and propose an allocating method based on the theory of comparative advantage, which is a framework to minimize makespan including the opportunity cost.

2.2. Task Allocation Based on the Theory of Comparative Advantage

2.2.1. Theory of Comparative Advantage. To increase the efficiency of tasks executed by heterogeneous robots through task allocation, we focus on "comparative advantage." Comparative advantage is an economic theory referring to the ability of any given economic actor to decrease the total economic cost. According to this theory, an appropriate allocation of tasks among robots is more efficient than each robot executing all tasks. We adapt this theory to the ability of robots and reduce the cost of multirobot decision.

Let e be the efficiency (the productivity per unit labor input) when the robot i executes task m. It is possible to compare the comparative advantage of robot i with robot j by

$$\frac{{}^{i}e_{m}}{{}^{i}e_{n}} > \frac{{}^{j}e_{m}}{{}^{j}e_{n}}. \tag{1}$$

If robot j has the comparative advantage in task n compared with robot i,

$$\frac{{}^{i}e_{n}}{{}^{i}e_{m}} < \frac{{}^{j}e_{n}}{{}^{j}e_{m}}. \tag{2}$$

We define ${}^{i}p_{m}$ as the "comparative advantage between robots" and ${}^{i}q_{m}$ as the "degree of comparative advantage between tasks":

$$\begin{aligned} {}^{i}p_{m} &= \frac{{}^{i}e_{m}}{{}^{j}e_{m}}, \\ {}^{i}q_{m} &= \frac{{}^{i}e_{m}}{{}^{i}e_{n}}. \end{aligned} \tag{3}$$

Each comparative advantage between each robot with each task becomes an index to allocate tasks. If robot i is inferior to robot j (${}^{i}e_{m} < {}^{i}e_{m}$) for task m and has the comparative advantage (see (1)), we should allocate robot i to task m. The total productivity of task m is given by

$$S_{m} = \sum_{i=1} {}^{i}e_{m} {}^{i}w_{m}, \quad \left(\sum_{i=1} {}^{i}w_{m} = 1 \right), \tag{4}$$

where ${}^{i}w_{m}$ is the working ratio (the labor input rate), which is the cost of robot i completing task m per the total cost of robot i. The Pareto front (Figure 2) gives a solution to fix the value of

$S_1 \; (= {}^0e_1 {}^0w_1 + {}^1e_1 {}^1w_1)$

total product of task 1

constraints

$$\begin{cases} {}^0w_0 + {}^0w_1 = 1 \\ {}^1w_0 + {}^1w_1 = 1 \\ 0 < {}^iw_m < 1 \end{cases}$$

Pareto front

feasible region

total product of task 0
$S_0 \; (= {}^0e_0 {}^0w_0 + {}^1e_0 {}^1w_0)$

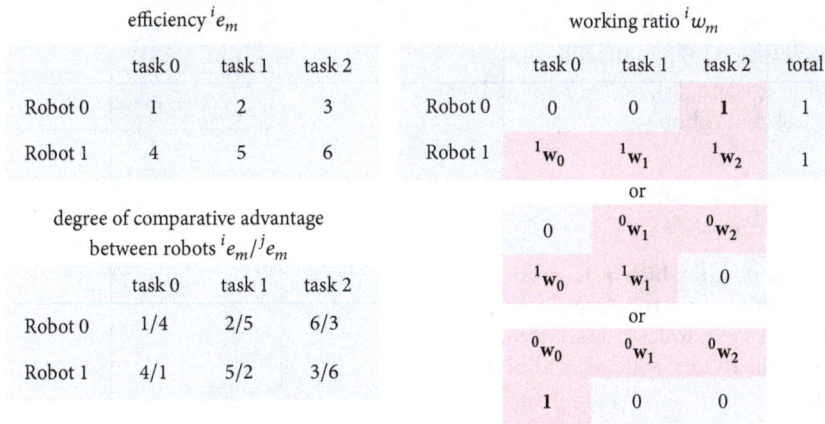

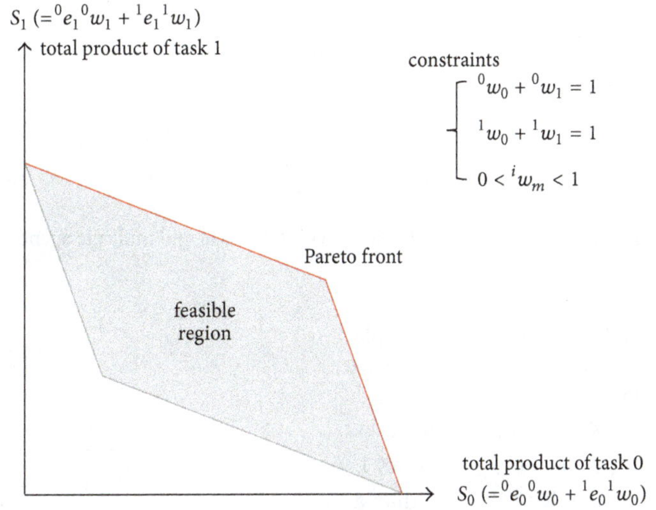

FIGURE 2: Pareto front.

efficiency ${}^ie_m$

	task 0	task 1
Robot 0	**4**	1
Robot 1	10	**4**

working ratio ${}^iw_m$

	task 0	task 1
Robot 0	**1**	0
Robot 1	${}^1\mathbf{w_0}$	${}^1\mathbf{w_1}$

or

	${}^0\mathbf{w_0}$	${}^0\mathbf{w_1}$
	0	**1**

FIGURE 3: Example of an optimal allocation on the Pareto front.

efficiency ${}^ie_m$

	task 0	task 1	task 2
Robot 0	1	2	3
Robot 1	4	5	6

working ratio ${}^iw_m$

	task 0	task 1	task 2	total
Robot 0	0	0	**1**	1
Robot 1	${}^1\mathbf{w_0}$	${}^1\mathbf{w_1}$	${}^1\mathbf{w_2}$	1

or

	task 0	task 1	task 2
	0	${}^0\mathbf{w_1}$	${}^0\mathbf{w_2}$
	${}^1\mathbf{w_0}$	${}^1\mathbf{w_1}$	0

or

	${}^0\mathbf{w_0}$	${}^0\mathbf{w_1}$	${}^0\mathbf{w_2}$
	1	0	0

degree of comparative advantage
between robots ${}^ie_m / {}^je_m$

	task 0	task 1	task 2
Robot 0	1/4	2/5	6/3
Robot 1	4/1	5/2	3/6

FIGURE 4: Comparative advantage with multiple tasks and two agents.

the working ratio. In this theory, to obtain the solution of the Pareto front (${}^iw_m = 1$), a robot with a comparative advantage (Figure 3) must specialize in a task. We define the cost of robot completing the task as the makespan.

For example, when two robots are allocated to two tasks by the Pareto front (Figure 1), robot 0 executes task 0 and robot 1 executes task 1 at first. Throughout the execution time, one robot is perfectly specialized in one task. The other robot first executes its comparatively advantageous task and subsequently executes the other task. When two robots are allocated to multiple tasks by the Pareto front, each robot performs the tasks in order of their comparative advantages. By the Pareto front, one can argue that maximizing the sum of efficiencies is inefficient. For example, Figure 4 shows a set of robots and tasks. According to the Pareto front, robot 0 executes the tasks in the order $(2, 1, 0)$ and robot 1 executes the tasks in the order $(0, 1, 2)$. By maximizing the sum of efficiency, if robot 0 executes any task after finishing task 2, it would not be an optimum allocation by the Pareto front.

2.2.2. Comparative Advantage with Multiple Agents and Multiple Tasks. The theory of comparative advantage can be

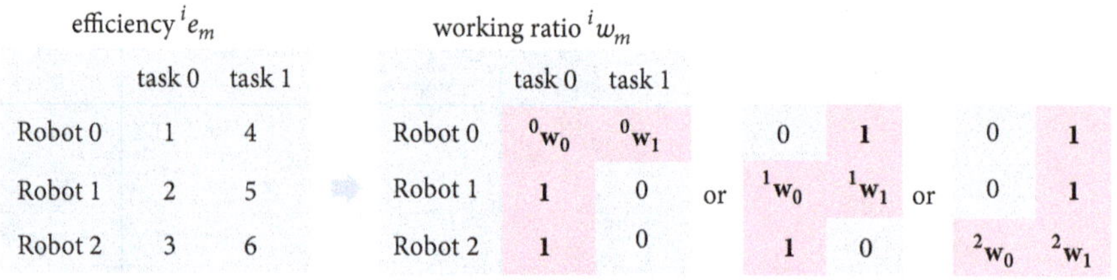

FIGURE 5: Comparative advantage with two tasks and multiple agents.

extended to allocation among multiple robots or multiple tasks [34]. Figure 4 shows an example of the allocation of two robots to multiple tasks. In Figure 4, tasks are arranged from the left in ascending order of ^ip_m of robot 0. The tasks are always arranged in descending order of ^ip_m of robot 1. Neither robot should perform tasks that have less ^ip_m than a benchmark task. We can choose any task as the benchmark task. If the benchmark is task 1, then robot 0 should first execute task 2 and robot 1 should first execute task 0 in the example in Figure 4. Figure 5 shows that each robot performs a task that has a larger ^iq_m than a benchmark agent. We can choose any agent as the benchmark agent.

In cases of multiple robots and multiple tasks, it will be complex to solve the problem. No general method for describing all the Pareto fronts has been established yet. In economics, several methods have been used to apply the theory of comparative advantage to cases of multiple agents and multiple commodities. There are methods of finding the solutions on the Pareto front or methods in which the number of agents or commodities is restricted [35–37]. In this study, we use a method proposed by Tian [37]. In the following, ^ir_m is the efficiency of robot i to execute task m (^ie_m) divided by the sum of efficiency per task and robot:

$$^ir_m = \frac{^ie_m}{\sum_k {}^ke_m \sum_l {}^ie_l}. \tag{5}$$

Robot i is allocated to task m for which r is maximum. Next, among the combinations of other robots and tasks, the robot and task combination that maximizes r is selected. This procedure is repeated for all the robots. A detailed allocation flowchart is given in Figure 6.

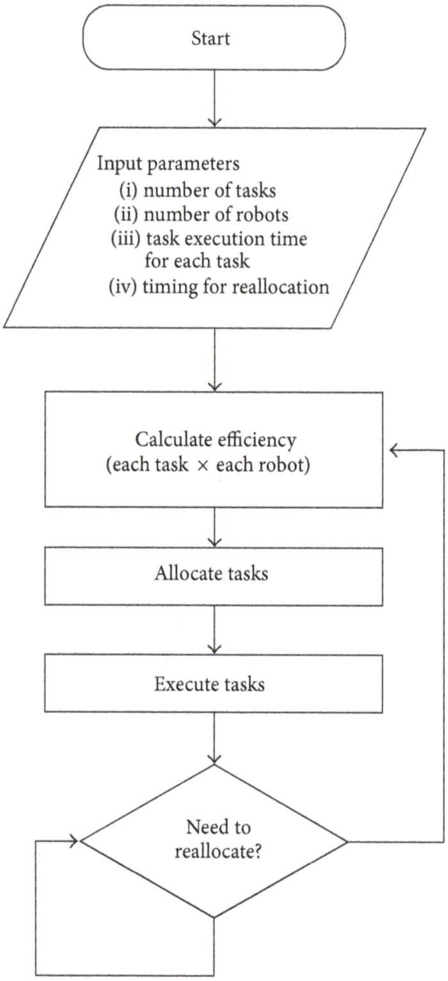

FIGURE 6: Allocation flowchart.

2.2.3. Optimal Assignment Problem. The optimal assignment problem [8] is a well-known problem in operations research. In this problem, M robots, N tasks, and nonnegative efficiencies that predict each robot's performance for each task are given. The goal is to assign tasks to robots to maximize the overall expected performance. In the task allocation problem, the goal is to find the appropriate $^m\alpha_n$ such that the sum of efficiency is maximized:

$$\underset{\alpha}{\operatorname{argmax}} \left(\sum_{m=1}^{M} \sum_{n=1}^{N} {}^me_n {}^m\alpha_n \right), \tag{6}$$

where e represents the task performance per time and $^m\alpha_n$ must be either 0 or 1; if robot m executes task n, then $^m\alpha_n = 1$. This is a method of instantaneous assignment which requires frequently solving the problem. It costs $O(MN^2)$ time using the Hungarian method [38]. However, maximizing the sum of efficiency does not always minimize the makespan. For example, in a case where each robot executes each different task (Table 1), robot 0 executes task 1 and robot 1 executes task 0 to maximize the sum of efficiency. Then, the total products are $S_0 = 1$ and $S_1 = 10$, where the solution is not on the Pareto front. In other instances, if robot 1 is specialized in task 1 and the working ratios of robots 0 and 1 are $^1w_0 = 0.25$ and $^1w_1 = 0.75$, respectively, the total products of task 0 are $S_0 = 1$ and $S_1 = 11.5$, where the solution is on the Pareto front.

TABLE 1: Example in which maximizing sum of efficiency does not minimize makespan.

	task 0	task 1
Robot 0	4	10
Robot 1	1	4

We can regard this multirobot task allocation problem as an extension of the traveling salesman problem (TSP) [39]. General TSPs with multiple agents are called multiagent traveling salesman problems [40], vehicle routing problems [41], or job-shop scheduling problems [6]. These problems are NP-hard [42]; they cannot be solved in polynomial time [43, 44]. Instead of listing all the orders (as in the full searching method), there are other methods to decrease the amount of calculation, e.g., by using dynamic programming of time computational quantity $O(2^N N^2)$ and the branch-and-cut algorithm [45]. General algorism, such as NN (Neural Network) [46] or ACO (Ant Colony Optimization) [47, 48], based approximate solutions are also given. In the study to solve vehicle routing problem, it is well known that the method clusters the area by the number of agents and calculates each shortest path [49]. Approximation algorithms and heuristics are, therefore, developed for these problems [48, 50, 51]. However, they require exponential time or longer to calculate the exact solution.

In the uncertain real environment, our proposed method should be able to solve large-scale problems with a large number of robots, tasks, people, animals, and so on. Our method should also be able to compare with different algorithms and the trajectories of robots. There are many things to clarify if our proposed method is feasible or not. But first and foremost, we think the full searching method is plausible to compare with the proposed method.

In this paper, we compare our method with existing methods and show the effectiveness of our method on situations: 2 task and 2 robots (Section 3.1), multiple tasks and 2 robots (Section 3.2), 2 robots and multiple tasks (Section 3.3), and multiple tasks and multiple robots (Section 3.4). As the reason for this, an optimal solution can be calculated for task allocation to the 2 robots situation, to the extent the number of tasks is not large. If our method in these situations is more effective than existing methods for the number of tasks that is more than a certain value, at least, the method has possibilities to solve large task allocation to multiple robots.

3. Task Allocation Method to Specific Problems

3.1. Allocating Two Types of Tasks to Two Types of Robots. To verify the effectiveness of the proposed method, we first simulate allocating two types of tasks to two types of robots (Figure 7). In this study, our purpose is to shorten the makespan of executing tasks in an unknown environment. Tasks in an unknown environment are divided into two main tasks: exploration and other tasks. We choose the carrying task as the basic other task in a field that consists of 20 × 20 areas. We define these two types of tasks as follows:

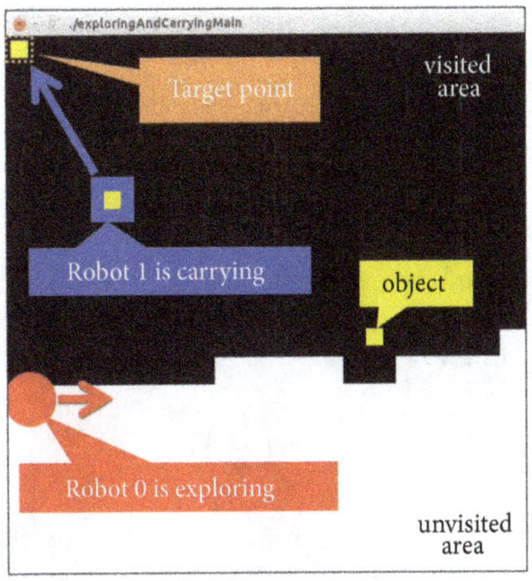

FIGURE 7: Simulation: two types of tasks by two types of robots. The field consists of 20 × 20 areas. The red circle and blue square represent the robots, the yellow squares represent objects to be carried, the white shading represents unvisited area, and the black shading represents visited area.

TABLE 2: Moving speeds of each robot.

	Not carrying (v_i)	Carrying ($v_{\mathrm{obj}(i)}$)
Agent 0	0.33	0.10
Agent 1	1.00	0.033

(i) Carrying task: carrying objects to a target point (Figure 7, dashed square). This task is finished once the robots have carried all the objects to the target point.

(ii) Exploring task: exploration of the unknown areas for mapping and discovering objects. Robots do not know where the objects are initially. Unknown areas, indicated by white, turn into known areas, indicated by black, when robots pass through the areas. This task is finished once the robots have passed through all the areas.

The two types of robots explore unknown areas and carry discovered objects. Table 2 shows the differences between the two types of robots. Here, v_i is the moving velocity when robot i does not carry any object, and $v_{\mathrm{obj}(i)}$ is the moving velocity when robot i carries an object. We define the efficiency (performance per unit work) of the carrying task, e_{car}. In this simulation, the efficiency is the inverse of the time taken for the robot to carry the nearest object to the target point:

$$ {}^{i}e_{\mathrm{car}} = \left(\frac{l_0}{v_i} + \frac{l_1}{v_{\mathrm{obj}(i)}} \right)^{-1}, \tag{7} $$

where l_0 is the distance from the robot to the object and l_1 is the distance from the object to the target point. We also define

TABLE 3: Moving velocity.

	Not carrying	Carrying
Robot 0	0.4	0.1
Robot 1	1.0	0.4

the efficiency of the exploring task, e_{exp}. In this situation, it refers to the inverse of the time taken for the robot to go to the nearest unknown area:

$$ {}^{i}e_{\text{exp}} = \left(\frac{l_2}{v_i}\right)^{-1}, \quad (8) $$

where l_2 is the distance from the robot to the nearest unknown area. In this simulation, the robots are allocated to tasks in the following situations: when one robot finishes carrying an object or exploring an area, when both robots try to carry the same object, and when the comparative advantages of the robots switch. If the allocated task is changed, while a robot is carrying an object, the robot leaves the object.

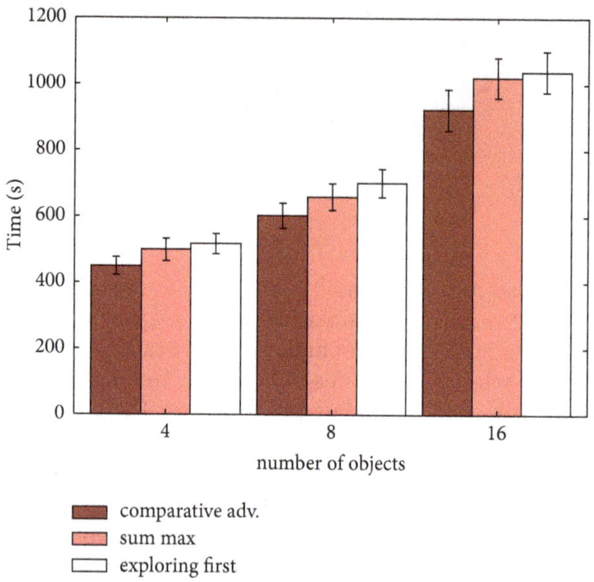

FIGURE 8: Results of proposed and conventional methods.

3.1.1. Comparison with Conventional Method. In this section, we compare the proposed method with conventional methods, i.e., maximizing the sum of efficiency [9] and exploring first method. In maximizing the sum of efficiency, new task is allocated to a robot which has best efficiency of the task as follows: the robot velocities are listed in Table 3. We defined the efficiency of exploring as the not-carrying velocity. We investigated the following three allocation methods:

(i) Proposed method: the theory of comparative advantage (Pareto front).

(ii) Maximizing the sum of efficiency: each robot executes each different task; if ${}^{0}e_0 + {}^{1}e_1$ is more than ${}^{0}e_1 + {}^{1}e_0$, then robot 0 executes task 0.

(iii) Exploring first method: both robots execute the exploring task at first and then execute the carrying task after the exploring task is finished.

We simulate three conditions of the number of objects (4, 8, 16). Under the proposed method, robot 0 executed the exploring task and robot 1 executed the carrying task. By maximizing the sum of efficiency, each robot executed opposite tasks. Figure 8 shows the results. Allocation based on our method minimized the makespan of the three types of allocation methods under all conditions.

3.2. Allocating Multiple Types of Tasks in Unknown Environment to Two Types of Robots. In unknown environments, robots have to execute many tasks which are manifold and uncertain. We propose a task allocation method based on the theory of comparative advantage in the case where there are multiple tasks in unknown environments. This allocation method is expected to minimize the makespan by executing in order of the comparative advantage between robots. There are many studies about situations consisting of multiple tasks (e.g., restaurants [52] and disaster mitigation [53]). We feature carrying multiple types of objects tasks as multiple types of tasks. We compare our method with the basic allocation method that gives all the task assignments of the entire period (we refer to this method as the full searching method). We also simulate allocation in unknown environments with possible accidents: later, adding, unreasonable, and speed-up.

3.2.1. Method to Allocate Tasks of Carrying Multiple Types of Objects. As noted above, we simulated the single-type carrying task allocated to heterogeneous robots by the Pareto front. We evaluate the Pareto front for allocation tasks of carrying multiple types of objects to heterogeneous robots. We define the efficiency of the carrying task ${}^{i}e_m$ as the time taken for the robot to carry the objects.

$$ {}^{i}e_m = \left(\frac{l_0}{v_i} + \frac{l_1}{v_{\text{obj}(i,m)}}\right)^{-1}, \quad (9) $$

where $v_{\text{obj}(i,m)}$ is the moving velocity when robot i is carrying object m. The carrying velocity $v_{\text{obj}(i,m)}$ depends on the carried object. The comparative advantage between robots is determined by ${}^{i}e_m / {}^{j}e_m$ in (3). When any robot finishes carrying an object, both robots redetermine the carrying objects.

We compare the Pareto front with an allocation method based on the full searching of carrying orders. The full searching method calculates the task-execution times of all the carrying orders in the case of M robots and N objects and calculates an order of minimal time. The number of possible combinations is

$$ \frac{(N + M - 1)!}{(M - 1)!}. \quad (10) $$

3.2.2. Setting of Simulation with Possible Accidents. We did a simulation to evaluate the proposed method by comparing

TABLE 4: Moving velocity.

	Not carry	Carry (0)	Carry (1)	⋯
Robot 0	1.0	0.10	0.11	⋯
Robot 1	1.0	0.20	0.21	⋯

The numbers in brackets represent the object IDs.

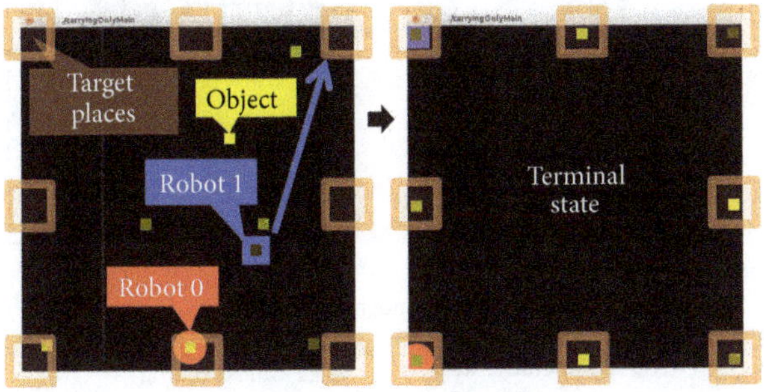

FIGURE 9: Simulation: heterogeneous robots and multiple carrying tasks.

it with the full searching method. Table 4 shows the moving velocities for types of robots and types of carrying objects. The field in the simulator consists of 20 × 20 areas (Figure 9). At first, the robots are in the center of the field, and the objects are in random positions. The robots know the positions of all the objects in the known area. The target place that each object should be carried to is different. Each robot carries an object that maximizes the comparative advantage between robots.

3.2.3. Conditions. We set simulation conditions defining unexpected events that may occur, including the following conditions:

(i) Later condition: the time required for a robot to carry is later than predicted.

(ii) Adding condition: new objects appear.

(iii) Unreasonable condition: some robots are disobedient to allocation orders.

(iv) Speed-up condition: the time to reach object positions is shorter than the carrying time.

We hypothesize that the proposed method based on the theory of comparative advantage performs better under these conditions than the full searching methods. In the proposed method, robots execute tasks that are given high priority according to the execution time difference among robots. Thus, the remaining tasks have shorter execution time difference among robots. If unexpected events occur, our method can avoid the situation in which an inappropriate robot executes a task and increases the total task-execution time.

Later Condition. This condition is the situation in which it takes more time to carry objects than predicted. For example, such situations occur when a robot carries an object that needs an unknown time to carry, accidents occur, and so

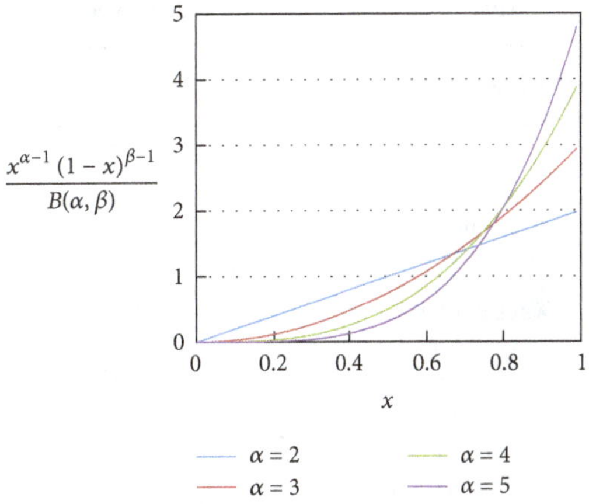

FIGURE 10: Beta distribution ($\beta = 1$).

forth. We simulated the delay by multiplying the estimated carrying velocity $v_{\text{obj}(i,m)}$ by noise x:

$$v_{\text{objreal}(i,m)} = x v_{\text{obj}(i,m)} \quad (0 < x < 1), \tag{11}$$

where $v_{\text{obj}(i,m)}$ is determined by the allocation and $v_{\text{objreal}(i,m)}$ is the carrying velocity when robots actually carry objects. We define noise x as a random number drawn on a beta distribution.

$$\frac{x^{\alpha-1}(1-x)^{\beta-1}}{B(\alpha,\beta)} \quad (0 \le x \le 1), \tag{12}$$

where x represents a random variable and $B(\alpha, \beta)$ is a beta function. At $x = 1$, the time is the same as predicted. At $x = 1/2$, it takes twice as much time as predicted. x is defined for each combination of robots and objects. The parameter α controls the noise variance. If the value α is smaller, the noise variance increases. Figure 10 shows a plot of the beta distribution with $\beta = 1$.

We simulated two types of full searching methods: (1) full searching (static), in which the robots carry objects in the order that is decided at first and (2) full searching (dynamic), in which the robots redetermine the order every time a robot finishes carrying an object. We ran simulations to compare the proposed method and the two types of full searching methods for certain patterns of α.

Adding Condition. This condition is the situation in which the number of objects increases. At first, there are seven objects, and some objects are added later. Table 5 lists the carrying velocity and time of added objects. The robots redetermine executing tasks in all the methods every time a robot finishes carrying an object, or an object is added.

Unreasonable Condition. This condition is the situation in which some robots are unreasonable agents (UAs). A human or agent is often disobedient to allocation orders because they are in an inaccessible place, they think of a new way, or for other reasons. In these situations, it is necessary to reallocate tasks for the shortest execution time that the other agents are not executing. The full searching method redetermines the

Table 5: Carrying velocity of added objects.

Appearing time [s]	1st 100	2nd 200	3rd 300	4th 400
Robot 0	0.11	0.19	0.13	0.17
Robot 1	0.21	0.29	0.23	0.27

Initial objects in Table 4.

Table 6: Moving velocity (object ID in brackets).

	Not carry v_i	Carry (0) $v_{obj(i,0)}$	Carry (1) $v_{obj(i,1)}$	...
Robot 0	0.6	0.010	0.011	...
Robot 1	1.0	0.020	0.021	...

Table 7: Moving velocity of added objects.

Appearing time [s]	1st 1000	2nd 1500	3rd 2000	4th 2500
Robot 0	0.011	0.019	0.013	0.017
Robot 1	0.021	0.029	0.023	0.027

order for the robots every time an agent finishes a task. In this simulation, Robot 1 is a UA and the robot velocities are given in Table 4.

Speed-Up Condition. In this condition, the carrying velocities (Table 6) are increased to 1/10 of the velocities in previous experiments (Table 4). The time to move is also shorter than the carrying time. In addition, to evaluate this condition, we combined it with adding a condition. The carrying velocity and time when objects are added are shown in Table 7.

Comparing with Allocation Method of Full Searching of Carrying Orders. The calculation complexity of the proposed method is $O(MN)$. Figure 11 shows the average task-execution and calculation time for 100 experiments, where there are initially 7, 8, 9, or 10 objects in random places. The task-execution time for the proposed method is slightly longer than that of the full searching method. However, the calculation time in full searching increases exponentially with an increasing number of objects. From this result, if the executing time of the proposed method is slightly longer than the full searching method, it means that the proposed method has a shorter makespan than the full searching method. Our proposed method is effective for the task allocation to multiple robots in uncertain environment. The calculation time in the proposed method is negligibly short compared with the task-execution time: only 0.51 s for an experiment with 10 objects. Our hardware platform was a ThinkPad T530 with a Core i7-3520M processor, 16 GB memory, and HDD running Ubuntu 14.04.

Maximize the Sum of Efficiency. We also compared our method with methods that maximize the sum of efficiency. Table 4 lists the robot velocities. Figure 12 shows the experimental results of 100 time simulations. There were significant

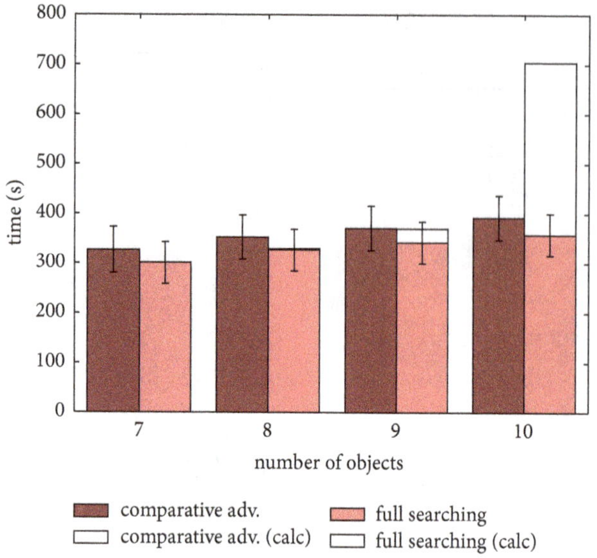

Figure 11: Makespan and calculation time in 100 simulations. There are 7, 8, 9, and 10 objects.

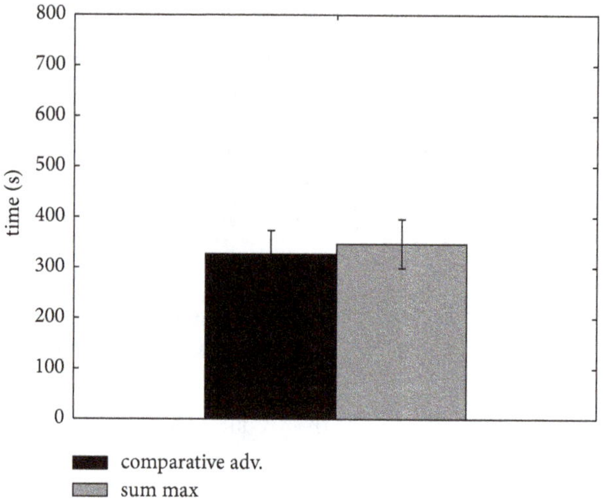

Figure 12: Makespan in 100 simulations.

differences: $t(99) = 9.08$; $p <: 05$. This is similar to the results in Section 3.1.

3.2.4. Results of Simulations with Possible Accidents

Results: Later Condition. Figure 13 shows the results. Compared with the full searching (static) method, the allocation based on our method reduced the makespan when the noise variance is large (e.g., $\alpha = 3$). Compared with full searching (dynamic) method, the Bonferroni method shows no significant difference between the methods for noises $\alpha = 5, 4, 3$, except for a significant difference between the method that maximizes the sum of efficiency with the full searching method at $\alpha = 5$ ($p < .01$). In the full searching (dynamic) method, the calculation time should be treated as idle time whenever the robots begin working.

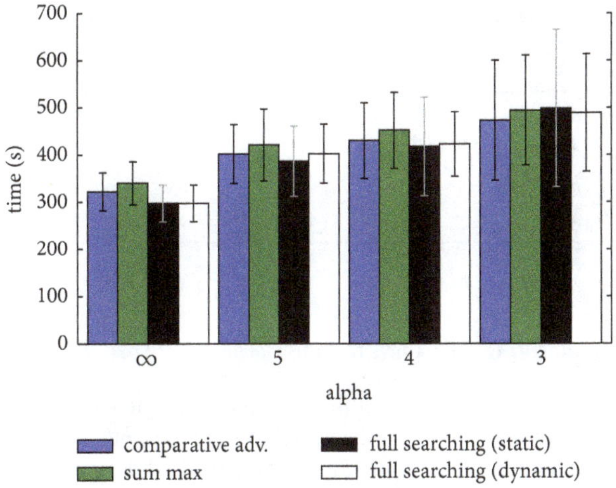

FIGURE 13: Makespan in the later condition.

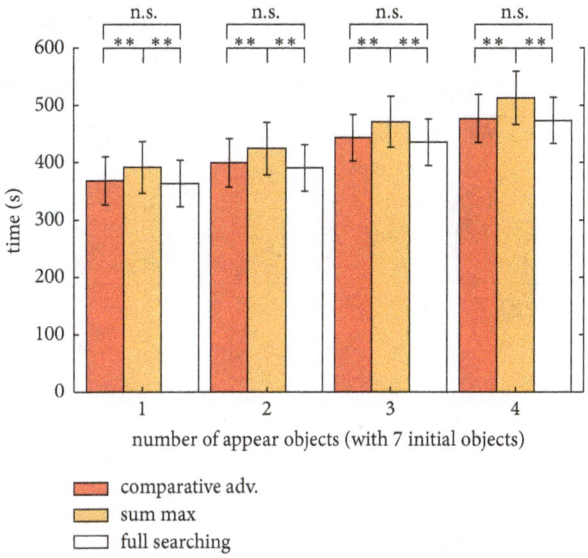

FIGURE 14: Makespan in the adding condition. $^{**}p < .01$, $^{*}p < .05$, and $^{+}p < .1$. A significant p value indicates that a significant difference in prescribing exists between the conditions. The lower p value indicates a strong significant difference. n.s. means no significant difference between the conditions.

The substantial task-execution time includes the working and calculating time. Our method shortens the substantial task-execution time compared to that in the full searching (dynamic) method, depending on the number of carrying objects. Compared with methods that maximize the sum of efficiency, the allocation based on the proposed method reduces the makespan.

Results: Adding Condition. Figure 14 shows the experimental results. There are no obvious differences depending on the number of added objects. We confirmed that allocation based on the proposed method reduces the makespan compared with the methods that maximize the sum of efficiency. The Bonferroni method shows a significant difference between the proposed method and the methods that maximize the sum of efficiency under all conditions. However, the Bonferroni method shows no significant difference between the proposed method and the full searching method under all conditions.

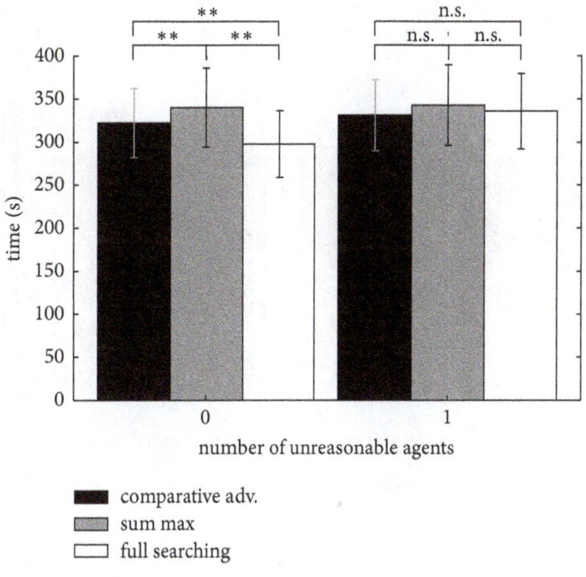

FIGURE 15: Makespan in the unreasonable condition. $^{**}p < .01$, $^{*}p < .05$, and $^{+}p < .1$. A significant p value indicates that a significant difference in prescribing exists between the conditions. The lower p value indicates a strong significant difference. n.s. means no significant difference between the conditions.

Results: Unreasonable Condition. Figure 15 shows the results. There are significant differences. The proposed method minimizes the makespan of the three methods, in comparison with the full searching method. The UA does not execute an allocated task. In these simulations, the UAs execute the shortest-makespan task of the tasks that other agents are not executing. In the full searching method, the task order for the robots is redetermined every time an agent completes a task. Robot 1 in Table 4 is a UA. Figure 15 shows the results. There are no significant differences.

Each cell in Tables 8 and 9 shows the method that reduced the makespan. The bars (-) in the tables indicate that the t-test shows no significant difference between the methods. As a result, in cases with a UA, the proposed method is more effective than the full searching method. In certain cases without a UA, the proposed method is as effective as the full searching method.

Results: Speed-Up Condition. We think the proposed method is more effective than the full searching method in cases where the differences in efficiency are smaller. Figure 16 shows the results for which the number of objects increases during the task execution. The allocation based on the proposed method reduces the makespan compared with the full searching method. The proposed method gives priority to the tasks whose carrying time differences are large. Therefore,

TABLE 8: Comparison between proposed method and full searching (with no unreasonable agents).

	$\alpha = \infty$	$\alpha = 5$	$\alpha = 4$	$\alpha = 3$
$n_{add} = 0$	full searching**	-	full searching$^+$	-
$n_{add} = 1$	full searching*	-	-	-
$n_{add} = 2$	full searching**	-	-	-
$n_{add} = 3$	full searching$^+$	-	-	-
$n_{add} = 4$	-	-	-	-

$^+p < .1$, $^*p < .05$, and $^{**}p < .01$; DOF = 199.

TABLE 9: Comparison between proposed method and full searching (with an unreasonable agent).

	$\alpha = \infty$	$\alpha = 5$	$\alpha = 4$	$\alpha = 3$
$n_{add} = 0$	proposed**	-	-	-
$n_{add} = 1$	proposed**	proposed*	proposed$^+$	-
$n_{add} = 2$	proposed**	proposed**	-	proposed*
$n_{add} = 3$	proposed**	proposed*	proposed**	-
$n_{add} = 4$	proposed**	proposed*	-	-

($^+p < .1$, $^*p < .05$, $^{**}p < .01$; DOF = 199).

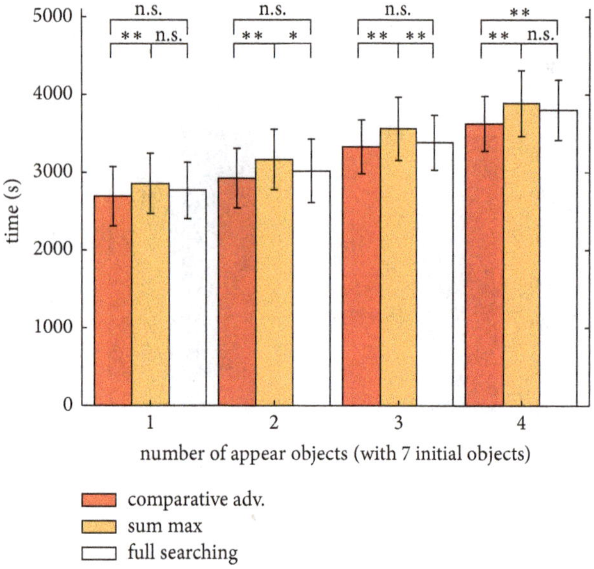

FIGURE 16: Makespan in the speed-up condition. (1/10 carrying speed condition). $^{**}p < .01$, $^*p < .05$, and $^+p < .1$. A significant p value indicates that a significant difference in prescribing exists between the conditions. The lower p value indicates a strong significant difference. n.s. means no significant difference between the conditions.

the opportunity cost, which is the difference in carrying times between robots, is unlikely to increase when compared with the full searching method. The efficiency of each task has less differences depending on the robot position, because the carrying time does not depend on it. The proposed method considers only the instantaneous robot position.

TABLE 10: Moving velocity (object ID in brackets).

	Not carrying	Carrying
Robot 0	0.2	0.1
Robot 1	0.3	0.08
Robot 2	0.4	0.06
Robot 3	0.5	0.04

3.3. Allocating Two Tasks to Many Types of Robots

3.3.1. Limitation of Allocating Method. We think the Pareto front can shorten the makespan of many robots executing two tasks (exploring and carrying). In this theory, we must nominate a benchmark robot to determine how many robots should be allocated to each task (exploring or carrying) and the benchmark robot is the only one that executes both tasks. For the adjustment to many robots, the following steps are added to the allocation method based on the theory of comparative advantage. To decide a benchmark robot, we performed a simulation.

(1) Nominate a benchmark robot in some way.

(2) Allocate one task to robots that have a comparatively greater degree of advantage than the benchmark robot, according to (3).

(3) Allocate the other task to the other robots.

The whole-period Pareto front has a limitation for allocating tasks to many types of robots. When one task is finished first, every agent will be executing the other task. This means that there are robots that execute both tasks and the whole-period Pareto front is unadaptable to this situation. The whole-period Pareto front can be adapted to the situation in which each robot executes a single task. Hence, to allocate two tasks to many types of robots, it is unadaptable to shorten the overall makespan. However, we think that the particular makespan (e.g., until one task is finished) will be shortened by the whole-period Pareto front. We propose a new method to shorten the overall makespan through the simulation results.

3.3.2. Simulation: Allocate a Carrying and an Exploring Task to Many Types of Robots. We performed simulation experiments to allocate a carrying and an exploring task to four types of robots by the theory of comparative advantage (Figure 17). We set the moving velocities of four robots as the comparative advantage (Table 10), and (8) gives the efficiency of the robots for each task. We performed the following steps to simulate allocation.

(1) Determine a benchmark robot in number order of robots (decide how many robots to allocate the exploring task).

(2) Allocate exploring task to robots that have a comparatively greater degree of advantage than the benchmark robot.

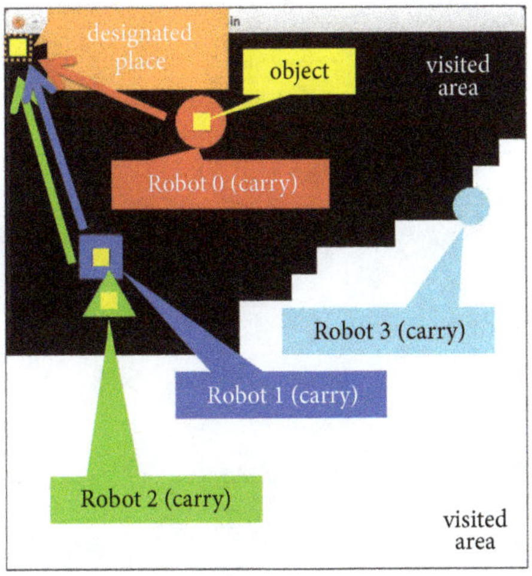

FIGURE 17: Simulation example: allocate the exploring and carrying tasks to four robots.

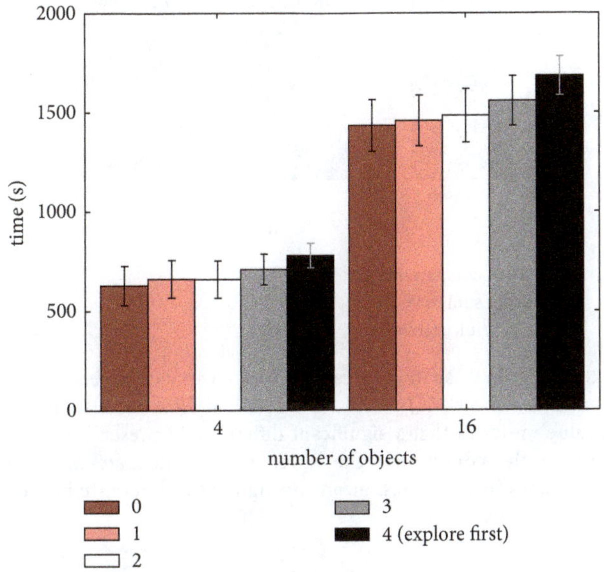

FIGURE 18: Results based on our method. The legend shows the benchmark agent id: minimal numbers of exploring agents.

(3) Allocate the carrying task to other robots in the order of the degree of comparative advantage.

(4) Allocate the exploring task to the remaining robots.

We simulated five robot conditions (benchmark robots are 1–4; exploring robots are 0–4) versus the object condition (number of objects are 4 or 16) on the instantaneous Pareto front. Figure 18 shows the results. In both object conditions, when the exploring robot was zero, the instantaneous Pareto front can shorten makespan the most. From these results, we can propose a method "explore until find." Under this method, if an object is found, it should be moved instantly. Accordingly, for few tasks, robots should give preference to other tasks over the exploring task. Thus, we propose a method in the case that there are many tasks in an unknown environment.

TABLE 11: Moving velocity.

	Not carry	Carry (0)	Carry (1)	···
	v_i	$v_{\text{obj}(i,0)}$	$v_{\text{obj}(i,1)}$	···
Robot 0	1.0	0.10	0.11	···
Robot 1	1.0	0.20	0.21	···
Robot 2	1.0	0.30	0.31	···
Robot 3	1.0	0.40	0.41	···

3.4. Allocating Many Types of Tasks to Many Types of Robots

3.4.1. Method. We propose a method to allocate many types of tasks to many types of robots, based on the theory of comparative advantage proposed by Tian [37]. The numbers of robots and objects are represented by M and N, respectively. Here, ${}^{i}r_m$ is the efficiency for robot i to execute task m (${}^{i}e_m$) divided by the sum of the efficiencies per task and per robot by (5). Robot i executes task m, for which r is maximum. Next, among the combinations of other robots and tasks, the combination for which r is maximum is selected. This procedure is repeated M times. In case of $M = 2$, the task selected by this method is always the same as the method of Section 3.1. Figure 19 shows an example of allocation based on this method. The red-colored values indicate the selected tasks.

3.4.2. Verification. We verify the proposed method to allocate carrying tasks to many robots by a simulation, as shown in Figure 20. The simulation settings used here are the same as those in Section 3.1. We employed the following conditions for verification.

(1) Robot condition: the robots have different comparative advantages. We set the moving velocities of four robots as the comparative advantage (Table 11).

(2) Environment condition: the system knows or does not know about the environment.

(3) Unreasonable condition: this is the same as in Section 3.2: task-execution time, added objects, and unreasonable agent.

We compare the proposed method with the full searching method and the method that maximizes the sum of efficiency. The full searching method in this section refers to the "dynamic full searching" method: robots are reallocated to each task every time a robot finishes carrying an object.

Figure 21 shows the makespan in these three methods in a known environment. Multiple comparisons with the Bonferroni methods show significant difference between these three methods. The method that maximizes the sum of efficiency is the highest in the known environment.

We show the following results of the conditions in an unknown environment.

Task-Execution Time. We performed simulation experiments in situations where it takes an unknown time to carry objects. Figures 22 and 23 show the experimental results. Multiple comparisons with the Bonferroni method revealed that there efficiency ${}^{i}e_m$

	task 0	task 1	task 2	task 3	task 4
Robot 0	1	2	3	4	5
Robot 1	6	7	8	9	10
Robot 2	11	12	13	14	15
Robot 3	16	17	18	19	20

$\frac{{}^{i}e_m}{(\Sigma_i {}^{i}e_m)(\Sigma_m {}^{i}e_m)} \times 10^{-3}$

1.96	3.51	4.76	5.80	**6.67**
4.41	4.61	4.76	**4.89**	5.00
4.98	**4.86**	4.76	4.68	4.62
5.23	4.97	4.76	4.59	4.44

red: selected tasks

FIGURE 19: Example of allocation based on the proposed method.

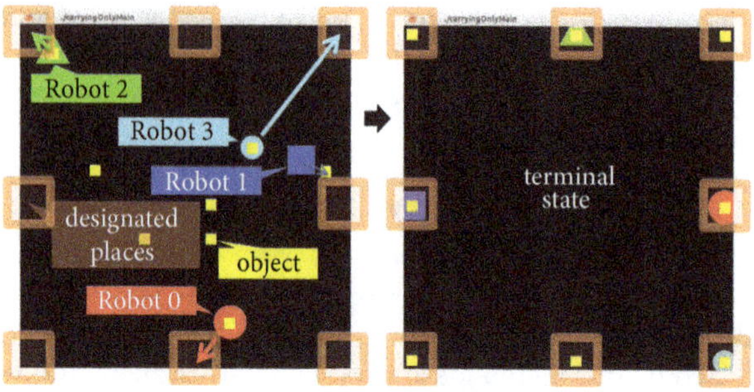

FIGURE 20: Simulation of carrying objects with four robots.

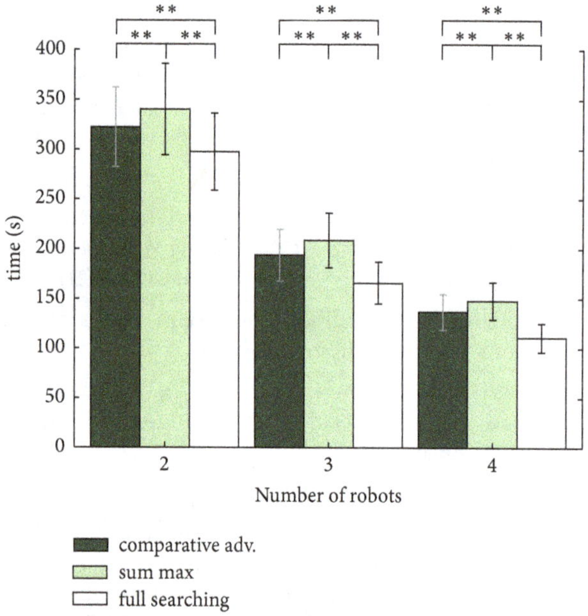

FIGURE 21: Makespan of some number of robots. $^{**}p < .01$, $^{*}p < .05$, and $^{+}p < .1$. A significant p value indicates that a significant difference in prescribing exists between the conditions. The lower p value indicates a strong significant difference. n.s. means no significant difference between the conditions.

are significant differences ($p < .001$) except for $\alpha = 4$, comparative adv versus sum max (n.s.). Compared with methods that maximize the sum of efficiency, the proposed method reduces the makespan under any condition. Compared with the full searching method, the makespan ratio decreased with

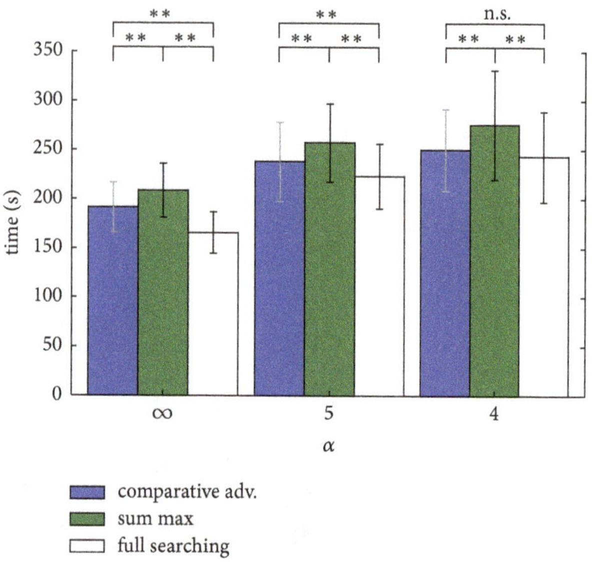

FIGURE 22: Makespan in situations where carrying objects can be late (three robots). $^{**}p < .01$, $^{*}p < .05$, and $^{+}p < .1$. A significant p value indicates that a significant difference in prescribing exists between the conditions. The lower p value indicates a strong significant difference. n.s. means no significant difference between the conditions.

increasing uncertainty. The Bonferroni method shows no significant differences ($p = .05$ level for $\alpha = 4$).

Adding Objects. Figures 24 and 25 show the experimental results. Compared with the methods that maximize the sum of efficiency, the proposed method reduces the makespan under any condition. Compared with the full searching method, the makespan ratio decreased with increasing uncertainty. The Bonferroni method shows no significant differences ($p = .05$ level for $n_{\text{add}} = 2, 3,$ or 4).

Unreasonable Robots. In this experiment, robots with large index numbers are unreasonable. For example, for two UAs and one reasonable agent (RA), robots 1 and 2 are unreasonable and robot 0 is reasonable. Figures 26 and 27 show the experimental results. Compared with the methods that maximize the sum of efficiency, the proposed method reduces the makespan under any condition. The Bonferroni method shows no significant differences ($p = .05$).

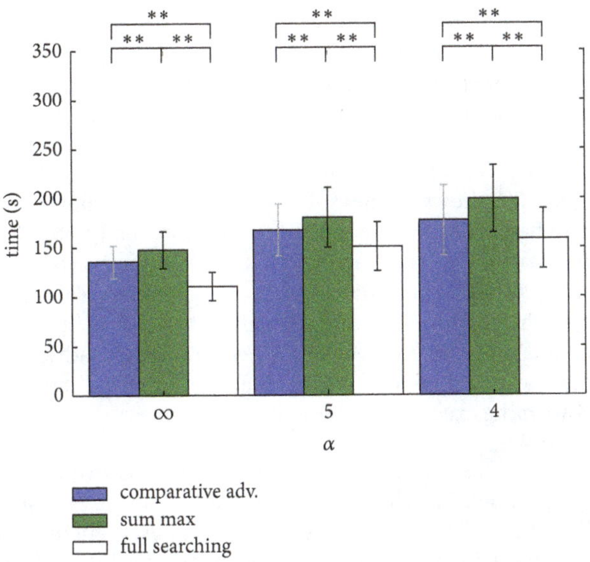

Figure 23: Makespan in situations where carrying objects can be late (four robots). $^{**}p < .01$, $^{*}p < .05$, and $^{+}p < .1$. A significant p value indicates that a significant difference in prescribing exists between the conditions. The lower p value indicates a strong significant difference. n.s. means no significant difference between the conditions.

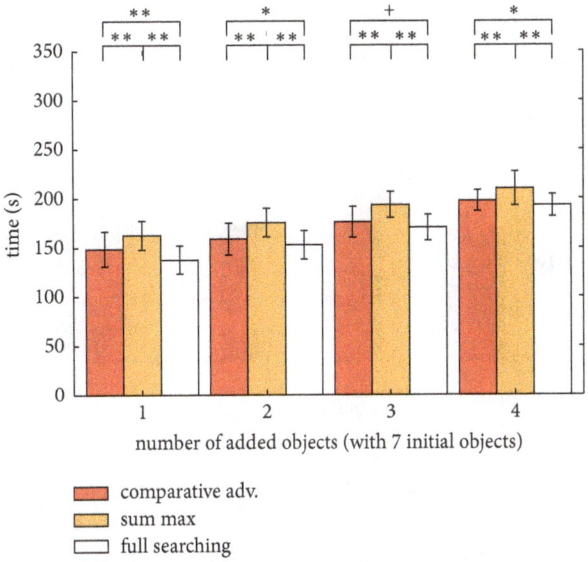

Figure 25: Makespan in situations where carrying objects are added later (four robots). $^{**}p < .01$, $^{*}p < .05$, and $^{+}p < .1$. A significant p value indicates that a significant difference in prescribing exists between the conditions. The lower p value indicates a strong significant difference. n.s. means no significant difference between the conditions.

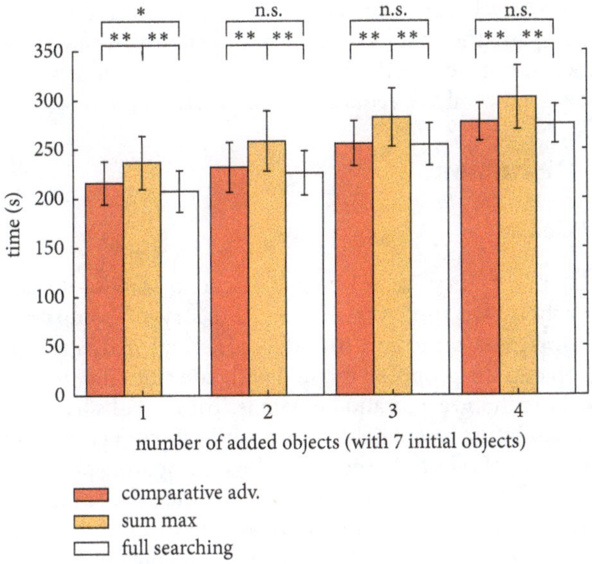

Figure 24: Makespan in situations where carrying objects are added later (three robots). $^{**}p < .01$, $^{*}p < .05$, and $^{+}p < .1$. A significant p value indicates that a significant difference in prescribing exists between the conditions. The lower p value indicates a strong significant difference. n.s. means no significant difference between the conditions.

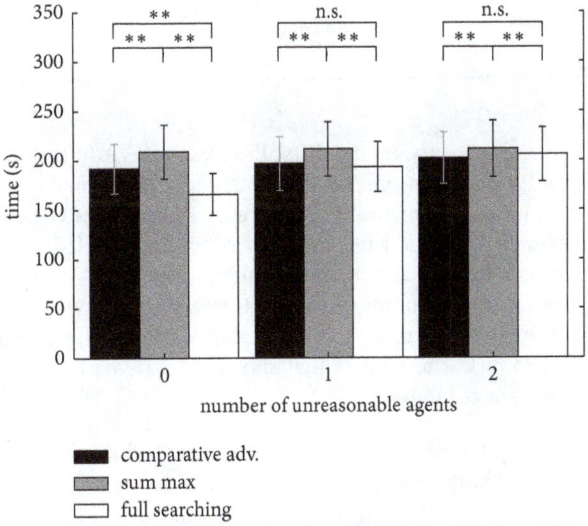

Figure 26: Makespan in situations where there is an unreasonable agent (three robots). $^{**}p < .01$, $^{*}p < .05$, and $^{+}p < .1$. A significant p value indicates that a significant difference in prescribing exists between the conditions. The lower p value indicates a strong significant difference. n.s. means no significant difference between the conditions.

The task-execution time for the proposed method is slightly longer than that of the full searching method from Figures 22–27. However, the calculation time in full searching increases exponentially with an increasing number of objects. We expect that our method will become shorter than full searching with an increase in the number of agents.

These results are similar to those in Figure 8, where the allocation based on our method reduces the makespan under any condition compared with methods that maximize the sum of efficiency. Compared with the full searching method, the makespan ratio decreased with increasing uncertainty.

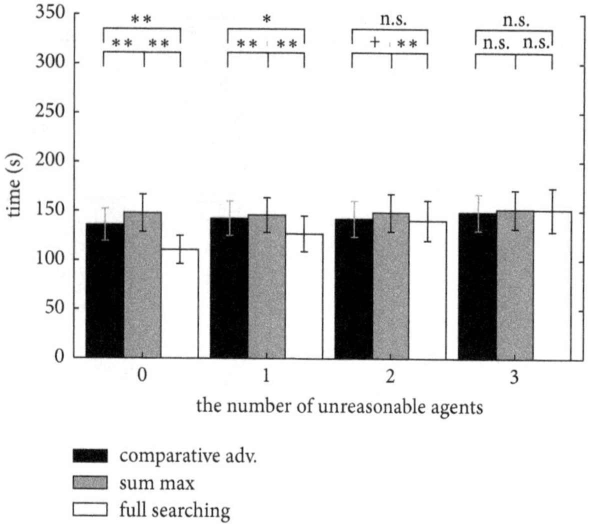

FIGURE 27: Makespan in situations where there is an unreasonable agent (four robots). $^{**}p < .01$, $^{*}p < .05$, and $^{+}p < .1$. A significant p value indicates that a significant difference in prescribing exists between the conditions. The lower p value indicates a strong significant difference. n.s. means no significant difference between the conditions.

TABLE 12: Moving ability of each robot.

	Proceeding	Rotating
Roomba	100 mm/s	100°/s
NAO	60 mm/s	60°/s

3.4.3. Experiments in the Real Environment. Additionally, beyond the simulations, we conducted an allocation experiment in a real environment. Figure 28 shows scenes of this experiment. We used two types of robots: a wheeled robot Roomba (iRobot) and a small humanoid NAO (Aldebaran Robotics). These robots can move forward or backward and rotate in their current position. Table 12 shows the speeds of the robots. The objects that should be carried and their features are as follows:

(i) Box: both robots can easily carry this object by pushing Roomba.

(ii) Ball on the wall: it is difficult for Roomba to push this object. Roomba cannot move the object to the target area directly, unless Roomba pushes it off the wall. Roomba needs to go around it and move it from behind. NAO can scrape objects from the wall using its hand, which costs approximately 5 s.

(iii) Table: it has four legs. Roomba is not wide enough to span the legs of the long side. Therefore, Roomba must alternately push the left and right legs in order to move the object. NAO has a comparative advantage to carry the table.

We calculated the carrying time using the moving path and moving speeds (Table 12). The robot controller gave a list of points as the moving path to the robots. Figure 28 shows the flow of the experiments. We regard the failure to carry as a factor of the delay of the carrying time. The proposed method is effective in allocating carrying tasks in an unknown environment as well as in the simulation.

Allocating Carrying Tasks to Wheeled Robot and Humanoid. Figure 28 shows scenes of experiments. (3) shows that Roomba failed to carry the box. If Roomba had not failed, Roomba would have carried the ball. We performed real-world experiments of allocating carrying tasks to real robots using the proposed method. We tried to confirm how our method works in the real world. As a result, the advantage of the proposed method will gradually reduce the differences between the execution times of remaining tasks.

Allocating Tasks to Robot and Human. We also conducted an experiment in which a robot and a human carry objects to confirm the effectiveness of the proposed method in allocation with an unreasonable agent. We used Roomba as the robot. Figure 29 shows the scene of this experiment, which included four types of objects (two boxes, a ball, and a table). Roomba is good at carrying the two boxes and is poor at carrying the table and ball. The human carries an object that the robot is not carrying. The results indicated that the human carried the table and ball, and the robot carried the two boxes. The robot saved carrying time for comparatively disadvantageous objects, thereby reducing the makespan. The robot executed comparatively advantageous tasks and prevented the UA from executing such tasks. The makespan would be reduced by the proposed method.

4. Discussion

4.1. Limitation

4.1.1. Computing Power. In this research, to allocate tasks to robots, we used the economic theory of comparative advantage as a method to reduce the total makespan. We compared the proposed method with conventional methods used in robotic exploration missions. These results have been enabled by the recent developments in computing power and the ability of robots. In the future, high computing power will reduce the total makespan, particularly the calculating time, and standardized robots that are able to execute multitasks will be developed. However, our method contributes to the efficiency of robots' labor output. Monofunctional robots can practically execute tasks other than their main task (e.g., cleaning task for the Roomba). Real robots can execute tasks that they should not or that their developers did not consider, but only at a low level (e.g., carrying task for the Roomba). In real environments, we cannot perceive everything that occurs. In such cases, our method can allocate the tasks to robots that include these low-level performances on a real-time basis according to the comparative advantage.

4.1.2. Simulation Setting. We conducted simulation in a field that is free (without obstacles) and small (20 × 20 grid), by robots without sensors and motions. Obstacles increase uncertainty and sensing capability relate closely to a robot's

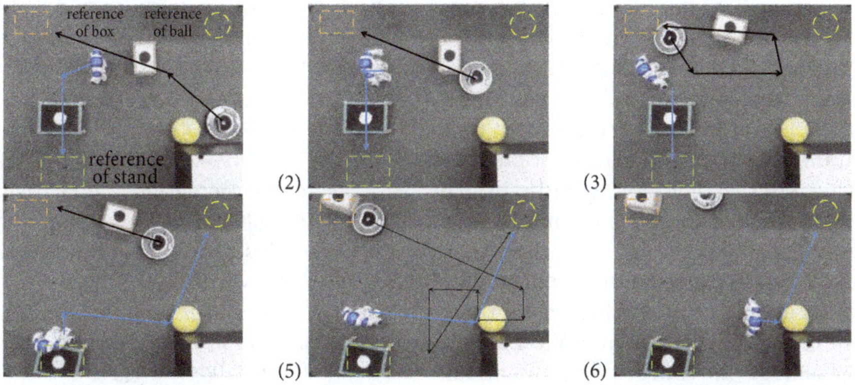

FIGURE 28: Experiment of real robots based on the proposed method. (1) Initial state. The robot decided the carrying object based on the list of carrying time. (2) Trying to move each object. (3) Roomba fails to carry the box and try to do again. (4) NAO finishes carrying the stand and begins to carry the ball. (5) Roomba finishes carrying the box. The method then determines which robot carries the ball, and NAO continues carrying the ball. (6) Roomba finished tidying, and NAO continues carrying the ball.

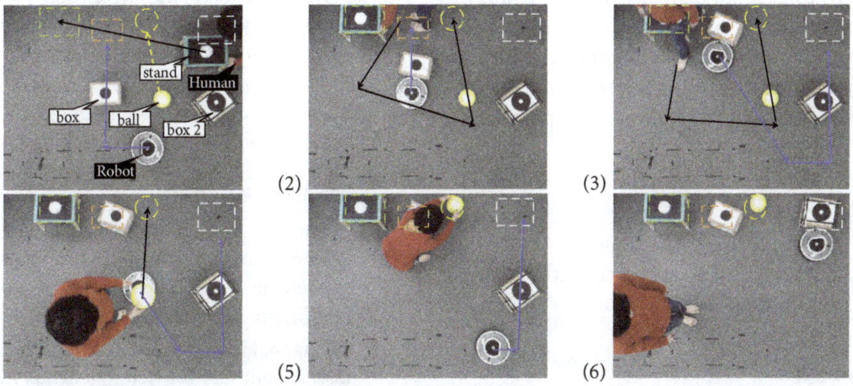

FIGURE 29: Experiment of the robot with a human.

performance. This study is a first step to investigate how to reduce the total makespan by using comparative advantage. We must unite the proposed method with existing algorithms of obstacle avoidance, exploration, and safe wandering. If it is possible to evaluate the performance of agents by real time, our method can be dynamically adaptive.

4.2. Future Work. There have been sparse improvements to the methods of allocating many tasks to many robots. We have a forward-looking approach in which the allocation of next tasks is preliminarily based on the theory of trade with the comparative advantage [54]. In the traveling salesman problem or the job-shop scheduling problem, there are many studies to reduce the calculation time. The proposed method should be applied to these studies and our method should be comparatively verified with them. In this study, one robot executes only one task (single-task robots). In the uncertain real environment, it is important to solve complex large-scale problems with a large number of robots, tasks, people, animals, and so on. Our method should also be able to be compared with different algorithms and the trajectories of robots. It is necessary to expand this to include the condition under which one robot executes certain tasks at the same time (multitask robots). In addition, the condition under which some robots execute one task at the same time should be considered. We supposed a centralized robot system in this study; thus, we should expand the proposed method to include the distributed autonomous systems.

This study investigated the allocation methods using simulations. We must comparatively verify these results in a real environment with other conventional methods such as market-based techniques inclusive of TraderBots [55, 56] and Hoplites [57]. The uncertainty instances with humans have numerous possible causes. Dealing with true "uncertainty" requires consideration of human-robot collaborations. Performance of a human is not certain quantitatively. Our method observes and calculates the performance of human and allocates tasks to robots with competence dynamically.

5. Conclusion

We investigated a method of using the theory of comparative advantage to allocate tasks to robots with uncertainty including humans. The proposed method is a dynamic sharing algorithm to allocate uncertainty tasks in an unknown environment, assuming timely reallocation. First, we confirmed that the proposed method reduces the total makespan (the total task-execution time) compared with conventional methods used in robotic exploration missions. We expect that our method is also effective in terms of calculation time when compared with the time-extended allocation method. We simulated carrying tasks and exploring tasks, which include uncertainty conditions of the work in an unknown environment. The proposed method is also more effective in dealing with uncertainty in task-execution time, uncertainty in the increasing number of tasks during task-execution, and uncertainty agents who are disobedient to allocation orders, compared to existing methods (the sum of efficiency and full searching methods). Finally, through experiments in a real environment, we confirmed that the proposed method can reduce the makespan.

This paper makes several contributions to human-robot interaction. First, the effectiveness of a new economic theory was shown for heterogeneous robots. In robot-robot collaboration, it is important to execute tasks even if the robot has inferior ability to accomplish a task. Second, the theory was effective in the uncertain environments including a human. Human-robot collaboration is receiving a lot of attention. The reallocation corresponding to uncertainness of people is a critical issue.

Disclosure

The full text is not published; it was accepted and presented in ROBOMECH2015 conference [58].

Conflicts of Interest

The authors declare that there are no conflicts of interest regarding the publication of this article.

Acknowledgments

The authors would like to thank Editage (https://www.editage.jp) for English language editing.

References

[1] D. Ricardo, *Principles of Political Economy and Taxation*, John Murray, 1817.

[2] A. Farinelli, L. Iocchi, and D. Nardi, "Multirobot systems: A classification focused on coordination," *IEEE Transactions on Systems, Man, and Cybernetics, Part B: Cybernetics*, vol. 34, no. 5, pp. 2015–2028, 2004.

[3] Z. Yan, N. Jouandeau, and A. A. Cherif, "A survey and analysis of multi-robot coordination," *International Journal of Advanced Robotic Systems*, vol. 10, article no. A399, 2013.

[4] B. Scassellati and B. Hayes, "Human-robot collaboration," *AI Matters*, vol. 1, no. 2, pp. 22-23, 2014.

[5] B. P. Gerkey and M. J. Matarić, "A formal analysis and taxonomy of task allocation in multi-robot systems," *International Journal of Robotics Research*, vol. 23, no. 9, pp. 939–954, 2004.

[6] R. L. Graham, "Bounds for certain multiprocessing anomalies," *Bell System Technical Journal*, vol. 45, no. 9, pp. 1563–1581, 1966.

[7] S. R. Lawrence and E. C. Sewell, "Heuristic, optimal, static, and dynamic schedules when processing times are uncertain," *Journal of Operations Management*, vol. 15, no. 1, pp. 71–82, 1997.

[8] D. Gale, *The Theory of Linear Economic Models*, McGraw-Hill Book Co, New York, NY, USA, 1960.

[9] B. P. Gerkey and M. J. Matarić, "Sold!: auction methods for multirobot coordination," *IEEE Transactions on Robotics and Automation*, vol. 18, no. 5, pp. 758–768, 2002.

[10] F. von Wieser and A. Ford Hinrichs, *Social Economics*, Adelphi, New York, NY, USA, 1927.

[11] L. E. Parker, "The effect of heterogeneity in teams of 100+ mobile robots," in *MultiRobot Systems Volume II: From Swarms to Intelligent Automata*, pp. 205–215, Kluwer, 2003.

[12] B. P. Gerkey, *Gerkey. on multi-robot task allocation [Ph.D. thesis]*, University of Southern California, 2003.

[13] A. Khamis, A. Hussein, and A. Elmogy, "Multi-robot task allocation: A review of the state-of-the-art," *Cooperative Robots and Sensor Networks*, vol. 604, Springer International Publishing, pp. 31–51, 2015.

[14] S. Liemhetcharat and M. Veloso, "Weighted synergy graphs for effective team formation with heterogeneous ad hoc agents," *Artificial Intelligence*, vol. 208, pp. 41–65, 2014.

[15] M. Scheutz, P. Schermerhorn, and P. Bauer, "The utility of heterogeneous swarms of simple UAVS with limited sensory capacity in detection and tracking tasks," in *Proceedings of the 2005 IEEE Swarm Intelligence Symposium, SIS 2005*, pp. 265–272, USA, June 2005.

[16] E. King, Y. Kuwata, M. Alighanbari, L. Bertuccelli, and J. How, "Coordination and control experiments on a multi-vehicle testbed," in *Proceedings of the 2004 American Control Conference (AAC)*, pp. 5315–5320, usa, July 2004.

[17] L. Zhang, H. Zhong, and S. Y. Nof, "Adaptive fuzzy collaborative task assignment for heterogeneous multirobot systems," *International Journal of Intelligent Systems*, vol. 30, no. 6, pp. 731–762, 2015.

[18] P. Brucker, *Scheduling Algorithms*, Springer, 2007.

[19] B. Gerkey and M. Mataric, "A framework for studying multi-robot task allocation," in *Proceedings of the In Proceedings of the Second International Naval Research Laboratory Workshop on Multi-Robot Systems*, 2003.

[20] R. T. Nelson, C. A. Holloway, and R. Mei-Lun Wong, "Centralized scheduling and priority implementation heuristics for a dynamic job shop model," *IIE Transactions*, vol. 9, no. 1, pp. 95–102, 1977.

[21] X. Zhang, Y. Deng, F. T. S. Chan, P. Xu, S. Mahadevan, and Y. Hu, "IFSJSP: a novel methodology for the Job-Shop Scheduling Problem based on intuitionistic fuzzy sets," *International Journal of Production Research*, vol. 51, no. 17, pp. 5100–5119, 2013.

[22] A. J. Davenport, G. Christophe, and J. Christopher Beck, "Slack-based Techniques for Robust Scheduling," in *Proceedings of the 6th European Conference on Planning ECP-01*, pp. 181–192, Toledo, Spain, 2001.

[23] J. Błażewicz, K. H. Ecker, E. Pesch, G. Schmidt, and J. Węglarz, *Scheduling Computer and Manufacturing Processes*, Springer, Berlin, Germany, 2nd edition, 2001.

[24] A. Edward, F. Silver David, and R. Peterson, *Inventory Management and Production Planning and Scheduling*, Wiley, 1998.

[25] B. B. Werger and M. J. Mataric, "Broadcast of local eligibility for multi-target observation," in *Distributed Autonomous Robotic Systems*, pp. 347–356, 2000.

[26] L. Vig and J. A. Adams, "Multi-robot coalition formation," *IEEE Transactions on Robotics*, vol. 22, no. 4, pp. 637–649, 2006.

[27] L. E. Parker and F. Tang, "Building multirobot coalitions through automated task solution synthesis," *Proceedings of the IEEE*, vol. 94, no. 7, pp. 1289–1304, 2006.

[28] L. Fanelli, A. Farinelli, L. Iocchi, D. Nardi, and G. P. Settembre, "Ontology-based coalition formation in heterogeneous MRS," in *Proceedings of the International Symposium on Practical Cognitive Agents and Robots, PCAR 2006*, pp. 105–116, Australia, November 2006.

[29] Y. Zhang and L. E. Parker, "Considering inter-task resource constraints in task allocation," *Autonomous Agents and Multi-Agent Systems*, vol. 26, no. 3, pp. 389–419, 2013.

[30] C.-H. Fua and S. S. Ge, "COBOS: Cooperative backoff adaptive scheme for multirobot task allocation," *IEEE Transactions on Robotics*, vol. 21, no. 6, pp. 1168–1178, 2005.

[31] R. Zlot, *An auction-based approach to complex task allocation for multirobot teams [PhD. thesis]*, Carnegie Mellon University, 2006.

[32] A. Gautam, J. K. Murthy, G. Kumar, S. P. A. Ram, B. Jha, and S. Mohan, "Cluster, Allocate, Cover: An Efficient Approach for Multi-robot Coverage," in *Proceedings of the IEEE International Conference on Systems, Man, and Cybernetics, SMC 2015*, pp. 197–203, Hong Kong, October 2015.

[33] D. Wu, G. Zeng, L. Meng, W. Zhou, and L. Li, "Gini coefficient-based task allocation for multi-robot systems with limited energy resources," *IEEE/CAA Journal of Automatica Sinica*, vol. 5, no. 1, pp. 155–168, 2018.

[34] R. Dornbusch, S. Fischer, and P. A. Samuelson, "Comparative advantage, trade, and payments in a Ricardian model with a continuum of goods," *American Economic Review*, vol. 67, no. 5, pp. 823–839, 1977.

[35] L. W. McKenzie, "Specialisation and efficiency in world production," *The Review of Economic Studies*, vol. 21, no. 3, pp. 165–180, 1954.

[36] R. W. Jones, "Comparative advantage and the theory of tariffs: A multi-country, multi-commodity model," *Review of Economic Studies*, vol. 28, no. 3, pp. 161–175, 1961.

[37] Y. Tian, "A New Idea about Ricardo's Comparative Advantage Theory on Condition of Multi-Commodity and Multi-Country," *International Journal of Business and Management*, vol. 3, no. 12, pp. 155–160, 2009.

[38] H. W. Kuhn, "The Hungarian method for the assignment problem," *Naval Research Logistics Quarterly*, vol. 2, pp. 83–97, 1955.

[39] M. M. Flood, "The traveling-salesman problem," *Operations Research*, vol. 4, pp. 61–75, 1956.

[40] T. Bektas, "The multiple traveling salesman problem: an overview of formulations and solution procedures," *Omega* , vol. 34, no. 3, pp. 209–219, 2006.

[41] G. B. Dantzig and J. H. Ramser, "The truck dispatching problem," *Management Science*, vol. 6, no. 1, pp. 80–91, 1959.

[42] C. H. Papadimitriou, *Computational Complexity*, Addison-Wesley, 1994.

[43] J. K. Lenstra and A. H. G. R. Kan, "Complexity of vehicle routing and scheduling problems," *Networks*, vol. 11, no. 2, pp. 221–227, 1981.

[44] M. R. Garey, D. S. Johnson, and R. Sethi, "The complexity of flowshop and jobshop scheduling," *Mathematics of Operations Research*, vol. 1, no. 2, pp. 117–129, 1976.

[45] A. H. Land and A. G. Doig, "An automatic method of solving discrete programming problems," *Econometrica*, vol. 28, pp. 497–520, 1960.

[46] D. E. Rumelhart, G. E. Hinton, and R. J. Williams, "Learning representations by back-propagating errors," *Nature*, vol. 323, no. 6088, pp. 533–536, 1986.

[47] M. Dorigo, *Optimization, learning and natural algorithms [Ph.D. thesis]*, Politecnico di Milano, Milano, Italy, 1992.

[48] M. Dorigo and L. M. Gambardella, "Ant colony system: a cooperative learning approach to the traveling salesman problem," *IEEE Transactions on Evolutionary Computation*, vol. 1, no. 1, pp. 53–66, 1997.

[49] J. Bramel and D. Simchi-Levi, "A location based heuristic for general routing problems," *Operations Research*, vol. 43, no. 4, pp. 649–660, 1995.

[50] G. Clarke and J. W. Wright, "Scheduling of vehicles from a central depot to a number of delivery points," *Operations Research*, vol. 12, no. 4, pp. 568–581, 1964.

[51] G. A. Croes, "A method for solving traveling-salesman problems," *Operations Research*, vol. 6, pp. 791–812, 1958.

[52] T. Van Der Zant and L. Iocchi, "Robocup@home: Adaptive benchmarking of robot bodies and minds," in *Proceedings of the International Conference on Simulation Modeling and Programming for Autonomous Robots*, vol. 7072, pp. 171–182, 2010.

[53] S. Tadokoro, H. Kitano, T. Takahashi et al., "RoboCup-Rescue project: a robotic approach to the disaster mitigation problem," in *Proceedings of the ICRA 2000: IEEE International Conference on Robotics and Automation*, pp. 4089–4094, April 2000.

[54] Y. Shiozawa, "Final resolution of the ricardian problem on international values," in *Proceedings of the International Conference on Structural Economic Dynamics*, 2012.

[55] R. Zlot, A. T. Stentz, M. B. Dias, and S. Thayer, "Multi-robot exploration controlled by a market economy," in *Proceedings of the 2002 IEEE International Conference on Robotics adn Automation*, pp. 3016–3023, usa, May 2002.

[56] R. Zlot and A. Stentz, "Market-based multirobot coordination for complex tasks," *International Journal of Robotics Research*, vol. 25, no. 1, pp. 73–101, 2006.

[57] N. Kalra, D. Ferguson, and A. Stentz, "Hoplites: A market-based framework for planned tight coordination in multirobot teams," in *Proceedings of the 2005 IEEE International Conference on Robotics and Automation*, pp. 1170–1177, Spain, April 2005.

[58] T. Morisawa and I. Mizuuchi, "2P1-V06 Allocating Multiple Tasks to Heterogeneous Robots Based on the Theory of Comparative Advantage," *The Proceedings of JSME annual Conference on Robotics and Mechatronics (Robomec)*, vol. 2015, no. 0, pp. _2P1-V06_1-_2P1-V06_4, 2015.

12

MapFuse: Complete and Realistic 3D Modelling

Michiel Aernouts, Ben Bellekens, and Maarten Weyn

IMEC, IDLab, Faculty of Applied Engineering, University of Antwerp, Groenenborgerlaan 171, 2000 Antwerp, Belgium

Correspondence should be addressed to Michiel Aernouts; michiel.aernouts@uantwerpen.be

Academic Editor: L. Fortuna

Validating a 3D indoor radio propagation model that simulates the signal strength of a wireless device can be a challenging task due to an incomplete or a faulty environment model. In this paper, we present a novel method to simulate a complete indoor environment that can be used for evaluating a radio propagation model efficiently. In order to obtain a realistic and robust model of the full environment, the OctoMap framework is applied. The system combines the result of a SLAM algorithm and secondly a simple initial model of the same environment in a probabilistic way. Due to this approach, sensor noise and accumulated registration errors are minimised. Furthermore, in this article, we evaluate the merging approach with two SLAM algorithms, three vision sensors, and four datasets, of which one is publicly available. As a result, we have created a complete volumetric model by merging an initial model of the environment with the result of RGB-D SLAM based on real sensor measurements.

1. Introduction

Due to the recently increased demands on location-based services, localisation of wireless devices based on received signal strength has become a significant research topic [1]. Within this research, simulation based approaches are used to evaluate localisation algorithms. These approaches use wave propagation models, which can predict the attenuation of a signal as it propagates through space. Moreover, such a propagation model will improve the localisation accuracy and precision [2].

Many different types of propagation models have been defined [3]. Previous research focussed on the validation of an indoor ray launching propagation model for sub-GHz frequencies [4–7]. In [4], Bellekens et al. utilised a sub-1 GHz measurement device that was programmed with the LPWAN (Low Power Wide Area Network) standard DASH7 [8]. A Pioneer 3DX robot was used to collect RF measurements from 6 stationary transmitters, laser range measurements, and the wheel odometry. Next, a laser based online SLAM algorithm (Gmapping) is being used to obtain a 2D map of the environment as well as the robots' estimated trajectory. By combining map and trajectory information with the RF measurements, the researchers were able to validate their 2D propagation simulation with an accuracy of 2.8 dB and a precision of 7.8 dB [4].

Evaluating a 3D propagation model based on a real environment model makes it possible to evaluate signal strengths at different heights as well as the specific multipath introduced by the environments. This will enable the evaluation of localisation algorithms that are highly influenced by multipath effects, that is, Angle of Arrival localisation. Therefore, a UAV (Unmanned Aerial Vehicle) can be utilised instead of a driving robot to receive first the RF measurements and secondly to capture the environment in 3D. Due to noisy depth image measurements and accumulated registration errors, the result of a 3D-SLAM does not cover the entire environment where ceilings, floor, walls, and objects are included.

In this paper, we aim to create a complete model of an environment of which we have foreknowledge. In order to benefit the completeness of our result, we used a probabilistic, volumetric mapping approach called OctoMap [9]. In contrast to using point clouds, OctoMap allows us to render a model which contains information about occupied spaces, free spaces, and unknown spaces. Also, we benefit from the probabilistic nature of an OctoMap, as it allows us to update

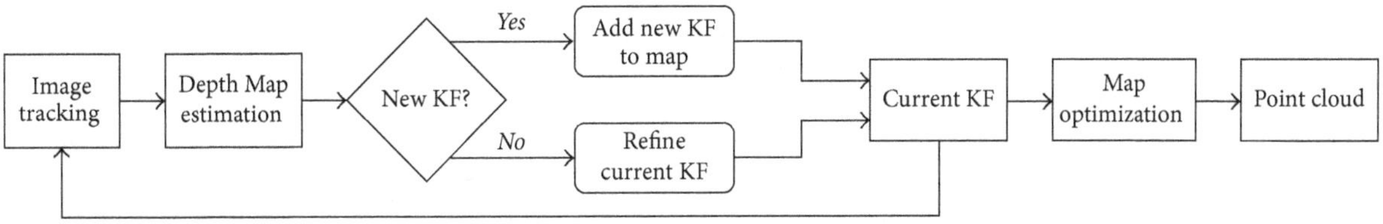

FIGURE 1: This figure illustrates a simplistic schematic of the LSD SLAM algorithm.

an initial guess model of the environment with real sensor measurements.

Current research about completing an environment model based on the surface of a captured point cloud has rather been limited [11, 12]. Breckon and Fisher presented a method to complete partially observed objects by deriving plausible data from known portions of the object. However, this method is time-consuming and involves complex calculations. Furthermore, the completions are not an accurate reconstruction of reality, as they are only meant to be visually acceptable for the viewer [11]. In [12], laser range data is used to create a 2D floor plan, which can be extruded to a 2.5D model. By aligning this simplified model with a complex octree of the environment, the researchers were able to build a final model which includes previously hidden surfaces. However, automatic generation of a 2.5D model is challenging when a flawed dataset is used as input. Also, this approach assumes floors and ceilings to be horizontal and to have fixed heights. Moreover, model merging is not done in a probabilistic fashion.

The main contributions of MapFuse regarding the state of the art are two different approaches for merging an initial environment model with real measurements. First, the initial model can be merged iteratively with the final SLAM result. With this technique, the accuracy can be regulated by changing the amount and sequence of both models. Secondly, an online merging process, which updates the initial model while SLAM is processing, can be applied. Both methods use OctoMap to probabilistically build a complete volumetric model.

In order to create and evaluate MapFuse, we have compared three different camera types in a simulated environment: a basic monocular camera, a wide field-of-view camera, and a depth-sense camera. With these cameras, datasets were recorded to be used as input for a visual SLAM algorithm such as Large-Scale Direct SLAM (LSD SLAM) or feature-based RGB-D SLAM. Afterwards, the simulated SLAM results were validated in four real environments.

The remainder of the paper is structured as follows: Section 2 lists techniques that were tested and implemented in our system. Section 3 describes the three main blocks of which our method consists. In Section 4, results of our approach are discussed. Finally, Section 5 concludes these results.

2. Related Work

For many years, researchers have come up with solutions to the SLAM problem. This problem occurs when a robot is placed in an unknown environment. Without a map or information about its own location, the robot has to be able to build a map of the environment while determining its own position at the same time [13]. When a SLAM algorithm is implemented, a robots' location can be determined in a probabilistic fashion by combining sensor measurements with odometry information. Bayes filters such as Kalman filters are well suited for this purpose, as they predict the robots' state based on its previous state and received motion commands [14]. Afterwards, the predicted state is corrected with the obtained sensor measurements [15].

Laser range sensors are commonly used for building 2D and 3D maps [4, 16, 17]. However, cameras have become increasingly more popular for building 3D maps. This is due to their low cost compared to laser range sensors and the possibility of obtaining 3D data when using depth-sense cameras or stereo cameras [18]. Also, compared to laser range sensors, depth-sense cameras allow us to record an entire depth image that contains a collection of points that are registered with an RGB image. For this reason, we will research visual SLAM (VSLAM) algorithms to map an indoor environment. VSLAM can be subdivided into two subclasses: *Feature-Based Methods* and *Direct Methods*. For our research, we evaluated the two merging processes with both of these VSLAM methods by comparing LSD SLAM to RGB-D SLAM.

With OctoMap, volumetric models can be created by calculating the occupancy probability of all nodes in the map. As information about occupied spaces, free spaces, and unknown spaces is obtained, this mapping approach can be used for autonomous exploration of various environments.

2.1. Large-Scale Direct SLAM. LSD SLAM is a direct VSLAM algorithm that can be implemented with monocular cameras as well as stereo cameras [19, 20]. Direct VSLAM methods profit from pixel intensity values of the entire image. Due to the fact that the complete image is used as input data, these methods result in a high accuracy and provide a lot of information about the geometry of the environment [19, 21].

Figure 1 illustrates the LSD SLAM workflow in a basic schematic. Firstly, the tracking component estimates the rigid body pose with $se(3)$ [22]. Secondly, the current KF will be refined or a new KF will be created from the most recent tracked image, depending on the estimated transformation between two images. Finally, the map optimisation component inserts keyframes into the global map when they are replaced by a new reference frame.

2.2. RGB-D SLAM. Contrary to LSD SLAM, RGB-D SLAM is a feature-based VSLAM algorithm. Feature-based VSLAM

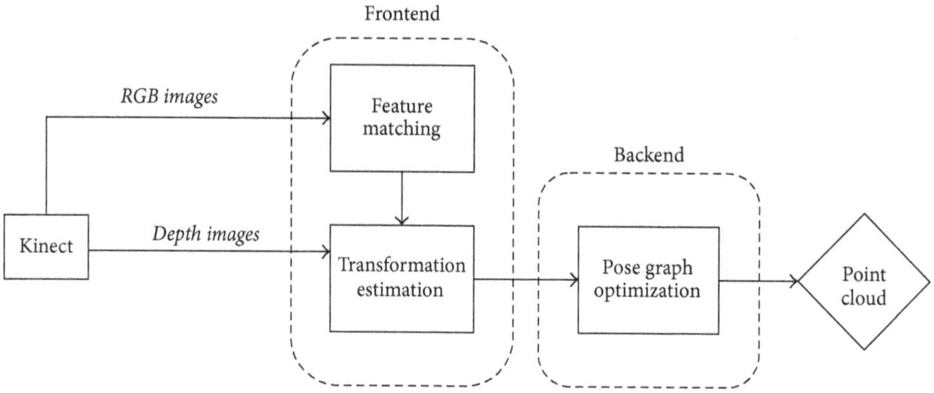

FIGURE 2: This figure illustrates a simplistic schematic of the RGB-D SLAM algorithm.

collects feature observations from the camera image and then compares these features to the previous camera image. Numerous feature detectors can be implemented for this purpose, for example, Oriented FAST and Rotated BRIEF (ORB), Scale Invariant Feature Transform (SIFT), and Speeded Up Robust Features (SURF) [23, 24].

In the RGB-D SLAM frontend, feature locations are visualised in three dimensions by overlaying the RGB image with its respective depth image. With *se*(3), a transformation estimation between two subsequent frames can be calculated [22]. However, this estimation cannot be considered accurate due to false positives in feature detection and the fact that RGB images can be inconsistent with depth images. Therefore, a Random Sample Consensus (RANSAC) algorithm is applied to abolish this effect [25].

In the SLAM backend, frames are added as a node to the pose graph. When a frame matches one of the previous frames, it will be connected to the existing pose graph of the matching frame. Otherwise, the new frame is connected to the previous node in the pose graph. After obtaining spatial relations via the SLAM frontend, RGB-D SLAM implements the g^2o framework for pose graph optimisation [26]. With pose graph optimisation, a trajectory is estimated using the robots' relative pose measurements, that is, the robots' current and previous poses. Figure 2 provides a simplified overview of the RGB-D SLAM workflow.

2.3. OctoMap. OctoMap is a volumetric mapping framework based on an octree data structure and probabilistic occupancy estimation [9]. When it comes to mapping arbitrary 3D environments, OctoMap has numerous advantages over other mapping approaches. Octrees are highly memory efficient; they consist of an octant which can be divided into eight leaf nodes. Subsequently, these leaf nodes can be seen as new octants, which in their turn can be divided into leaf nodes again. The desired resolution of the 3D model is determined by the depth of the octree. For example, large adjacent volumes can be represented by a single leaf node to save memory. The octree data structure is shown in Figure 3.

Figure 3 also displays that an octant node n can have different states: free, occupied, or unknown. This state can be derived from the calculated probability according to

$$P(n \mid z_{1:t}) = \left[1 + \frac{1 - P(n \mid z_t)}{P(n \mid z_t)} \frac{1 - P(n \mid z_{1:t-1})}{P(n \mid z_{1:t-1})} \frac{P(n)}{1 - P(n)}\right]^{-1}. \quad (1)$$

Equation (1) states that the probability of whether a node is occupied or free is determined by the current sensor measurement z_t, the previous estimate $P(n \mid z_{1:t-1})$, and a prior probability $P(n)$, which is assumed to equal 0.5. In order to rewrite the equation, we will apply the log-odds notation:

$$L(n) = \log\left[\frac{P(n)}{1 - P(n)}\right]. \quad (2)$$

Thus, (1) can be rewritten as

$$L(n \mid z_{1:t}) = L(n \mid z_{1:t-1}) + L(n \mid z_t). \quad (3)$$

With log-odds, probabilities of 0% to 100% are mapped to $-\infty$ dB and $+\infty$ dB. A main advantage of this notation is that small differences at the outer edges of the range have the strongest influence on the probability. For example, 50.00% and 50.01% are mapped to 0 dB and 0.0017 dB, while 99.98% and 99.99% are mapped to 37 dB and 40 dB. As (3) makes use of additions instead of multiplications, the probability of a leaf node can be updated faster than in (1). Faulty measurements due to noise or reflections are cancelled out by the update formula.

Furthermore, the map can be extended at any time when the robot explores new unknown areas. As the OctoMap holds information about unmapped space, the robot knows which areas it has to avoid for safety reasons, or which areas are yet to be explored.

3. System Approach

In pursuance of building a complete model, we propose a system that consists of three main steps. Figure 4 displays a basic schematic that represents the MapFuse workflow. Firstly, visual information from the camera is recorded into a dataset, and an initial guess is modelled. Secondly, the dataset that was gathered in the first step is used as input for a SLAM algorithm. Finally, OctoMap is used to merge the SLAM point cloud with the initial guess.

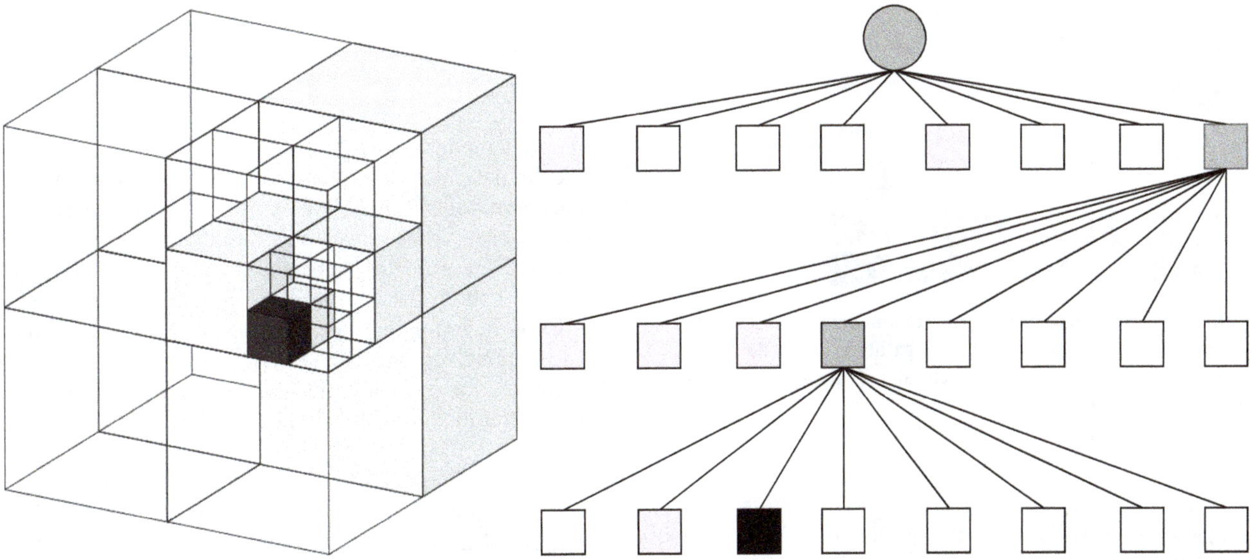

FIGURE 3: A visual representation of the octree data structure [9]. The black leaf node represents occupied space, whereas grey nodes indicate free space. Unknown space is marked by transparent nodes.

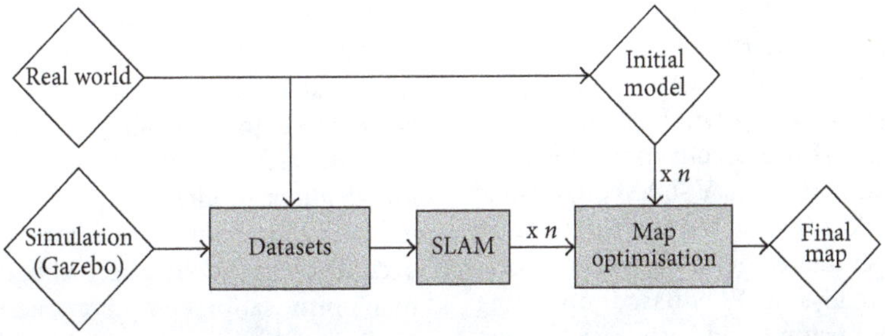

FIGURE 4: With MapFuse, a dataset that is recorded in a simulated or real environment is used as input for a SLAM algorithm. In the map optimisation component, the resulting SLAM point cloud is merged with an initial model which was modelled based on exact dimensions of the environment. The final MapFuse result is a complete volumetric model of the environment.

(a) (b)

FIGURE 5: An initial guess point cloud will be used so as to complete the unfinished SLAM point cloud.

3.1. Dataset. Since we have foreknowledge of the environment, an initial guess model can be created. In order to do so, we resort to OpenSCAD. This 3D modelling software allows us to build a model that matches the exact dimensions of the real environment. The most important goal for such a model is to provide an incomplete SLAM map with complementary information about the environment. The amount of detail that has to be included in the initial guess mostly depends on the quality of the dataset. With an admissible dataset, SLAM will provide a lot of details, so that the initial guess can be limited to a bounding box of the environment.

Figure 5 displays two different initial guess models. In Figure 5(a), a bounding box of an indoor environment with doors and windows is shown. For visualisation purposes,

Figure 6: With Gazebo, we are able to simulate quadcopter flight and sensor measurements in order to gather an ideal dataset. This dataset was used to evaluate which SLAM algorithm was most suitable for our approach.

ceilings were not included in this model. In Figure 5(b), we created a simplified model of an industrial train cart.

An additional benefit is that the model can be imported in a simulator, which induces numerous advantages. Above all, simulation is time-saving and abates the risk of crashing the quadcopter. It has allowed us to experiment with multiple camera types and algorithms in order to design an optimal work flow. Therefore, our system was assessed by employing Gazebo. This software allows us to spawn a quadcopter as well as our initial guess model. However, our model needs to be extended with colour and objects, as VSLAM algorithms require visual features to build a map of the environment. In Figure 6, an example of the simulated environment is shown.

As our quadcopter employs a ROS-based operating system, trajectories can be scripted and tested in Gazebo before real world tests are conducted. By doing so, the quadcopter will always follow the same trajectory. Hence, a better comparison between camera types can be made. Figure 7 shows which cameras we have evaluated in our experiments.

However, scripting a trajectory requires some form of ground truth such as GPS. Since we are operating the quadcopter indoors, we cannot rely on GPS communication. An accurate indoor ground truth pose estimation system would have to be implemented in order to use these trajectory scripts in reality, which is an expensive and time-consuming process [27]. Therefore, our real world implementation of the system will control the quadcopter via a remote controller.

After setting up the simulator with an environment, a flying quadcopter, and a camera, datasets can be recorded via ROS topics. Such a topic can hold camera images, odometry, or information about the relationship between all coordinate frames. With the latter, it is possible to deduce the camera pose relative to the quadcopter. Consecutively, we can deduce the initial pose of the quadcopter relative to the map. A VSLAM algorithm combines all this information with visual odometry of the camera, with the purpose of obtaining a more accurate trajectory estimate.

Datasets that were recorded in Gazebo were used as input for several VSLAM algorithms in order to determine which camera and which algorithm are most suitable for our method. In order to validate our simulation results, we implemented the same process to gather datasets in real environments.

3.2. SLAM. The second step adopts the dataset as input for a ROS implementation of a visual SLAM algorithm. By playing back the datasets, camera images will be published to ROS topics required by the SLAM algorithm. The playback speed of the dataset can be slowed down, so that the applied SLAM algorithm has more time to detect and process visual features. We have experimented with LSD SLAM as well as RGB-D SLAM in order to analyse which of these algorithms is most suitable for our method. As discussed in Section 4, parameters for both algorithms were changed empirically until we found an optimal result.

3.3. Map Optimisation. The final step in our system combines the initial guess point cloud of Figure 5 and the SLAM point cloud into a single OctoMap. In order to obtain an accurate OctoMap, these point clouds have to be aligned as well as possible. Point cloud alignment is achieved by empirically transforming the initial guess coordinate frame to the SLAM coordinate frame. After the transformation is regulated correctly, both clouds are sent into an OctoMap server node. This way, the initial guess model will be updated with real measurements from a camera. Because SLAM is not able to map all elements in the environment, for example, ceilings or walls that are blocked by furniture, our initial guess model will provide the OctoMap server with information about these missing elements and updates the occupancy probability accordingly. A basic schematic of our map optimisation component is shown in Figure 8.

Two options can be considered to merge point clouds. On the one hand, the final SLAM result can be merged iteratively with the initial guess. Occupancy estimations can be altered by changing the initial OctoMap occupancy probability or by inserting both point clouds multiple times. When using this method, a balance between map completeness and detail has to be mediated. On the other hand, the merging process can be affected while SLAM is building a point cloud. After sending the initial guess point cloud to the OctoMap server a single time, node probabilities will be updated iteratively as the SLAM algorithm refines its point cloud based on current and previous measurements. Due to the fact that multiple measurements are taken into account, the occupancy probability will be more conclusive. Figure 9 demonstrates the difference between both merging methods.

4. Results

MapFuse was evaluated using four different datasets, three of which we have recorded ourselves. Dataset 4 is publicly available via the RGB-D benchmark dataset [27]. Dataset 1 was recorded with a wide field-of-view camera; all other datasets were recorded with a Microsoft Kinect. For all datasets, we have built an initial guess model with OpenSCAD.

(i) *Dataset 1*: room V329 at the University of Antwerp. This meeting room contains many empty tables and closets

(a) Logitech C615 (b) Genius WideCam (c) Microsoft Kinect

FIGURE 7: In both simulation and reality, we conducted tests with a common webcamera (a), a wide field-of-view webcamera (b), and a Microsoft Kinect (c).

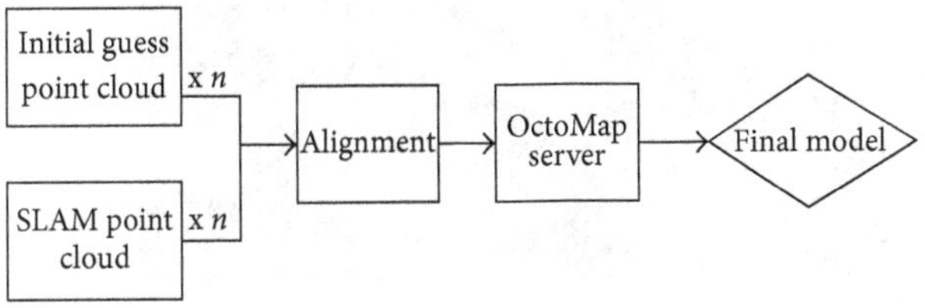

FIGURE 8: Basic schematic of the optimisation block of our system.

(ii) *Dataset 2*: room V315 at the University of Antwerp (6.67 m × 7.02 m × 3.77 m). The adjacent room V317 was also included in this dataset (4.12 m × 3.42 m × 3.77 m). Both rooms contain desks and closets with a high amount of clutter

(iii) *Dataset 3*: an industrial tank car located in a small hangar at the port of Antwerp (9.2 m × 2.45 m × 3.75)

(iv) *Dataset 4*: the "freiburg1_room" dataset provided by Sturm et al. This dataset is recorded in a small office environment.

These datasets were used as input for the two VSLAM algorithms that were discussed in Section 2: LSD SLAM and RGB-D SLAM.

Finally, the MapFuse optimisation step of Section 3.3 was evaluated by applying iterative merging and online merging on the initial model and the SLAM output.

All tests were performed on a Dell Inspiron 15 5548 laptop, which is provided with an Intel i7 5500U 2.4 GHz, 8 GB RAM, and an AMD Radeon R7 M265 graphics card. Ubuntu 14.04.5 LTS was used as the operating system.

4.1. LSD SLAM. Our first tests were conducted with LSD SLAM. We connected a wide field-of-view web camera (120°) to a laptop and downsampled the image to a 640 × 480 resolution in order to evaluate the algorithm. With this setup, we recorded dataset 1 by walking around the room in a sideways motion. This was necessary to ensure sufficient camera translation, which is required for LSD SLAM. In order to optimise the map with loop closures, the same trajectory was repeated multiple times.

When running LSD SLAM, a few important parameters have to be reckoned with. First, a pixel noise threshold is set to handle faulty sensor measurements. Second, the amount of keyframes to be saved is defined. This amount is based on the image overlap and the distance between two consecutive keyframes. A large number of keyframes will result in an accurate trajectory, but also induces more noise in the map.

Figure 10(b) illustrates the point cloud and trajectory estimate of LSD SLAM. Empirical comparison with the real environment of Figure 10(a) leads us to conclude that LSD SLAM produces an accurate trajectory estimate. However, the point cloud holds a high amount of noise. As our approach requires a dense and detailed SLAM point cloud with little noise, we will not pursue LSD SLAM in our research any further.

4.2. RGB-D SLAM. For our tests with RGB-D SLAM, we mounted a Kinect camera to an Erle-Copter as shown in Figure 11 [10]. The Kinect was slightly tilted downwards to capture as many visual features as possible. The camera was connected to a laptop which ran the camera driver correctly. Contrary to LSD SLAM, RGB-D SLAM can handle camera translation as well as camera rotation. We found that the best trajectory for this algorithm is to rotate the camera 360 degrees at the centre of the room and then apply coastal navigation. This process should be repeated for every new room that is entered.

RGB-D SLAM allowed us to configure numerous parameters. First, a feature extractor had to be chosen. Our tests indicated that the SIFTGPU extractor, combined with FLANN feature matching, induces a satisfying result. Second, we filtered the depth image by implementing a minimum and maximum processing depth. As a result, noisy measurements outside the valid range were diminished. The optimal values for these parameters depend on the environment that was recorded. For example, for Figure 12, we have set these parameters to 0.5 metres and 7 metres, respectively. Lastly, the computed point cloud was downsampled, as we noticed that RGB-D SLAM failed to process new visual features when the CPU is overloaded. Downsampling the point cloud with a factor n significantly decreases CPU usage, while maintaining

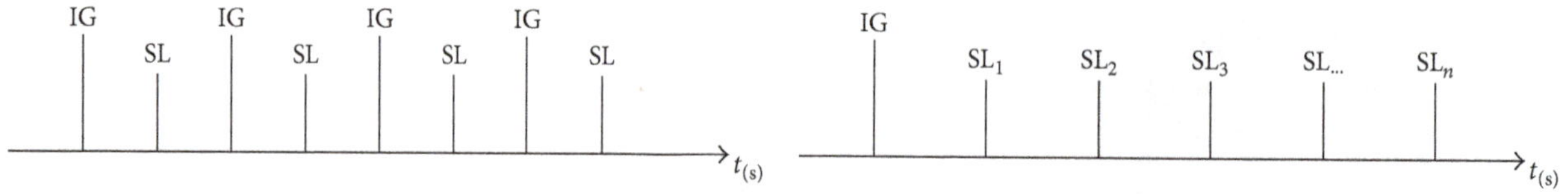

(a) Iterative merging

(b) Online merging

FIGURE 9: In (a), the initial guess (IG) is iteratively merged with the complete SLAM point cloud (SL). A balance between map completeness and detail is regulated by the amount of IG or SL point clouds we merge. (b) illustrates another option, where a single IG is merged with partial online SLAM clouds (SL_n). The online merging process is finished when SLAM has completely processed the dataset. The difference between both merging methods is discussed in detail in Section 4.3.

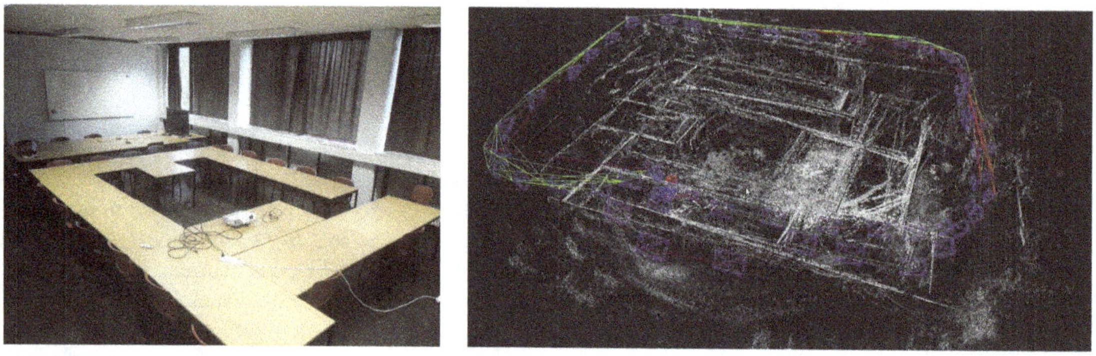

(a) The meeting room that was used to record dataset 1

(b) Resulting point cloud

FIGURE 10: LSD SLAM result.

FIGURE 11: For our research, we mounted a Kinect camera to an Erle-Copter [10].

an acceptable point cloud density. Normally, the Kinect outputs a 640 × 480 array (307200 entries). By downsampling this array, RGB-D SLAM keeps every nth entry in the Kinect array. For example, if n equals 4, only 76800 entries (25%) are kept to be processed by RGB-D SLAM.

After configuring the RGB-D SLAM parameters, we managed to build the point clouds shown in Figure 12. When we compare these results with the LSD SLAM result in Figure 10, it becomes clear that RGB-D SLAM builds point clouds with higher density and less noise. This is mainly due to the fact that RGB-D SLAM inserts all visual information into the point cloud, whereas LSD SLAM creates a point cloud which merely consists of pixels that were used for depth map estimation. However, the RGB-D SLAM point clouds still contain gaps due to registration errors and limited observations. For example, the ceilings of dataset 2 are not visible in Figure 12(b) and the industrial train car of dataset 3 is incomplete in Figure 12(e).

Inquiring the accuracy of a trajectory requires the implementation of a ground truth estimation system. As mentioned in Section 3.1, implementing such a system does not lie within the scope of our research, so we adopt the accuracy measurements of Endres et al. [25]. In their research, the authors state that the RGB-D SLAM trajectory estimate has an average root mean square error (RMSE) of 9.7 cm and 3.95 degrees if SIFTGPU is used as feature extractor. This number was obtained by testing the SLAM algorithm with datasets which include ground truth information [27].

We conducted our own tests in order to determine the xyz precision of the trajectory estimate. RGB-D SLAM was launched several times, each time with the same parameters. One reference trajectory estimate and ten test trajectory estimates were extracted from these tests. For every trajectory, we plotted the error relative to the reference trajectory. Boxplots for the x-, y-, and z-axis error can be found in Figures 13, 14, and 15.

In Figure 13(a), we can observe that the test trajectories correspond well to our reference trajectory, as well as to each other. Errors relative to the reference remain very limited for all trajectories. Nonetheless, we also detect outliers with a difference of up to 80 cm relative to the reference trajectory. Figure 13(b) illustrates when these outliers occur in time. This plot shows high precision until a certain point where the trajectories start to spread out. At this point, RGB-D SLAM was not able to process visual features. Thus, the trajectory

(a) Room V315 at the University of Antwerp, where we recorded dataset 2

(b) RGB-D SLAM result for dataset 2

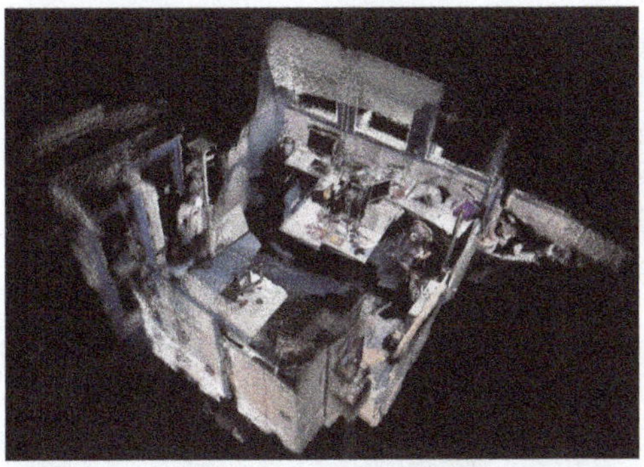

(c) RGB-D SLAM result for dataset 4

(d) Dataset 3 was recorded with the purpose of modelling an industrial tank car at the port of Antwerp

(e) RGB-D SLAM result for dataset 3

FIGURE 12: We assessed the RGB-D SLAM algorithm in several environments. First, we tested indoor environments as shown in (a), (b), and (c). Second, we applied the algorithm to map an industrial train cart ((d) and (e)).

estimate could not be calculated correctly until visual features were tracked again. Compared to the y- and z-axis, the x-axis trajectory accumulated more errors due to the fact that the camera mainly travelled along the x-axis.

Similarly to the x-axis trajectory, Figure 14 demonstrates a high precision on the y-axis. As this axis contains fewer translations than the x-axis, outliers are less distinct.

Finally, the z-axis boxplot in Figure 15 exhibits precision results that are comparable with the x and y precision plots.

In general, we can conclude that the RGB-D SLAM algorithm results in precise trajectory estimates, as long as visual features are continuously detected while mapping an environment. In order to ensure continuous feature tracking, the dataset can be played out at a lower speed. By doing so, RGB-D SLAM will have more time to process new images which is beneficial for feature extraction.

4.3. Optimisation Results. Although RGB-D SLAM has provided us with an accurate and dense point cloud, Figure 12 has shown us that the map merely contains information about all environmental elements that were visible in the dataset images. For example, ceilings were not recorded, so they will not be included in the resulting map. When we use this point cloud to render an OctoMap, only a partial volumetric model of the environment is obtained, as seen in Figure 16.

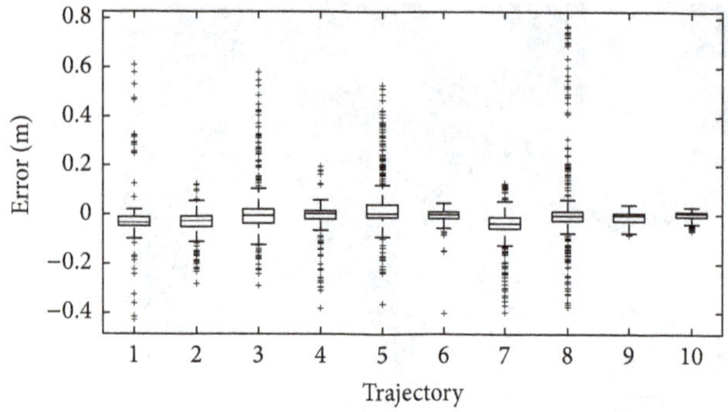

(a) This boxplot demonstrates the x-axis precision error of all trajectories relative to a test trajectory

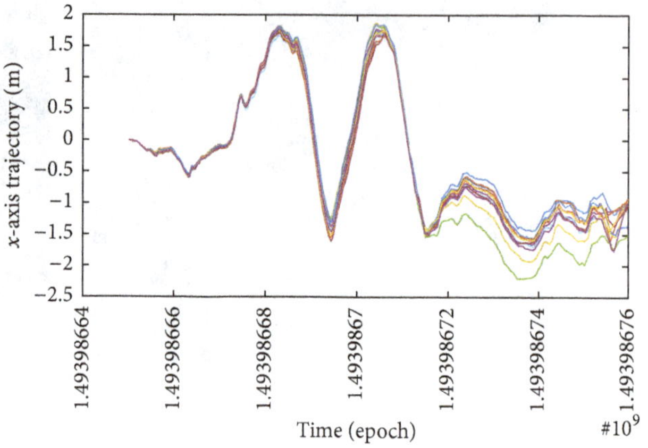

(b) This plot shows the x-axis precision error over time of all trajectories relative to a test trajectory

FIGURE 13: RGB-D SLAM x-axis precision.

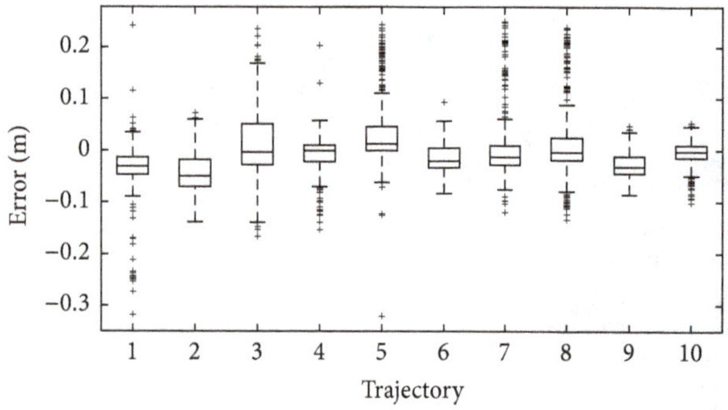

FIGURE 14: RGB-D SLAM y-axis precision.

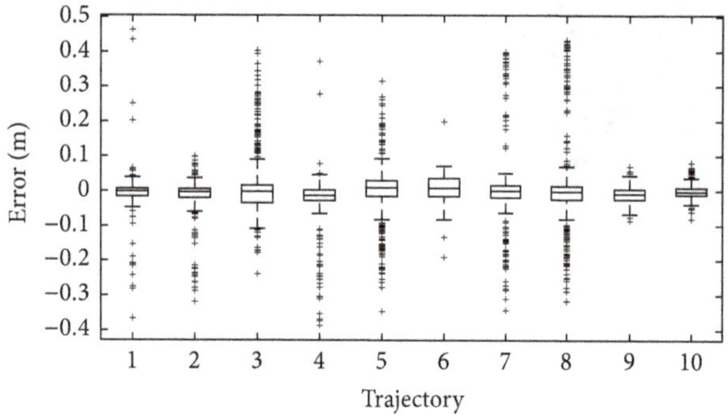

FIGURE 15: RGB-D SLAM z-axis precision.

Our approach resolves this issue by combining the SLAM point cloud with an initial guess. Merging these clouds can be achieved in two ways: iterative and online. The first method builds an OctoMap using the initial guess and the final SLAM result as input. Occupancy probability is steered by iteratively inserting the point clouds multiple times. However, this form of map completion also updates valid measurements with free space, causing the map to lose some of its detail. Figure 17 demonstrates this problem. In Figure 17(a), both point clouds were inserted once. The initial

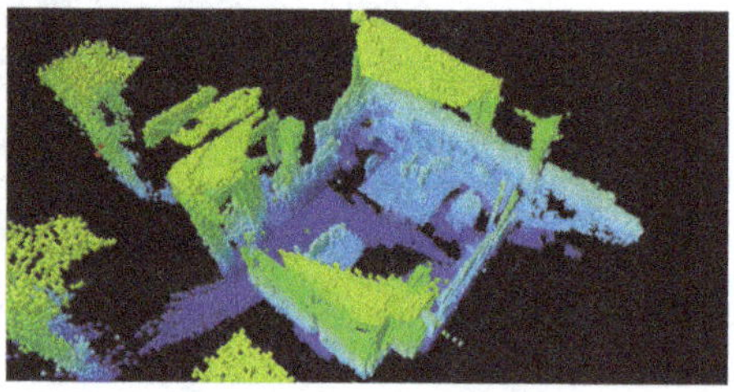

FIGURE 16: An OctoMap created from our RGB-D SLAM result of Figure 12(c).

guess has successfully filled in gaps that were present in the SLAM result of Figure 16, although it has also caused doors and windows to disappear. By adding another instance of the SLAM result (Figure 17(c)), doors and windows started to reappear along with unwanted gaps in the floor. Another factor that has to be taken into account is the order in which point clouds are being merged. As can be seen in Figures 17(e) and 17(g), inverting the merging sequence has a significant effect on the occupancy probability calculation. Concisely, balancing the amount of point clouds and uncovering an appropriate merging sequence are a troublesome task.

A second method to merge point clouds was mentioned in Section 3.3. With this method, point clouds are already being merged while SLAM is running. Instead of using a single SLAM point cloud, RGB-D SLAM constantly pushes its current online point cloud. The main advantage of this method is that OctoMap can now render a volumetric model based on previous and current observations, which leads to a more conclusive probability calculation. Contrary to iterative merging, online merging allows us to obtain an adequate balance between map completeness and detail. This is demonstrated in Figure 18(a): undesirable gaps were completed by the initial guess model, without completely closing up doors and windows.

Both optimisation methods were evaluated using our own datasets as well. First, we take a look in Figure 19(a). An initial

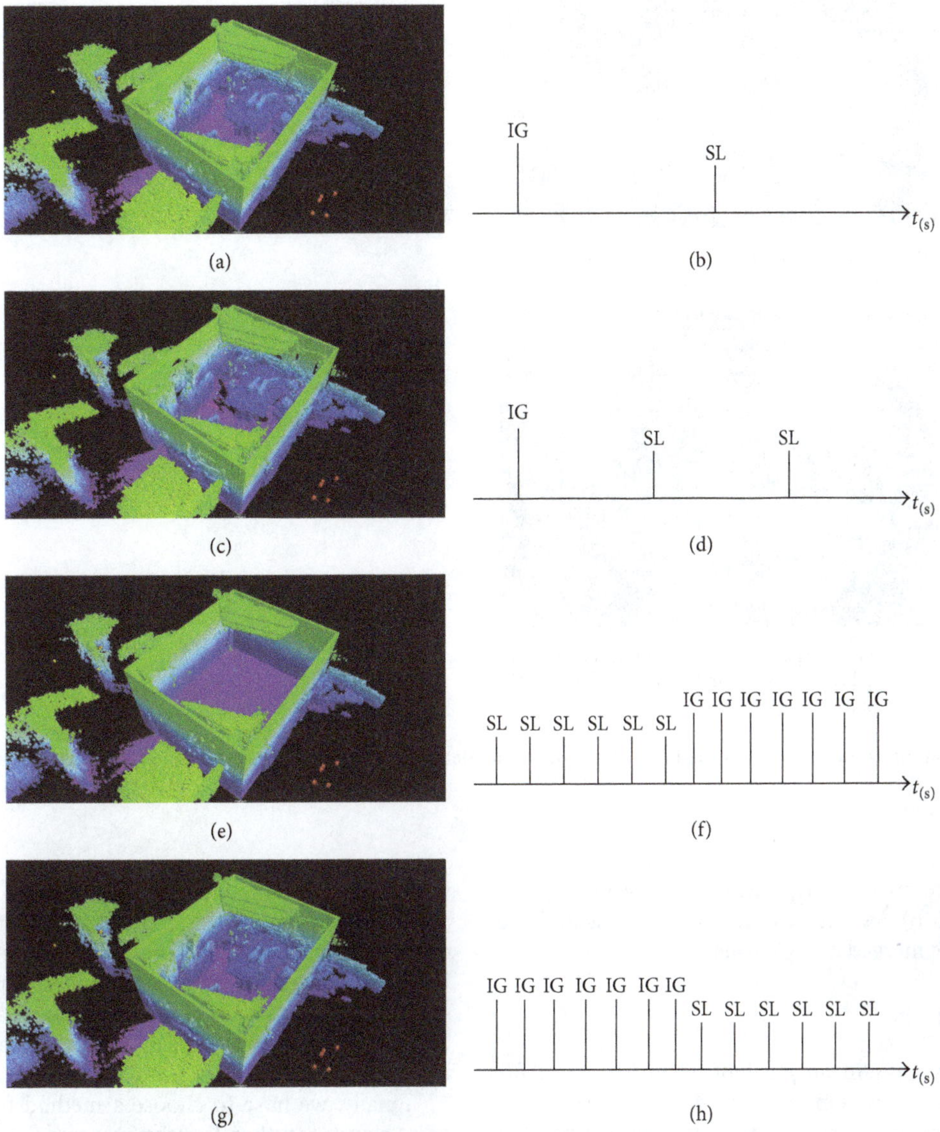

FIGURE 17: Iterative merging of our initial guess model (IG) with the complete SLAM point cloud (SL).

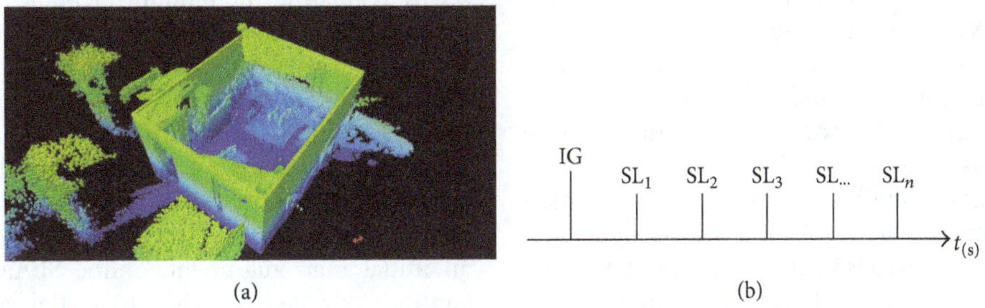

FIGURE 18: Online merging.

guess model was created with OpenSCAD and converted to a point cloud (Figure 5(a)). Also, we recorded a dataset in our indoor environment (Figure 12(a)) to generate online point clouds via RGB-D SLAM. Through an OctoMap server, these online point clouds were constantly merged with our initial guess until the entire dataset was played.

Secondly, we discuss Figure 19(c). In order to build this model, RGB-D SLAM was used to create a point cloud of an industrial environment, as can be seen in Figures 12(d) and 12(e). Before merging the SLAM point cloud with our initial guess, we removed all unnecessary data, as we only wished to obtain a model of the train cart. Next, this point cloud

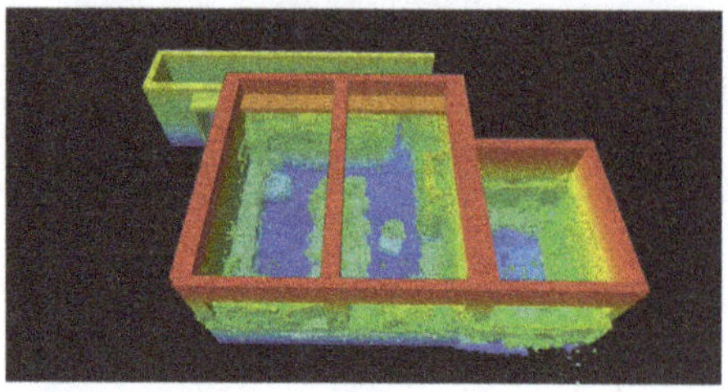

(a) Online merging of an initial guess model (Figure 5(a)) with live RGB-D SLAM output

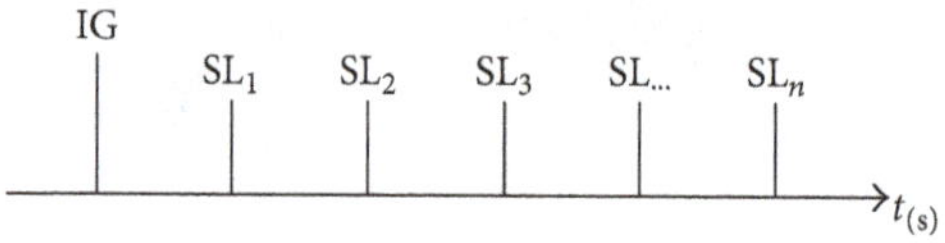

(b)

(c) Iterative merging of Figure 5(b) with Figure 12(e). The model consists of one initial guess and one SLAM point cloud

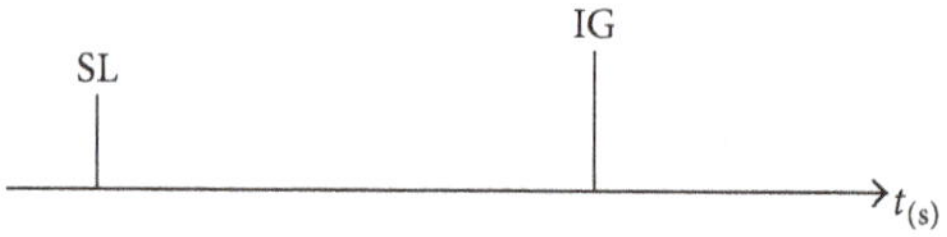

(d)

FIGURE 19: Optimisation results for our own datasets. For (a), online merging was applied. In (c), we conducted iterative merging of 2 point clouds.

was aligned and iteratively merged with the initial guess point cloud of Figure 5(b). In this case, the initial guess and the SLAM result were merged a single time.

5. Conclusion

In this paper, we present an efficient, robust method for completion and optimisation of 3D models using MapFuse. In a simulator as well as in reality, we have evaluated combinations of proven open-source technologies in order to attain a realistic map optimisation technique. Apart from these technologies, our method does not require additional complex calculations for map optimisation.

Several aspects affect the quality of our final result. Firstly, the accuracy of the initial guess model has to be considered. For known environments, the accuracy is assumed to be 100%, as exact measurements can be collected. In other situations, the user has to speculate about dimensions based on visual observations or the result of a SLAM algorithm. Also, the amount of detail that is included in the initial guess, for example, windows, doors, and furniture, will affect the occupancy probability for those elements within the model. In general, outlines of the filtered SLAM environment are sufficient to serve as initial guess, as detail will be provided by SLAM. Future work could involve automatic generation of the initial model from the SLAM output.

Secondly, MapFuse requires an accurate RGB-D SLAM point cloud in order to update the initial guess correctly. For this purpose, the SLAM algorithm has to be provided with a valid dataset that contains a significant amount of visual information about the environment. SLAM accuracy and completeness are directly related to the amount and quality of visual observations in the dataset. Improvements for the SLAM result can be made by altering parameters of the algorithm, or by lowering the playback speed of the dataset. The latter measure allows RGB-D SLAM more time per frame to process visual features.

Finally, we have to choose a method for bringing both point clouds together. Iterative merging fuses complete point clouds by aligning them and sending them to an OctoMap server. A balance between map completeness and detail is set by regulating the amount of point clouds that is being forwarded, as well as implementing an appropriate merging sequence. However, obtaining this balance has proven to be a difficult exercise. A main advantage of the iterative merging method is that the SLAM point cloud can be edited before using it in the merging process. Online merging starts by sending a single initial guess to an OctoMap server and continues with running the RGB-D SLAM algorithm. After an initial map alignment, online SLAM point clouds are continuously merged with the initial guess, leading to an improved balance between map completeness and detail. For both merging methods, OctoMap parameters such as initial probability and resolution can be altered in order to influence the final result. Also, MapFuse could cope with dynamic environments by setting an occupancy probability threshold which cancels out moving objects.

As initially intended, MapFuse is suitable for creating 3D models of various environments for the purpose of validating wireless propagation models. Furthermore, the proposed approach can be applied in other application domains such as the optimisation of dynamic control algorithms. Researchers would be able to model realistic 3D objects which makes it possible to validate complex control simulations [28]. Due to the realistic nature of our approach, such validations could improve control systems which work with complex 3D objects. Additionally, the accuracy and precision of the validation will be affected by the OctoMap resolution. Hence, a performance trade-off for the control system could be analysed.

Conflicts of Interest

The authors declare that there are no conflicts of interest regarding the publication of this article.

References

[1] F. Xia, L. T. Yang, L. Wang, and A. Vinel, "Internet of things," *International Journal of Communication Systems*, vol. 25, no. 9, pp. 1101-1102, 2012.

[2] M. F. Iskander and Z. Yun, "Propagation prediction models for wireless communication systems," *IEEE Transactions on Microwave Theory and Techniques*, vol. 50, no. 3, pp. 662–673, 2002.

[3] C. Phillips, D. Sicker, and D. Grunwald, "A Survey of wireless path loss prediction and coverage mapping methods," *IEEE Communications Surveys & Tutorials*, vol. 15, no. 1, pp. 255–270, 2013.

[4] B. Bellekens, R. Penne, and M. Weyn, "Validation of an indoor ray launching RF propagation model," in *Proceedings of the 6th IEEE-APS Topical Conference on Antennas and Propagation in Wireless Communications, IEEE APWC 2016*, pp. 74–77, Australia, September 2016.

[5] Z. Yun and M. F. Iskander, "Ray tracing for radio propagation modeling: Principles and applications," *IEEE Access*, vol. 3, pp. 1089–1100, 2015.

[6] Z. Lai, G. De La Roche, N. Bessis et al., "Intelligent ray launching algorithm for indoor scenarios," *Radioengineering*, vol. 20, no. 2, pp. 398–408, 2011.

[7] J. Chan, C. Zheng, and X. Zhou, "3D printing your wireless coverage," in *Proceedings of the 2nd ACM International Workshop on Hot Topics in Wireless, HotWireless 2015*, pp. 1–5, France.

[8] M. Weyn, G. Ergeerts, R. Berkvens, B. Wojciechowski, and Y. Tabakov, "DASH7 alliance protocol 1.0: Low-power, mid-range sensor and actuator communication," in *Proceedings of the IEEE Conference on Standards for Communications and Networking, CSCN 2015*, pp. 54–59, Japan, October 2015.

[9] A. Hornung, K. M. Wurm, M. Bennewitz, C. Stachniss, and W. Burgard, "OctoMap: An efficient probabilistic 3D mapping framework based on octrees," *Autonomous Robots*, vol. 34, no. 3, pp. 189–206, 2013.

[10] E. Robotics, "Erle-Copter — Erle Robotics," http://erlerobotics.com/blog/erle-copter/.

[11] T. P. Breckon and R. B. Fisher, "Non-parametric 3D surface completion," in *Proceedings of the 5th International Conference on 3-D Digital Imaging and Modeling, 3DIM 2005*, pp. 573–580, Canada, June 2005.

[12] E. Turner and A. Zakhor, "Automatic Indoor 3D Surface Reconstruction with Segmented Building and Object Elements," in *proceedings of the 2015 International Conference on 3D Vision*, vol. 10, Institute of Electrical and Electronics Engineers (IEEE), Lyon, France, 2015.

[13] H. Durrant-Whyte and T. Bailey, "Simultaneous localization and mapping: part I," *IEEE Robotics & Automation Magazine*, vol. 13, no. 2, pp. 99–110, 2006.

[14] L. Zhang, R. Zapata, and P. Lepinay, "Self-adaptive monte carlo localization for mobile robots using range sensors," in *Proceedings of the IEEE/RSJ International Conference on Intelligent Robots and Systems (IROS '9)*, pp. 1541–1546, St. Louis, Mo, USA, October 2009.

[15] D. Fox, J. Hightower, L. Liao, D. Schulz, and G. Bordello, "Bayesian filtering for location estimation," *IEEE Pervasive Computing*, vol. 2, no. 3, pp. 24–33, 2003.

[16] D. M. Cole and P. M. Newman, "Using laser range data for 3D SLAM in outdoor environments," in *Proceedings of the 2006 IEEE International Conference on Robotics and Automation*, IEEE, Orlando, FL, USA, 2006, http://www.robots.ox.ac.uk/mobile/Papers/3DScanMacthingCole_ICRA2006.pdf.

[17] A. Aghamohammadi, A. H. Tamjidi, and H. D. Taghirad, "SLAM Using Single Laser Range Finder," *IFAC Proceedings Volumes*, vol. 41, no. 2, pp. 14657–14662, 2008, http://linkinghub.elsevier.com/retrieve/pii/S1474667016413479.

[18] A. Gil, O. M. Mozos, M. Ballesta, and O. Reinoso, "A comparative evaluation of interest point detectors and local descriptors for visual SLAM," *Machine Vision and Applications*, vol. 21, no. 6, pp. 905–920, 2010.

[19] J. Engel, T. Schöps, and D. Cremers, "LSD-SLAM: Large-Scale Direct monocular SLAM," *Lecture Notes in Computer Science (including subseries Lecture Notes in Artificial Intelligence and Lecture Notes in Bioinformatics): Preface*, vol. 8690, no. 2, pp. 834–849, 2014.

[20] J. Engel, J. Stuckler, and D. Cremers, "Large-scale direct SLAM with stereo cameras," in *proceedings of the 2015 IEEE/RSJ International Conference on Intelligent Robots and Systems (IROS)*, vol. 9, IEEE, Hamburg, Germany, 2015.

[21] G. Silveira, E. Malis, and P. Rives, "An Efficient Direct Approach to Visual SLAM," in *Proceedings of the IEEE Transactions on Robotics*, vol. 24, IEEE, Roma, Italy, 2007, https://www.researchgate.net/profile/Patrick_Rives/publication/224330808_An_efficient_direct_approach_to_visual_SLAM/links/00b7d52983db5ed842000000/An-efficient-direct-approach-to-visual-SLAM.pdf.

[22] S. Umeyama, "Least-Squares Estimation of Transformation Parameters Between Two Point Patterns," *IEEE Transactions on Pattern Analysis and Machine Intelligence*, vol. 13, no. 4, pp. 376–380, 1991.

[23] E. Rublee, V. Rabaud, K. Konolige, and G. Bradski, "ORB: an efficient alternative to SIFT or SURF," in *Proceedings of the IEEE International Conference on Computer Vision (ICCV '11)*, pp. 2564–2571, Barcelona, Spain, November 2011.

[24] P. M. Panchal, S. R. Panchal, and S. K. Shah, "A comparison of SIFT and SURF," *International Journal of Innovative Research in Computer and Communication Engineering*, vol. 1, no. 2, pp. 323–327, 2013.

[25] F. Endres, J. Hess, N. Engelhard, J. Sturm, D. Cremers, and W. Burgard, "An evaluation of the RGB-D SLAM system," in *Proceedings of the IEEE International Conference on Robotics and Automation (ICRA '12)*, pp. 1691–1696, 2012.

[26] R. Kümmerle, G. Grisetti, H. Strasdat, K. Konolige, and W. Burgard, "G^2o: a general framework for graph optimization," in *Proceedings of the IEEE International Conference on Robotics and*

Automation (ICRA '11), pp. 3607–3613, Shanghai, China, May 2011.

[27] J. Sturm, N. Engelhard, F. Endres, W. Burgard, and D. Cremers, "A benchmark for the evaluation of RGB-D SLAM systems," in *Proceedings of the 25th IEEE/RSJ International Conference on Robotics and Intelligent Systems (IROS '12)*, pp. 573–580, October 2012.

[28] L. Fortuna and G. Muscato, "A roll stabilization system for a monohull ship: modeling, identification, and adaptive control," *IEEE Transactions on Control Systems Technology*, vol. 4, no. 1, pp. 18–28, 1996.

13

Decentralized Cooperative Localization Approach for Autonomous Multirobot Systems

Thumeera R. Wanasinghe, George K. I. Mann, and Raymond G. Gosine

IS Lab, Faculty of Engineering and Applied Science, Memorial University of Newfoundland, St. John's, NL, Canada A1B 3X5

Correspondence should be addressed to Thumeera R. Wanasinghe; ruwansiriwat@gmail.com

Academic Editor: Giovanni Muscato

This study proposes the use of a split covariance intersection algorithm (Split-CI) for decentralized multirobot cooperative localization. In the proposed method, each robot maintains a local cubature Kalman filter to estimate its own pose in a predefined coordinate frame. When a robot receives pose information from neighbouring robots, it employs a Split-CI based approach to fuse this received measurement with its local belief. The computational and communicative complexities of the proposed algorithm increase linearly with the number of robots in the multirobot systems (MRS). The proposed method does not require fully connected synchronous communication channels between robots; in fact, it is applicable for MRS with asynchronous and partially connected communication networks. The pose estimation error of the proposed method is bounded. As the proposed method is capable of handling independent and interdependent information of the estimations separately, it does not generate overconfidence state estimations. The performance of the proposed method is compared with several multirobot localization approaches. The simulation and experiment results demonstrate that the proposed algorithm outperforms the single-robot localization algorithms and achieves approximately the same estimation accuracy as the centralized cooperative localization approach, but with reduced computational and communicative cost.

1. Introduction

With the advancement of information and communication technology, robots from different vendors, different domains of operation, and various perception and operational capabilities are unified into a single framework in order to perform a collaborative task. These heterogeneous multirobot systems (MRS) offer greater coverage in exploration and searching tasks, robustness against individual failures, improved productivity through parallel tasking, and implementation of more complex operations than a single robot [1]. Integration of robots from different domains of operation, such as a robot team with unmanned ground vehicles (UGV) and unmanned aerial vehicles (UAV), enhances team perception while increasing accessibility in a cluttered environment [2].

Robots in a heterogeneous MRS may host different exteroceptive and proprioceptive sensory systems resulting in a significant variation in self-localization across teammates. Interrobot observations and flow of information between teammates can establish a sensor sharing technique such that the localization accuracy of each member improves over the localization approaches that solely depend on the robot's onboard sensors. This sensor sharing technique is known as cooperative localization and was initially developed to improve the localization accuracy of MRS that are navigating in global positioning system- (GPS-) denied areas or areas without preinstalled (known) landmarks [3, 4]. However, it is reported that a minimum of one agent in a cooperative localization team should possess absolute positioning capabilities in order to have a bounded estimation error and uncertainty [5].

Available cooperative localization algorithms can be classified into three groups: centralized/multicentralized approaches, distributed approaches, and decentralized approaches. Although the centralized/multicentralized approaches generate consistent pose estimate with high accuracy, these approaches entail considerably higher computational and communicative cost. Distributed cooperative localization algorithms attempt to reduce the communication cost. However, the computational cost of most of the distributed algorithms remains as high as the centralized/multicentralized approaches. Decentralized schemes have reduced both the computational and communication cost of multirobot cooperative localization tasks. However, available decentralized cooperative localization (DCL) schemes sometime generate overconfidence pose estimates for robots in an MRS.

Therefore, computing a nonoverconfidence state estimate with bounded error while reducing computation and communication cost is a key requirement for successful implementation of multirobot collaborative missions that rely on cooperative localization. In this paper, we extend our previous work presented in [6] and propose a scalable DCL approach for heterogeneous MRS which is guarantee for nonoverconfidence pose estimate with the bounded estimation error. The proposed method applies the split covariance intersection (Split-CI) algorithm to accurately track independencies and interdependencies between teammates' local pose estimates. The recently developed cubature Kalman filter (CKF) is exploited for sensor fusion. Each robot periodically measures the relative pose of its neighbours. These measurements and associated measurement uncertainties are first transformed into a common reference frame and then communicated to corresponding neighbour robots. Once a robot receives pose measurements from a neighbour, it uses the Split-CI algorithm to fuse received measurements with its local belief. The work presented in this paper offers the following key contributions: (i) a novel DCL approach that combines the properties of the Split-CI algorithm and CKF in order to avoid double counting of the information while generating pose estimation with bounded estimation error, (ii) an extension to the general CKF to accurately calculate and maintain both the independent and the dependent covariances, and (iii) a consistent and debiased algorithm to convert information between two Cartesian coordinate systems. The proposed algorithm has the following properties: (i) per-measurement computational and communication cost of the proposed DCL approach is constant, (ii) there is no requirement for a synchronous or fully connected communication network, (iii) the algorithm is robust against the single point of failure, and (iv) the algorithm is capable of generating a nonoverconfidence pose estimate with the bounded estimation error for each member of the MRS.

2. Background

In 1994, Kurazume et al. proposed the first cooperative localization algorithm [3]. Ever since, numerous approaches for cooperative localization have been reported. These implementations can be categorized into three main groups: (1) centralized/multicentralized cooperative localization, (2) distributed cooperative localization, and (3) decentralized cooperative localization.

Centralized cooperative localization approaches augment each robot's pose into a single state vector and perform state estimation tasks at a central processing unit [3, 4, 8–12]. The computational cost of the centralized algorithms scales to $\mathcal{O}(|N|^4)$, where $|N|$ is the number of robots in the MRS. Since these implementations require each robot to communicate its high frequency egocentric data and interrobot observations to a central processing unit, communication links should have wider bandwidth. Multicentralized approaches have been reported to improve the robustness of the general centralized approaches against the single point of failure by duplicating the state estimation process on a few or all robots in the team [12, 13]. This causes an increasing per measurement communicative cost linearly with respect to the number of independent processing units. A multicentralized approach that enables cooperative localization for sparsely communicating robot networks requires more communication bandwidth and onboard memory than traditional multicentralized approaches [14].

In the distributed cooperative localization, each robot locally runs a filter to fuse egocentric measurements and propagate the state over time. As a result, there is no requirement for exchanging high-frequency egocentric measurements with teammates or with a central server. This enables the implementation of distributed cooperative localization algorithms using communication channels with limited bandwidth. However, interrobot observations are still fused at a central processor leading to the same computational complexity as the centralized/multicentralized approaches (i.e., $\mathcal{O}(|N|^4)$). Work presented in [15] proposed distributed maximum a posteriori estimation for cooperative localization and achieved reduced computational complexity, that is, $\mathcal{O}(|N|^2)$, and also improved robustness against the single point of failure.

Decentralized algorithms focus on reducing both computational and communicative complexity. In general, each robot in a DCL team locally runs a filter (estimator) to fuse its ego-centric measurements as well as the interrobot observations received from its neighbours with its local belief. Available decentralized algorithms perform the sensor fusion either in a suboptimal manner [6, 16–21] or in an approximate manner [22–25]. Approximate approaches assume that a given robot local pose estimate is independent from the measurements sent by neighbours. This assumption leads to an overconfidence state estimation. Work presented in [17] uses a *dependency-tree* to track the recent interaction between robots. However, this approach maintains only the recent interdependencies of robot pose estimate; it tends to be overconfidence. An interlaced extended Kalman filter- (EKF-) based suboptimal filtering approach is presented in [18, 19] to avoid the possibility of generating an overconfidence state estimation. This approach requires each robot in the MRS to maintain a bank of EKFs representing the interaction between teammates. Although it produces a nonoverconfidence state estimation, this book-keeping approach is unscalable, as the number of EKF runs on a single robot increases exponentially with the number of robots in the MRS. A suboptimal filtering approach called *channel filtering* is presented in [16] which requires a communication network without loops. However, a communication network without loops is an unrealistic assumption for the practical implementation of cooperative localization. An extended information filter is used for implementing the DCL [20], in which each robot maintains the history of robot-to-robot relative measurement updates. CI-based approaches are also reported for the DCL [21]. However, the general CI algorithm neglects possible independencies between local estimates. This causes more conservative estimates and may produce an estimation error covariance which is larger than that of the best unfused estimate [26].

Available cooperative localization algorithms use four types of interrobot observations: relative range-only [27, 28], relative bearing-only [29, 30], both relative range and relative bearing [31–33], and full relative pose measurement [5, 6, 21, 24, 25, 34]. Full relative pose measurement-based cooperative localization schemes always generate more accurate pose estimate than range-only, bearing-only, and range-and-bearing measurement systems [35]. The lowest estimation accuracy was found with the bearing-only measurement system [35].

3. Problem Statement

To facilitate the mathematical formulation, let N represent the set that contains the unique identification indices of all robots in the MRS. The cardinality of this set, that is, $|N|$, corresponds to the total number of robots in the MRS. Further, it is assumed that the robots are represented by $\{\mathscr{R}_1, \mathscr{R}_2, \ldots, \mathscr{R}_{|N|}\}$ where the identification indices of the robots range from 1 to $|N|$. The set N and its cardinality $|N|$ may or may not be known to each agent in the MRS.

It is assumed that the robots are navigating in a flat, two-dimensional (2D) terrain. Each robot in the MRS hosts a communication device to exchange information between teammates; an interoceptive sensory system to measure ego-motion (linear and angular velocities); and an exteroceptive sensory system to measure the relative pose of neighbouring robots. Known data correspondence is assumed for the exteroceptive sensory system. Besides these sensory systems and communication modules, some of the robots in the MRS host a sensory system, such as a differential global positioning system (DGPS), to receive absolute positioning information periodically.

3.1. Robot's Motion Model. Robot navigation in a 2D space is modeled by a general three-degree-of-freedom (3-DOF) discrete-time kinematic model

$$\mathbf{x}_{q,k} = \mathbf{g}\left(\mathbf{x}_{q,k-1}, \overline{\mathbf{u}}_{q,k-1}\right) \quad \forall q \in N,$$
$$\begin{bmatrix} x \\ y \\ \phi \end{bmatrix}_{q,k} = \begin{bmatrix} x \\ y \\ \phi \end{bmatrix}_{q,k-1} + \delta t \begin{bmatrix} \overline{v}_x \cos(\phi) - \overline{v}_y \sin(\phi) \\ \overline{v}_x \sin(\phi) + \overline{v}_y \cos(\phi) \\ \overline{\omega}_z \end{bmatrix}_{q,k-1}, \tag{1}$$

where $\mathbf{x}_{q,k} \in \mathbb{R}^3$ is the robot's pose with respect to a given coordinate frame (or global coordinate frame) at discrete time k. The nonlinear state propagation function and the sampling time are represented by $\mathbf{g}(\cdot)$ and δt, respectively. $\overline{\mathbf{u}}_{q,k} \in \mathbb{R}^3$ is the system input and $\overline{\mathbf{u}}_{q,k} = \mathbf{u}_{q,k} + \boldsymbol{\nu}_{q,k}$, where $\mathbf{u}_{q,k} = [v_x \ v_y \ \omega_z]^T$. The parameters v_x and v_y are nominal linear velocities in x- and y-directions, respectively. ω_z is the nominal angular velocity. $\boldsymbol{\nu}_{q,k}$ represents the additive white Gaussian noise term with covariance $\mathbf{Q} \in \mathbb{R}^{3\times3}$. For nonholonomic robots, terms associated with linear velocities in y-direction, that is, $\overline{v}_y \sin(\phi)$ and $\overline{v}_y \cos(\phi)$, should be omitted from the system model.

3.2. Interrobot Relative Measurement Model. Consider the scenario where robot $\mathscr{R}_q$ measures relative pose of robot $\mathscr{R}_r$. This relative pose measurement is modeled as

$$\mathbf{y}_{q,k}^{r,q} = \mathbf{h}\left(\mathbf{x}_{q,k}, \mathbf{x}_{r,k}\right) + \mathbf{n}_{q,k}^{r,q}; \quad q \in N, \ r \in N_{q,k}, \ d_{q,k}^{r,q} \le d_m,$$
$$\begin{bmatrix} \delta x \\ \delta y \\ \delta \phi \end{bmatrix}_{q,k}^{r,q} = \boldsymbol{\Gamma}_{\mathbf{x}_{q,k}}^T \left(\mathbf{x}_{r,k} - \mathbf{x}_{q,k}\right) + \mathbf{n}_{q,k}^{r,q} \tag{2}$$

with

$$\boldsymbol{\Gamma}_{\mathbf{x}_{q,k}} = \begin{bmatrix} \mathbf{C}\left(\phi_{q,k}\right) & \mathbf{0}_{2\times1} \\ \mathbf{0}_{1\times2} & 1 \end{bmatrix},$$
$$\mathbf{C}\left(\phi_{q,k}\right) = \begin{bmatrix} \cos\left(\phi_{q,k}\right) & -\sin\left(\phi_{q,k}\right) \\ \sin\left(\phi_{q,k}\right) & \cos\left(\phi_{q,k}\right) \end{bmatrix}, \tag{3}$$

where $\mathbf{y}_{q,k}^{r,q} \in \mathbb{R}^3$ is the relative pose of robot $\mathscr{R}_r$ as measured by $\mathscr{R}_q$; that is, $\mathbf{y}_{q,k}^{r,q} = [\delta x_{q,k}^{r,q} \ \delta y_{q,k}^{r,q} \ \delta \phi_{q,k}^{r,q}]^T$, where δx, δy, and $\delta\phi$ are x-position, y-position, and the orientation of robot $\mathscr{R}_r$ with respect to robot $\mathscr{R}_q$. This pose measurement is on the body-fixed coordinate system of robot $\mathscr{R}_q$. The nonlinear measurement function is represented by $\mathbf{h}(\cdot)$. The measurement noise covariance, $\mathbf{n}_{q,k}^{r,q}$, is assumed to be an additive white Gaussian noise with covariance $\mathbf{R}_{q,k}^{r,q} \in \mathbb{R}^{3\times3}$. Parameters $d_{q,k}^{r,q}$ and d_m represent the distance between two robots and the sensing range of robot $\mathscr{R}_q$, respectively. It is assumed that the communication range of a given robot is greater than or equal to its sensing range. $N_{q,k}$ represents the set that contains the unique identification indices of robots which are within the sensing range of robot $\mathscr{R}_q$ at the discrete time k. The matrix transpose operation is represented by T. At a given time step, there may be no robots, one robot, or multiple robots operating within the sensing range of $\mathscr{R}_q$. The maximum number of robots that can operate within the sensing range of $\mathscr{R}_q$ is one robot less than the total number of robots in the MRS, that is, $|N| - 1$. These imply that $0 \le |N_{q,k}| \le |N| - 1$ for $k = 1, 2, \ldots, \infty$. Let

$$\mathscr{Y}_{q,k}^{q} = \left\{p\left(\mathbf{z}_{q,k}^{r,q}\right) \mid r \in N_{q,k}, \ d_{q,k}^{r,q} \le d_m\right\} \tag{4}$$

represent all relative pose measurements acquired by $\mathscr{R}_q$ at time step k, where $p(\mathbf{z}^{r,q}_{q,k}) = \mathcal{N}(\mathbf{y}^{r,q}_{q,k}, \mathbf{R}^{r,q}_{q,k})$. Symbol $\mathcal{N}(a, b)$ represents that the distribution is Gaussian with mean (actual measurement) a and covariance b. The cardinality of the set $\mathscr{Y}^q_{q,k}$ follows the same property as set $N_{q,k}$; that is, $0 \le |\mathscr{Y}^q_{q,k}| \le |N| - 1$ for $k = 1, 2, \ldots, \infty$.

4. Decentralized Cooperative Localization Algorithm

4.1. State Propagation. The objective of the state propagation step is to predict current pose and the associated estimation uncertainties of a given robot using both the robot's posterior state density and the odometry reading at the previous time step. In order to avoid the cyclic update (cyclic update is the process that uses the same information more than once), each robot maintains two covariance matrices: total covariance and independent covariance. Once the total and independent covariances are known, dependent covariance can be calculated as

$$P_{q_d,k} = P_{q,k} - P_{q_i,k}, \quad (5)$$

where $P_{q,k}$, $P_{q_i,k}$, and $P_{q_d,k}$ are total covariance, independent covariance, and dependent covariance of R_q's pose estimation at time step k, respectively.

This study employs CKF for sensor fusion. CKF is a recently developed suboptimal nonlinear filter which uses the spherical-radial cubature rule to solve the multidimensional integral associated with the Bayesian filter under the Gaussian approximation [36]. CKF is a Jacobian-free approach that is always guaranteed to have a positive definite covariance matrix and has demonstrated superior performance than the celebrity EKF and the unscented Kalman filter (UKF) [37–39].

For a system with n state variables, the third-order spherical-radial cubature rule selects $2n$ cubature points in order to compute the standard Gaussian integral, as follows:

$$I_{\mathcal{N}}(\mathbf{f}) = \int_{\mathbb{R}^{n_x}} \mathbf{f}(\mathbf{x}) \mathcal{N}\left(\mathbf{x}; \hat{\mathbf{x}}, \widehat{\mathbf{P}}\right) dx \approx \sum_{i=1}^{2n_x} \frac{1}{2n} \mathbf{f}\left(\hat{\mathbf{x}} + \mathbf{S}\xi_i\right), \quad (6)$$

where the square-root factor $\mathbf{S}$ of the covariance $\widehat{\mathbf{P}}$ satisfies the equality $\widehat{\mathbf{P}} = \mathbf{S}\mathbf{S}^T$. The cubature points ξ_i are given by

$$\xi_i = \begin{cases} \sqrt{n}e_i, & i = 1, 2, \ldots, n, \\ -\sqrt{n}e_{i-n}, & i = n+1, n+2, \ldots, 2n, \end{cases} \quad (7)$$

where $e_i \in \mathbb{R}^{n_x \times 1}$ represents the ith elementary column vector. In this paper, robot pose and odometry vectors are augmented into a single state vector leading to $n = n_x + n_c$, where n_x is the size of pose vector and n_c is the size of the odometry vector. Further, the general CKF algorithm is extended to accommodate independent covariance calculation capabilities.

The proposed state propagation approach is summarized as follows.

Algorithm 1 (state propagation).

Data. Assume at time k posterior density function of robot's pose estimation $p(\mathbf{x}_{q,k_+}) = \mathcal{N}(\hat{\mathbf{x}}_{q,k_+}, \mathbf{P}_{q,k_+})$, independent covariance matrix $\mathbf{P}_{q_i,k_+}$, and odometry reading $\overline{\mathbf{u}}_{q,k}$ are known.

Result. Calculate the predictive density function of robot's pose estimation $p(\mathbf{x}_{q,(k+1)_-}) = \mathcal{N}(\hat{\mathbf{x}}_{q,(k+1)_-}, \mathbf{P}_{q,(k+1)_-})$ and associated independent covariance matrix $\mathbf{P}_{q_i,(k+1)_-}$.

(1) Augment state end odometry reading into single vector:

$$\hat{\mathbf{x}}_{k_+} = \left[\hat{\mathbf{x}}^T_{q,k_+} \;\; \overline{\mathbf{u}}^T_{q,k}\right]^T. \quad (8)$$

(2) Compute the corresponding covariance matrix:

$$\mathbf{P}_{k_+} = \begin{bmatrix} \mathbf{P}_{q,k_+} & \mathbf{0}_{n_x \times n_c} \\ \mathbf{0}_{n_c \times n_x} & \mathbf{Q} \end{bmatrix}. \quad (9)$$

(3) Factorize

$$\mathbf{P}_{k_-} = \mathbf{S}_{k_+} \mathbf{S}^T_{k_+}. \quad (10)$$

(4) Generate cubature points ($j = 1, 2, \ldots, m$):

$$\mathbf{X}_{j,k_+} = \mathbf{S}_{k_+} \xi_j + \hat{\mathbf{x}}_{k_+}, \quad (11)$$

where $m = 2(n_x + n_c)$.

(5) Propagate each set of cubature points through nonlinear state propagation function given in (1) ($j = 1, 2, \ldots, m$):

$$\mathscr{X}_{j,(k+1)_-} = \mathbf{g}\left(\mathbf{X}_{j(1:n_x),k_+}, \mathbf{X}_{j(n_x+1:n_x+n_c),k_+}\right). \quad (12)$$

(6) Predict next state:

$$\hat{\mathbf{x}}_{q,(k+1)_-} = \frac{1}{m} \sum_{j=1}^{m} \mathscr{X}_{j,(k+1)_-}. \quad (13)$$

(7) Estimate the predictive error covariance:

$$\mathbf{P}_{q,(k+1)_-} = \frac{1}{m} \sum_{j=1}^{m} \mathscr{X}_{j,(k+1)_-} \mathscr{X}^T_{j,(k+1)_-} - \hat{\mathbf{x}}_{q,(k+1)_-} \hat{\mathbf{x}}^T_{q,(k+1)_-}. \quad (14)$$

(8) To calculate independent covariance, construct new block diagonalization covariance matrix as follows:

$$\mathbf{P}_{i,k_+} = \begin{bmatrix} \mathbf{P}_{q_i,k_+} & \mathbf{0}_{n_x \times n_c} \\ \mathbf{0}_{n_c \times n_x} & \mathbf{Q} \end{bmatrix}. \quad (15)$$

(9) Factorize $\mathbf{P}_{i,k_+}$, then generate a new set of cubature points, and propagate this new cubature point set through the nonlinear state propagation function (1) (refer to lines (3), (4), and (5) for equations).

(10) Predict new state using independent covariance:

$$\hat{\mathbf{x}}_{q_i,(k+1)_-} = \frac{1}{m}\sum_{j=1}^{m}\mathcal{X}_{j,(k+1)_-}. \quad (16)$$

(11) Estimate the independent predictive error covariance:

$$\mathbf{P}_{q_i,(k+1)_-} = \frac{1}{m}\sum_{j=1}^{m}\mathcal{X}_{j,(k+1)_-}\mathcal{X}_{j,(k+1)_-}^{T} - \hat{\mathbf{x}}_{q_i,(k+1)_-}\hat{\mathbf{x}}_{q_i,(k+1)_-}^{T}. \quad (17)$$

n_x is size of robot's pose vector, n_c is size of robot's odometry vector, $\mathbf{0}_{a\times b}$ is matrix with a rows and b columns and all entries are zeros, k_+ represents $k \mid k$, and $(k+1)_-$ represents $(k+1) \mid k$.

The algorithm is initialized with known prior density $p(\mathbf{x}_{q,k_+}) = \mathcal{N}(\hat{\mathbf{x}}_{q,k_+}, \mathbf{P}_{q,k_+})$, independent covariance matrix $\mathbf{P}_{q_i,k_+}$, and odometry reading $\bar{\mathbf{u}}_{q,k}$ at the previous time step (say, time step k). The algorithm predicts the robot pose for the next time step along with associated total and independent covariances. First, the algorithm augments the estimated pose vector $\hat{\mathbf{x}}_{q,k_+}$ with the odometry vector at time k (line (1)). The associated covariance matrix is then computed by block-diagonalization of the estimation and process covariance matrices (line (2)). In the CKF, a set of the cubature points is used to represent the current estimated pose and associated estimation uncertainties (line (4)). To generate these cubature points, the square-root factor of the covariance matrix is required. Any matrix decomposition approach that preserves the equality given in (10) can be exploited to compute the square-root factor of the covariance matrix (line (3)). The cubature points that represent current state and odometry measurements are evaluated on the nonlinear state propagation function (line (5)), which generates the cubature point distribution for a predicted state. The predicted pose (or state) of the robot is the average of the propagated cubature points (line (6)). Total predictive covariance is then computed from (14). Once the total predictive covariance is calculated, a new block-diagonalized covariance matrix, that is, $\mathbf{P}_{i,k_+}$, is generated using the independent covariance matrix of time k, that is, $\mathbf{P}_{q_i,k_+}$, and process covariance matrix, that is, $\mathbf{Q}$ (line (8)). After computing $\mathbf{P}_{i,k_+}$, its square-root factor is computed (similar to line (3)); then a set of cubature points are generated using the new square-root factor (similar to line (4)); and, finally, the newly generated cubature points are propagated through the nonlinear state propagation function (similar to line (5)) (line (9)). These steps are followed by the computation of prediction for independent propagated state (line (10)) and the associated covariance matrix (line (11)).

4.2. Computing Pose of Neighbours. At a relative pose measurement event each robot acquires the relative pose of its neighbours. These measurements are in the local coordinate system of the observing robot and are required to be transformed to the reference coordinate system prior to executing the sensor fusion at the neighbouring robot's local processor. Assume that, at time $(k+1)$, robot $\mathcal{R}_q$ measures the relative pose of robot $\mathcal{R}_r$. This nonlinear coordinate transformation can be modeled as

$$\begin{aligned}\mathbf{y}_{q,k+1}^{r,*} &= \mathbf{f}\left(\hat{\mathbf{x}}_{q,(k+1)_-}, \mathbf{y}_{q,k+1}^{r,q}\right) = \hat{\mathbf{x}}_{q,(k+1)_-} \oplus \mathbf{y}_{q,k+1}^{r,k} \\ &= \hat{\mathbf{x}}_{q,(k+1)_-} + \mathbf{\Gamma}_{x_{q,(k+1)_-}}\mathbf{y}_{q,k+1}^{r,k},\end{aligned} \quad (18)$$

where $\oplus$ is known as the pose composition operator and $\mathbf{y}_{q,k+1}^{r,*}$ is the global pose of $\mathcal{R}_r$ on the reference coordinate frame, as measured by $\mathcal{R}_q$. The superscript asterisk "$*$" is used to indicate that the measurement is on the reference coordinate system. Symbol $\mathbf{\Gamma}_{x_{q,(k+1)_-}}$ has the same meaning as (2). Since this Cartesian-to-Cartesian transformation is nonlinear, a cubature point-based approach, as summarized in Algorithm 2, is exploited to achieve consistent and unbiased coordinate transformation (see Appendix).

Algorithm 2 (relative to global conversion).

Data. Assume at time k the predictive density function of a robot's (say, $\mathcal{R}_q$) pose estimation $p(\mathbf{x}_{q,(k+1)_-}) = \mathcal{N}(\hat{\mathbf{x}}_{q,(k+1)_-}, \mathbf{P}_{q,(k+1)_-})$, independent covariance matrix $\mathbf{P}_{q_i,(k+1)_-}$, and relative pose measurement of a neighbour (say, $\mathcal{R}_r$) are available.

Result. Calculate global pose measurement of $\mathcal{R}_r$, that is, $\mathbf{y}_{q,k+1}^{r,*}$, and associated independent and dependent measurement covariances, i.e., $\mathbf{R}_{q_i,k+1}^{*}$ and $\mathbf{R}_{q_d,k+1}^{*}$.

(1) Augment the predictive state and relative pose into single vector:

$$\mathbf{y}_{k+1} = \left[\mathbf{x}_{q,(k+1)}^{T} \; \left(\mathbf{y}_{q,(k+1)}^{r,*}\right)^{T}\right]^{T}. \quad (19)$$

(2) Construct corresponding covariance matrix:

$$\overline{\mathbf{R}}_{k+1} = \begin{bmatrix}\mathbf{P}_{q,(k+1)_-} & \mathbf{0}_{n_x\times n_x} \\ \mathbf{0}_{n_x\times n_x} & \mathbf{R}_q\end{bmatrix}. \quad (20)$$

(3) Factorize

$$\overline{\mathbf{R}}_{k+1} = \overline{\mathbf{S}}_{k+1}\overline{\mathbf{S}}_{k+1}^{T}. \quad (21)$$

(4) Generate set of cubature points ($j = 1, 2, \ldots, m$):

$$\mathcal{Y}_{j,k+1}^{r,q} = \overline{\mathbf{S}}_{k+1}\xi_j + \mathbf{y}_{k+1}, \quad (22)$$

where $m = 2 * n_x$.

(5) Perform coordinate transform for each set of cubature points ($j = 1, 2, \ldots, m$):

$$\mathcal{Y}_{j,k+1}^{r,*} = \mathbf{f}\left(\mathcal{Y}_{j(1:n_x),k+1}^{r,q}, \mathcal{Y}_{j(n_x+1:2n_x),k+1}^{r,q}\right). \quad (23)$$

(6) Compute global pose of neighbour:

$$\mathbf{y}_{q,k+1}^{r,*} = \frac{1}{m}\sum_{j=1}^{m}\mathcal{Y}_{j,k+1}^{r,*}. \quad (24)$$

(7) Compute total noise (error) covariance:

$$\mathbf{R}^{r,*}_{q,k+1} = \frac{1}{m}\sum_{j=1}^{m}\left(\mathcal{Y}^{r,*}_{j,k+1}\right)\left(\mathcal{Y}^{r,*}_{j,k+1}\right)^{T} - \left(\mathbf{y}^{r,*}_{q,k+1}\right)\left(\mathbf{y}^{r,*}_{q,k+1}\right)^{T}. \quad (25)$$

(8) Construct a block-diagonalized matrix using independent predictive covariance and measurement noise covariance:

$$\overline{\mathbf{R}}_{i,k+1} = \begin{bmatrix} \mathbf{P}_{q_i,(k+1)_-} & \mathbf{0}_{n_x \times n_x} \\ \mathbf{0}_{n_x \times n_x} & \mathbf{R}_q \end{bmatrix}. \quad (26)$$

(9) Factorize $\overline{\mathbf{R}}_{i,k+1}$ and then generate a new set of cubature points followed by the coordination transformation for each cubature point (refer to lines (3), (4), and (5) for equations).

(10) Compute coordinate transformed measurement using independent covariance:

$$\mathbf{y}^{r,*}_{q_i,k+1} = \frac{1}{m}\sum_{j=1}^{m}\mathcal{Y}^{r,*}_{j,k+1}. \quad (27)$$

(11) Estimate independent covariance for the pose measurement:

$$\mathbf{R}^{r,*}_{q_i,k+1} = \frac{1}{m}\sum_{j=1}^{m}\left(\mathcal{Y}^{r,*}_{j,k+1}\right)\left(\mathcal{Y}^{r,*}_{j,k+1}\right)^{T} - \left(\mathbf{y}^{r,*}_{q_i,k+1}\right)\left(\mathbf{y}^{r,*}_{q_i,k+1}\right)^{T}. \quad (28)$$

(12) Estimate dependent covariance for the pose measurement:

$$\mathbf{R}^{r,*}_{q_d,k+1} = \mathbf{R}^{r,*}_{q,k+1} - \mathbf{R}^{r,*}_{q_i,k+1}. \quad (29)$$

The algorithm is initialized with a known predictive density of the pose estimation of the observing robot along with the predictive independent covariances. At an inter-robot-relative-pose measurement event, the observing robot augments its predictive pose and relative pose measurement into a single state vector (line (1)). The associated covariance matrix is obtained by block-diagonalization of the predictive total covariance ($\mathbf{P}_{q,k+1_-}$) and noise covariance of the relative pose measurement ($\mathbf{R}_q$) (line (2)). This block-diagonalized covariance matrix is then factorized and exploited for generating a set of cubature points to represent the state vector (lines (3) and (4)). The generated cubature points are evaluated on the nonlinear Cartesian-to-Cartesian coordinate transformation function, that is, (18), in order to compute the coordinate transformed cubature points (line (5)). This step is followed by the computation of the observed robot pose in the reference coordinate system (line (6)) and associated total noise (error) covariance matrix (line (7)). Once the total noise covariance is calculated, the algorithm constructs a new block-diagonalized covariance matrix, $\overline{\mathbf{R}}_{i,k+1}$, using the predictive independent covariance matrix and the relative pose measurement noise covariance matrix (line (8)). After computing $\overline{\mathbf{R}}_{i,k+1}$, its square-root factor is computed similar to line (3); then a set of cubature points are generated using the new square-root factor (similar to line (4)); and, finally, newly generated cubature points are transformed from the local coordinate system to the global coordinate system (similar to line (5)) (line (9)). These steps are followed by the computing of coordinate transformed measurement and associated independent noise covariance matrix (lines (10) and (11)). Finally, the dependent covariance of the coordinate transformed measurement is calculated as the difference between total and independent covariances (line (12)).

4.3. Update Local Pose Estimation Using the Pose Sent by Neighbours. When a robot receives a pose measurement from a neighbour, this measurement is fused with the observed robot's local estimate in order to improve its localization accuracy.

Algorithm 3 (state update with the measurement sent by neighbours).

Data. Assume predictive density of robot pose estimation $p(\mathbf{x}_{r,(k+1)_-} = \mathcal{N}(\hat{\mathbf{x}}_{r,(k+1)_-}, \mathbf{P}_{r,(k+1)_-}))$, the associated independent covariance matrix $\mathbf{P}_{q_i,(k+1)_-}$, and pose measurements from a neighbour $\mathbf{y}^{r,*}_{q,k+1}$ along with the associated independent and dependent covariances are available.

Result. Calculate posterior density of time $k+1$, that is, $p(\mathbf{x}_{r,(k+1)_+}) = \mathcal{N}(\hat{\mathbf{x}}_{r,(k+1)_+}, \mathbf{P}_{r,(k+1)_+})$, and the associated independent covariance matrix $\mathbf{P}_{r_i,(k+1)_+}$.

(1) Calculate the predictive dependant covariance:

$$\mathbf{P}_{q_d,(k+1)_-} = \mathbf{P}_{q,(k+1)_-} - \mathbf{P}_{q_i,(k+1)_-}. \quad (30)$$

(2) Compute the weighted predictive covariance

$$\mathbf{P}_1 = \frac{\mathbf{P}_{q_d,(k+1)_-}}{\alpha} + \mathbf{P}_{q_i,(k+1)_-}. \quad (31)$$

(3) Compute the weighted measurement covariance

$$\mathbf{P}_2 = \frac{\mathbf{R}^{r,*}_{q_d,k+1}}{1-\alpha} + \mathbf{R}^{r,*}_{q_i,k+1}. \quad (32)$$

(4) **if** measurement gate validated **then**

(5) Compute Kalman gain

$$\mathbf{K} = \mathbf{P}_1\left(\mathbf{P}_1 + \mathbf{P}_2\right)^{-1}. \quad (33)$$

(6) Update robot pose

$$\hat{\mathbf{x}}_{r,(k+1)_+} = \hat{\mathbf{x}}_{r,(k+1)_-} + \mathbf{K}\left(\mathbf{y}^{r,*}_{q,k+1} - \hat{\mathbf{x}}_{r,(k+1)_-}\right). \quad (34)$$

(7) Update total covariance

$$\mathbf{P}_{r,(k+1)_+} = \left(\mathbf{I}_{n_x} - \mathbf{K}\right)\mathbf{P}_1. \quad (35)$$

(8) Update independent covariance

$$\mathbf{P}_{r_i,(k+1)_+} = \left(\mathbf{I}_{n_x} - \mathbf{K}\right)\mathbf{P}_{r_i,(k+1)_-}\left(\mathbf{I}_{n_x} - \mathbf{K}\right)^T + \mathbf{K}\mathbf{R}^{r,*}_{q_i,k+1}\mathbf{K}^T. \tag{36}$$

(9) **else**

(10) Assign predictive state and covariances into posterior state and covariances

$$\begin{aligned}\widehat{\mathbf{x}}_{r,(k+1)_+} &\longleftarrow \widehat{\mathbf{x}}_{r,(k+1)_-}\\ \mathbf{P}_{r,(k+1)_+} &\longleftarrow \mathbf{P}_{r,(k+1)_-}\\ \mathbf{P}_{r_i,(k+1)_+} &\longleftarrow \mathbf{P}_{r_i,(k+1)_-}.\end{aligned} \tag{37}$$

(11) **end if**

$\mathbf{I}_{n_x}$ is identity matrix of $(n_x \times n_x)$ and α is weighting coefficient and belongs to the interval $[0, 1]$.

Algorithm 3 summarizes the steps involved in this measurement update. The algorithm is initialized by calculating the observed robot's predictive dependent covariance (line (1)). The weighted predictive covariance and weighted measurement covariance are then calculated as given in (31) and (32), respectively (lines (2) and (3)). Coefficient α belongs to the interval $[0, 1]$ and can be determined such that the trace or determinant of the updated total covariance is minimized. Detection and elimination of outliers are important for preventing the divergence of state estimates. This requirement is fulfilled by employing an ellipsoidal measurement validating gate [40] (line (4)). As the pose measurements from neighbours are in the reference coordinate frame, the measurement model of this sensor fusion becomes linear. Therefore, the linear Kalman filter can be exploited for sensor fusion. In this measurement update, the measurement matrix $\mathbf{H}$ of the traditional Kalman filter becomes an identity matrix, $(\mathbf{I}_{n_x})$, of $n_x \times n_x$. Using the multiplicative property of the identity matrix (i.e., $\mathbf{I}_m\mathbf{A} = \mathbf{A}\mathbf{I}_n = \mathbf{A}$, where $\mathbf{A}$ is $m \times n$) the Kalman gains, the updated robot pose, and the associated total and independent covariance matrices can be computed from (33), (34), (35), and (36), respectively (lines (5), (6), (7), and (8)). For outliers, measurements are discarded and the predictive pose and the associated total and independent covariance matrices are directly assigned to the corresponding posterior quantities (lines (9) and (10)).

4.4. Update Local Pose Estimation Using the Measurement Acquired by the Absolute Positioning System. It is assumed that some of the robots in the MRS host DGPS sensor in order to measure global position information. This position measurement at time $k + 1$ is modeled as

$$\mathbf{y}^A_{q,k+1} = \mathbf{H}^A_{q,k+1}\mathbf{x}_{q,k+1} + \boldsymbol{\nu}^A_{K+1}, \tag{38}$$

where $\mathbf{H}^A_{q,k+1} = [\mathbf{I}_2 \ \ \mathbf{0}_{2\times1}]$ is the measurement vector and $\boldsymbol{\nu}^A_{K+1}$ is the additive white Gaussian noise term with covariance $\mathbf{R}^A \in \mathbb{R}^{2\times2}$. This measurement is linear and independent from the robot's pose estimate. Thus, this measurement can be fused with the current state estimation using the general linear Kalman filter measurement update steps followed by

$$\mathbf{P}_{q_i,(k+1)_+} = \left(\mathbf{I} - \mathbf{K}\mathbf{H}^A_{q,k+1}\right)\mathbf{P}_{q_i,(k+1)_-}\left(\mathbf{I} - \mathbf{K}\mathbf{H}^A_{q,k+1}\right)^T + \mathbf{K}\mathbf{R}^A\mathbf{K}^T. \tag{39}$$

This equation computes the updated independent covariance matrix at the event of DGPS measurement update.

4.5. Sensor Fusion Architecture. This study assumes that each agent in the MRS initially knows its pose with respect to a given reference coordinate frame. The recursive state estimation framework of the proposed DCL algorithm is outlined in Algorithm 4 and is graphically illustrated in Figure 1.

Algorithm 4 (Split-CI based cooperative localization algorithm).

(1) Initialize known $\mathbf{x}_{q,\circ}$ and $\mathbf{P}_{q,\circ}$

(2) Set initial independent covariance: $\mathbf{P}_{q_i,\circ} \leftarrow \mathbf{P}_\circ$

(3) **for** $k \in (1, \ldots, \infty)$ **do**

(4) Read ego-motion sensor $\overline{\mathbf{u}}_{q,k}$

(5) Propagate state: Algorithm 1

(6) **if** $|\mathcal{Y}^q_{q,k+1}| > 0$ **then**

(7) **for** $\forall r \in \mathbf{N}_{q,k+1}$ **do**

(8) Read $\mathbf{y}^{r,q}_{q,k+1}$

(9) Transform relative pose measurement to reference coordinate frame: Algorithm 2

(10) Transmit $(\mathbf{y}^{r,*}_{q,k+1}, \mathbf{R}^{r,*}_{q_i,k+1}, \mathbf{R}^{r,*}_{q_d})$

(11) **end for**

(12) **end if**

(13) **if** pose measurement receives from neighbours **then**

(14) **for** $\forall r \in \overline{\mathbf{N}}_{q,k+1}$ **do**

(15) Collect $(\mathbf{y}^{q,*}_{r,k+1}, \mathbf{R}^{q,*}_{r_i,k+1}, \mathbf{R}^{q,*}_{r_d})$ from $\mathcal{R}_r$

(16) Perform Split-CI-based measurement update: Algorithm 3

(17) Enable recursive update

$$\begin{aligned}\widehat{\mathbf{x}}_{q,(k+1)_-} &\longleftarrow \widehat{\mathbf{x}}_{q,(k+1)_+}\\ \mathbf{P}_{q,(k+1)_-} &\longleftarrow \mathbf{P}_{q,(k+1)_+}\\ \mathbf{P}_{q_i,(k+1)_-} &\longleftarrow \mathbf{P}_{q_i,(k+1)_+}\end{aligned} \tag{40}$$

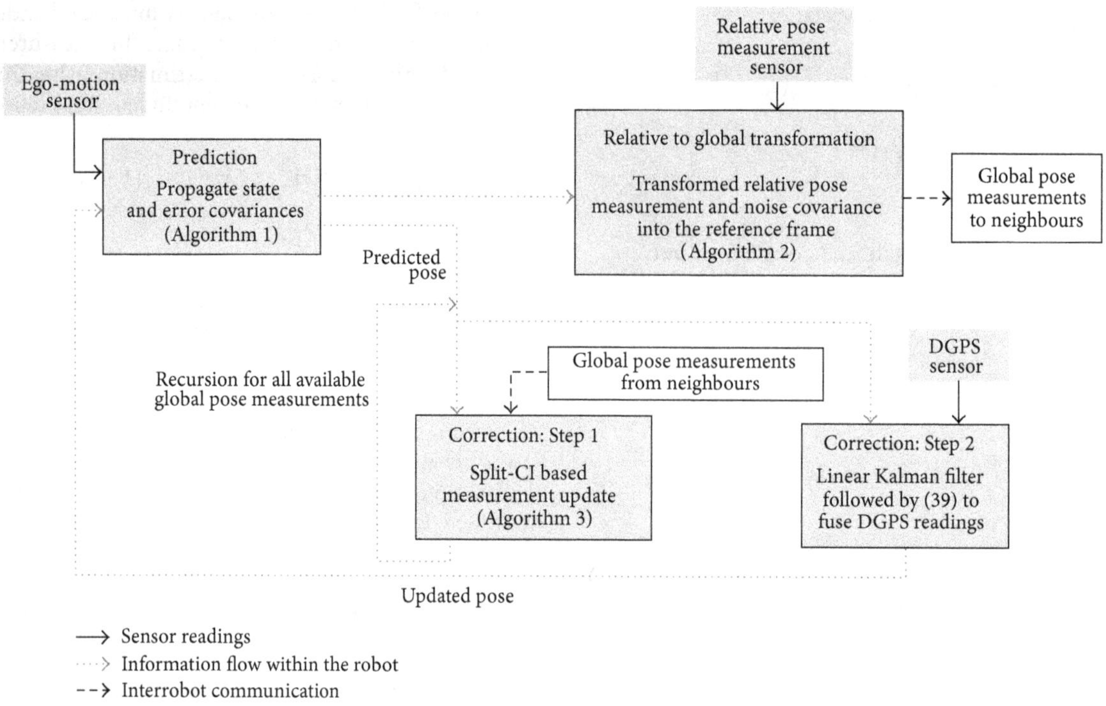

FIGURE 1: Sensor fusion architecture of the proposed decentralized multirobot cooperative localization scheme.

(18) **end for**

(19) Set independent covariance to zero: $\mathbf{P}_{q_i,(k+1)_-} \longleftarrow [\mathbf{0}_{3\times3}]$

(20) **end if**

(21) **if** DGPS measurement available **then**

(22) Read $\mathbf{y}^{A}_{q,k+1}$

(23) **if** measurement gate validated **then**

(24) Compute $\widehat{\mathbf{x}}_{q,(k+1)_+}$, $\mathbf{P}_{q,(k+1)_+}$, and $\mathbf{P}_{q_i,(k+1)_+}$ as detailed in Section 4.4

(25) **else**

(26) Assign predictive quantities to corresponding posterior quantities

$$\begin{aligned} \widehat{\mathbf{x}}_{q,(k+1)_+} &\longleftarrow \widehat{\mathbf{x}}_{q,(k+1)_-} \\ \mathbf{P}_{q,(k+1)_+} &\longleftarrow \mathbf{P}_{q,(k+1)_-} \\ \mathbf{P}_{q_i,(k+1)_+} &\longleftarrow \mathbf{P}_{q_i,(k+1)_-} \end{aligned} \quad (41)$$

(27) **end if**

(28) **else**

(29) Assign predictive quantities to corresponding posterior quantities: (41)

(30) **end if**

(31) **end for**

$\overline{\mathbf{N}}_{q,k+1}$ is the set containing unique identification indices of robots that communicate global pose measurements to $\mathscr{R}_q$ at time $k+1$.

The proposed DCL algorithm has four main steps.

Step 1 (propagate state (lines (4)-(5) in Algorithm 4)). At each time step, the robot acquires its ego-motion sensor reading (odometry). This measurement is fused with the previous time step's posterior estimate in order to compute the predicted pose and the associated total and independent error covariance matrices as detailed in Algorithm 1.

Step 2 (measure neighbours' pose (lines (6)–(12) in Algorithm 4)). At an interrobot relative pose measurement event, first, the robot reads its exteroceptive sensors and collects the relative poses of its neighbours. Then, each relative pose measurement is transformed into the reference coordinate frame as outlined in in Algorithm 2. Finally, the transformed global pose measurements and the associated independent and dependent covariance matrices are transmitted to the corresponding neighbouring robots.

Step 3 (update with pose measurements sent by neighbours (lines (13)–(20) in Algorithm 4)). At a given time step, a robot may receive pose measurements from one (or more) neighbour(s). First, the received pose measurement is fused with the local estimation using the Split-CI measurement update structure that is detailed in Algorithm 3. In order to enable the recursion for available pose measurements from multiple neighbours, the updated pose and associated total and independent covariances are assigned back to the corresponding predictive parameter (line (17)). The recursion is then continued until all the received pose measurements are considered. Work presented in [41] provides a complete theoretical analysis and simulation-based validation for the consistency of the Split-CI-based filtering. However, the simulation study presented in [21] surfaced that the estimated states using the Split-CI based DCL algorithm sometimes diverge. This may occur because the resulting pose estimation might be correlated partially or fully to subsequent pose measurements received from neighbours. To overcome this issue, the proposed DCL algorithm directly assigns the known-independent covariance component to the correlated component (line (19)). In other words, this study set the independent covariance component to zero after interrobot measurement update event, which is not included in the standard Split-CIF algorithm described in [42].

Step 4 (update with absolute position measurement (lines (21)–(30) in Algorithm 4)). The final step of the proposed DCL algorithm is to update the robot's local pose with the position measurement acquired from an absolute positioning system. When a new position measurement is available, it is evaluated through an ellipsoidal validation gate to identify whether the acquired measurement is a valid measurement or an outlier (line (23)). If it is a valid measurement then the measurement is fused with the local estimation (line (24)). Otherwise, the predictive quantities are directly assigned to the corresponding posterior quantities (line (26)). For the time steps where no absolute position measurements are available, the predictive quantities are directly assigned to the corresponding posterior quantities (line (29)).

5. Simulation Results

5.1. Setup. The performance of the proposed DCL algorithm was evaluated using a publicly available multirobot localization and mapping dataset [43]. This 2D indoor dataset was generated from five robots (which we refer to as $\mathcal{R}_1$, $\mathcal{R}_2$, $\mathcal{R}_3$, $\mathcal{R}_4$, and $\mathcal{R}_5$) that navigated in a $15\,\text{m} \times 8\,\text{m}$ indoor space. Although this dataset consists of odometry readings, ground truth measurements, and range and bearing measurements to neighbours and landmarks we only used the odometry readings and ground truth measurements of each robot in order to evaluate the proposed DCL algorithm. This simulation study assumed that all five robots would be equipped with lightweight sensory systems to uniquely identify and measure the relative pose of their neighbours. Further, it was assumed that only two members of the robot team (i.e., $\mathcal{R}_1$ and $\mathcal{R}_2$) would be capable of acquiring DGPS measurements periodically. Interrobot measurements and DGPS measurements were synthesized from the ground truth data. Simulation parameters and sensor characteristics related to this simulation setup are summarized in Tables 1 and 2, respectively.

5.2. Results. Figures 2 and 3 illustrate the mean estimation error and the associated 3-σ error boundaries of the proposed DCL algorithm for 20 Monte-Carlo runs. Figure 2 corresponds to a robot with absolute position measuring capabilities and Figure 3 corresponds to a robot without absolute position measuring capabilities. From these results, it can be seen that the estimation errors of the proposed DCL algorithm are always inside the corresponding 3-σ error boundaries. This observation verifies that the proposed DCL algorithm is capable of avoiding the cyclic update and generating nonoverconfidence state estimations. Additionally, it is clear that the robots with absolute position measuring capabilities can achieve a more accurate pose estimation than the robots without such capabilities (note that the y-axis of Figures 2 and 3 is presented in two different scales). Further, the results confirm that the estimation error of the proposed DCL algorithm is bounded.

Table 1: Simulation parameters.

Symbol	Parameter description	Value
$\lvert N \rvert$	Number of robots in the team	5
t	Total time period of the dataset	1500 s
d_m	Maximum sensing range	10 m
$d \times l$	Size of the navigation arena	$15\,\text{m} \times 8\,\text{m}$
MC	Number of Monte-Carlo runs	20

Table 2: Sensor characteristics.

Sensor type	Measure	Update rate	Noise σ
Odometry	Linear velocity	50 Hz	$\sqrt{5.075\bar{v}_{x_q,k}}$
	Angular velocity	50 Hz	$\sqrt{0.345}\,\text{rads}^{-1}$
Relative pose	Relative x-position	10 Hz	0.1 m
	Relative y-position	10 Hz	0.1 m
	Relative orientation	10 Hz	1 deg
DGPS	Global x-position	10 Hz	0.1 m
	Global y-position	10 Hz	0.1 m

Noise parameters for velocities were extracted from [7].

5.3. Comparison. The estimation accuracy of the proposed DCL algorithm is compared with the estimation accuracy that were obtained from the following localization schemes:

(1) *Single-Robot Localization (SL) Method.* Each robot continually integrates its odometry readings in order to estimate its pose in a given coordinate frame. This method is also known as dead-reckoning. Robots with DGPS measuring capability fuse their DGPS sensor readings with the local estimate in order to improve pose estimation accuracy.

(2) *DCL Using Naïve Block-Diagonal (NB) Method.* In this method, the pose measurements sent by neighbours are treated as an independent information and are fused directly with the local estimate. In other words, possible correlations between local estimate and pose measurements sent by neighbours are neglected at the sensor fusion step.

(3) *DCL Using Ellipsoidal Intersection (EI) Algorithm.* The EI algorithm always assumes there exist unknown correlations between each robot's local pose estimations and uses a set of explicit expressions to

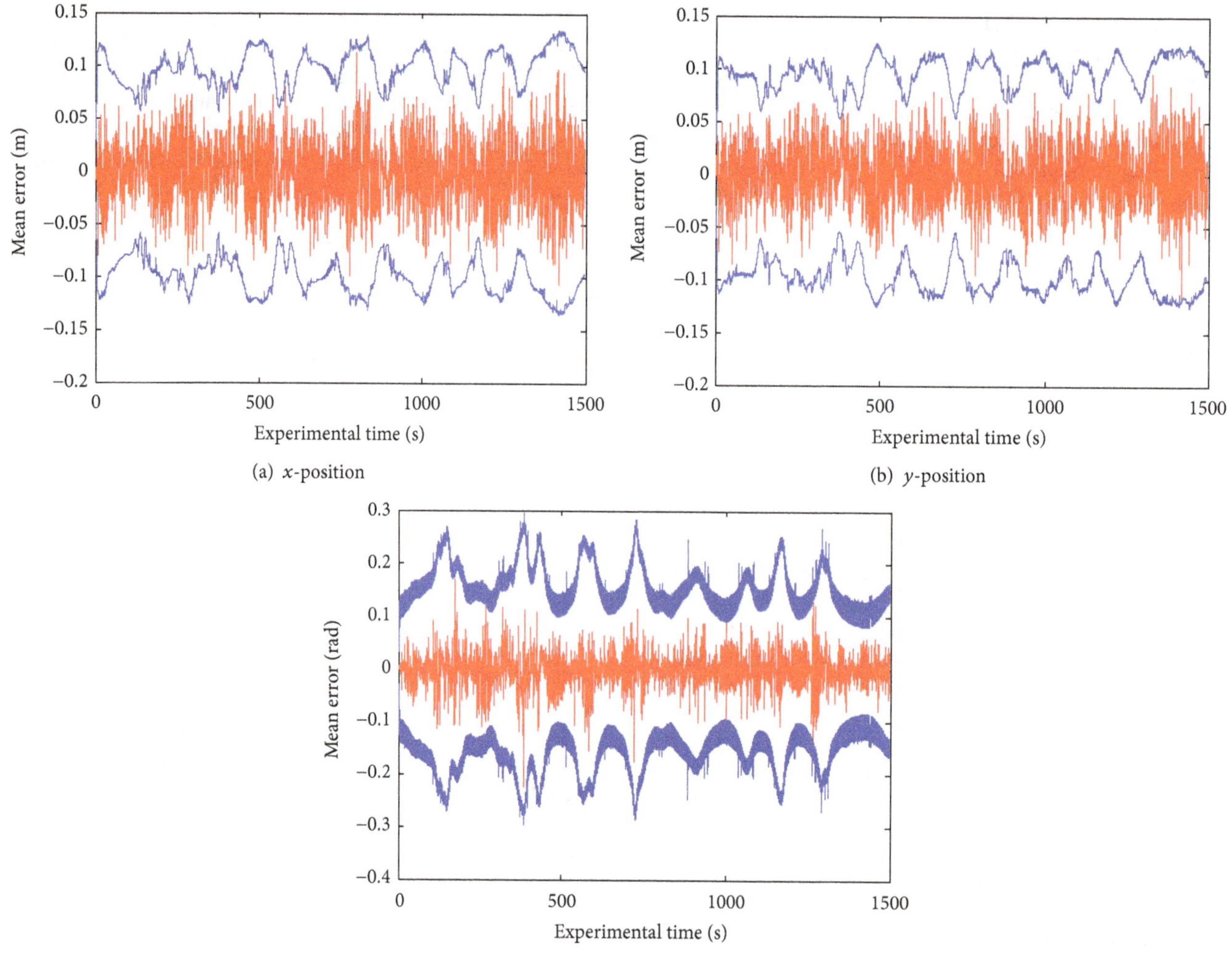

(a) x-position

(b) y-position

(c) ϕ-orientation

FIGURE 2: Mean estimation error of $\mathscr{R}_1$ for 20 Monte-Carlo simulations (a robot with absolute position measuring capabilities). In each graph, the solid red line indicates mean estimation error while the solid black lines indicate double-sided 3-σ error boundaries.

calculate these unknown correlations, that is, mutual-mean and mutual-covariance. When a robot receives a pose measurement(s) from its neighbour(s) EI algorithm first calculates these unknown correlations. In order to obtain the updated estimation the calculated mutual-mean and mutual-covariance are fused with the robot's local estimates and the pose measurements received from the robot's neighbours [44].

(4) *DCL Using Covariance Intersection (CI) Algorithm.* Each robot runs a local estimator to estimate its pose using onboard sensors and the pose measurement from neighbours. When a robot receives pose measurements from its neighbours, the covariance intersection algorithm is used to fuse these pose measurements with the robot's local estimate [21].

(5) *Centralized Cooperative Localization (CCL) Approach.* The pose of each robot is augmented into a single state vector. The ego-centric measurements of robots and interrobot observations are fused using EKF. This is a centralized approach which can accurately track the correlations between robots' pose estimations. Therefore, the results of this approach will serve as the benchmark for the performance evaluation of the proposed DCL algorithm.

We performed 20 Monte-Carlo simulations per each localization algorithm. Then the RMSE of position and orientation estimation for 20 Monte-Carlo simulations were computed. Finally, the time averaged RMSE values and associated standard deviations were calculated to perform the comparison between different localization schemes.

Consider robots without DGPS measuring capabilities (i.e., $\mathscr{R}_3$, $\mathscr{R}_4$, and $\mathscr{R}_5$). The pose estimations of these robots entirely rely on the odometry readings and the interrobot observations. Therefore, the time averaged RMSE and the associated standard deviation values of the pose estimation of these robots provide insight into the performance of each localization algorithm. The time averaged RMSE and the associated standard deviation values of the localization of $\mathscr{R}_5$ using the single-robot localization scheme were found as

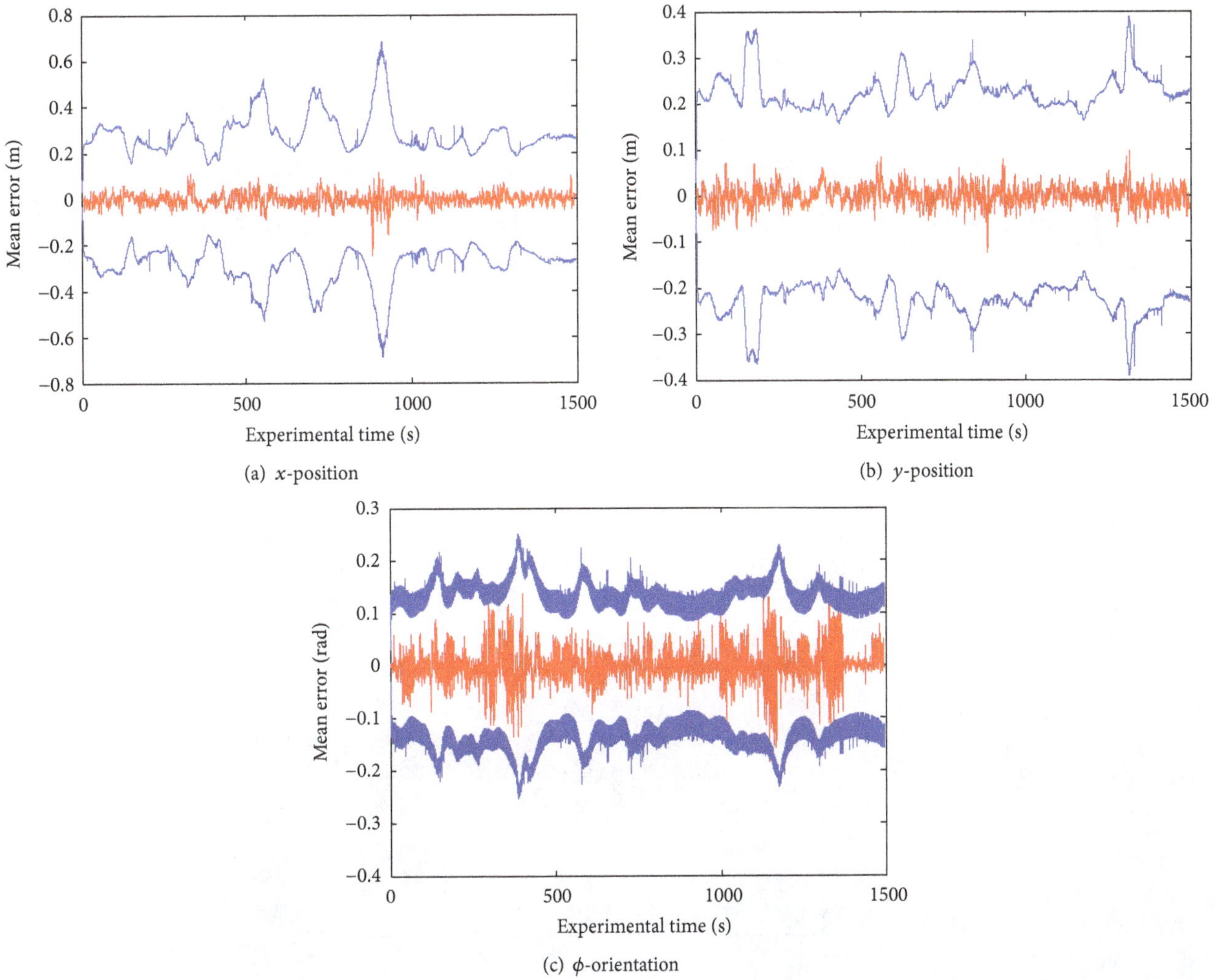

FIGURE 3: Mean estimation error of $\mathscr{R}_5$ for 20 Monte-Carlo simulations (a robot that does not have absolute position measuring capabilities). In each graph, the solid red line indicates mean estimation error while the solid black lines indicate double-sided 3-σ error boundaries.

3.1762 ± 2.3680 m in x-direction, 5.0073 ± 2.3339 m in y-direction, and 1.1776 ± 0.9015 rad in orientation estimation (the format of the listed estimation errors is mean ± standard deviation). The time averaged RMSE of the localization of $\mathscr{R}_5$ using any of the cooperative localization algorithms (NB, EI, CI, Split-CI, and CCL) were less than 10 cm in both x- and y-directions and less than 0.1 rad in orientation estimation. These observations imply that the cooperative localization approaches can significantly improve the accuracy of pose estimation of agents in a MRS.

Time averaged RMSE and the associated standard deviation values of x-position, y-position, and ϕ-orientation estimates of $\mathscr{R}_5$ using different cooperative localization schemes are compared in Figure 4. This comparison shows that the centralized cooperative localization algorithm outperforms all the other approaches. This was the expected result, as the centralized estimator maintained the joint-state and the associated dense covariance matrix in order to accurately represent the correlation between teammates pose estimates.

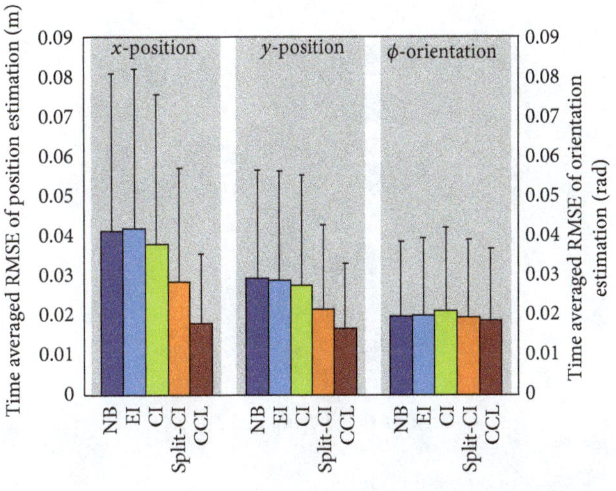

FIGURE 4: Comparison of estimation error of different cooperative localization algorithms. This result is for robot $\mathscr{R}_5$ (one of the robots without the DGPS measuring capabilities).

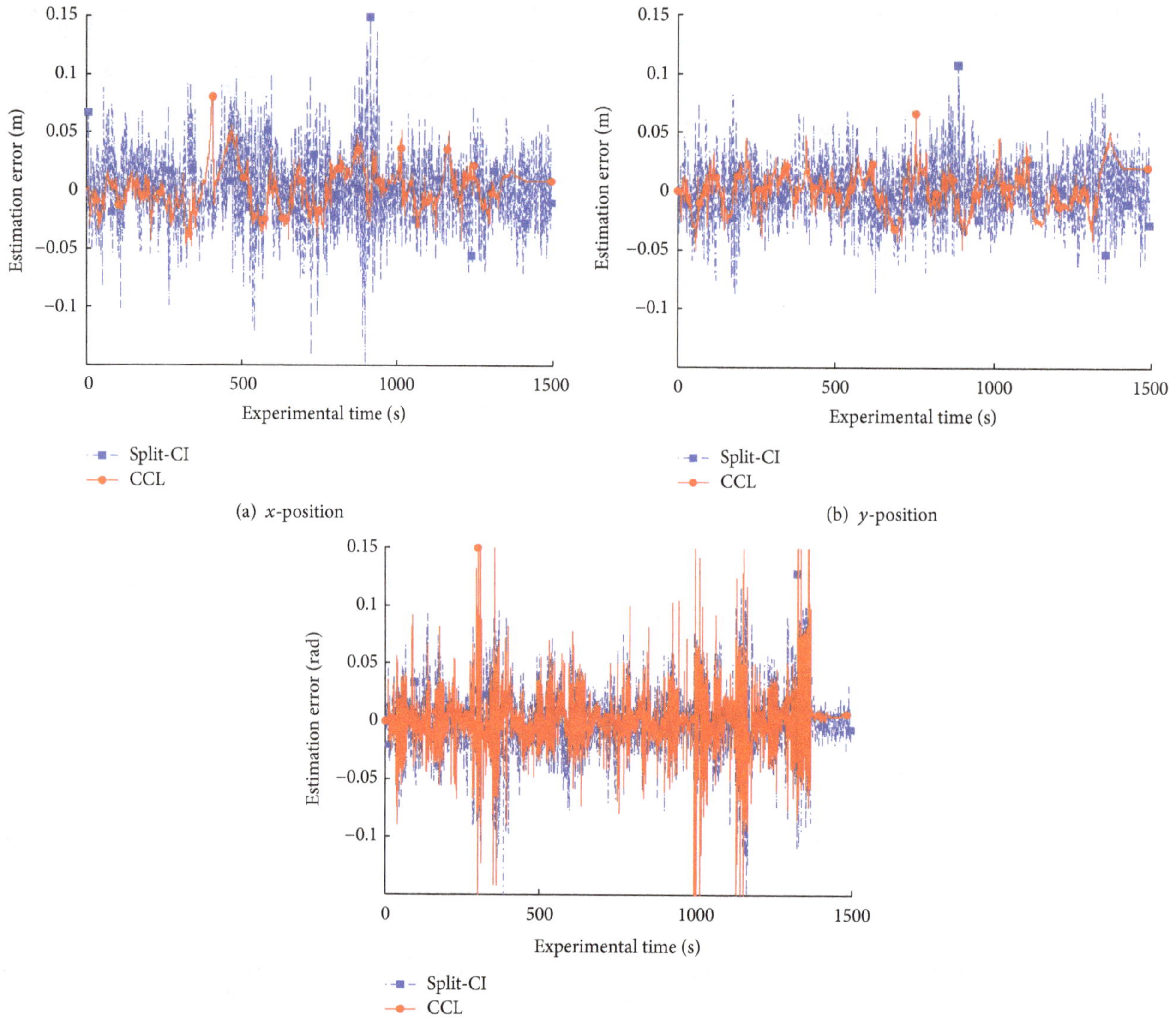

(a) x-position

(b) y-position

(c) ϕ-orientation

FIGURE 5: Estimation error comparison between the proposed Split-CI based approach and the centralized cooperative localization approach.

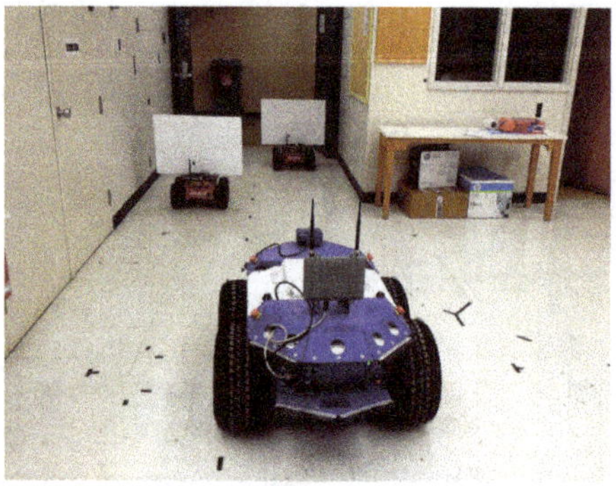

FIGURE 6: Experimental setup.

Although the estimated pose using the proposed Split-CI based DCL algorithm is less accurate than that of the centralized cooperative localization algorithm, it demonstrates better accuracy over all other DCL approaches that we have evaluated in this paper.

Figure 5 illustrates the estimation error comparison between the proposed Split-CI based DCL algorithm and the centralized cooperative localization algorithm. It indicates that the centralized approach has better accuracy; however, the estimation accuracy obtained from the proposed DCL algorithm is comparable with the estimation accuracy obtained from the centralized approach.

6. Experimental Results

6.1. Setup. The proposed DCL algorithm was experimentally evaluated on a team of three robots (see Figure 6): one Seekur Jr. (we will refer to this as platform A) and two Pioneer robots (we will refer to these as platforms B and C). Each robot is equipped with wheel encoders for odometry.

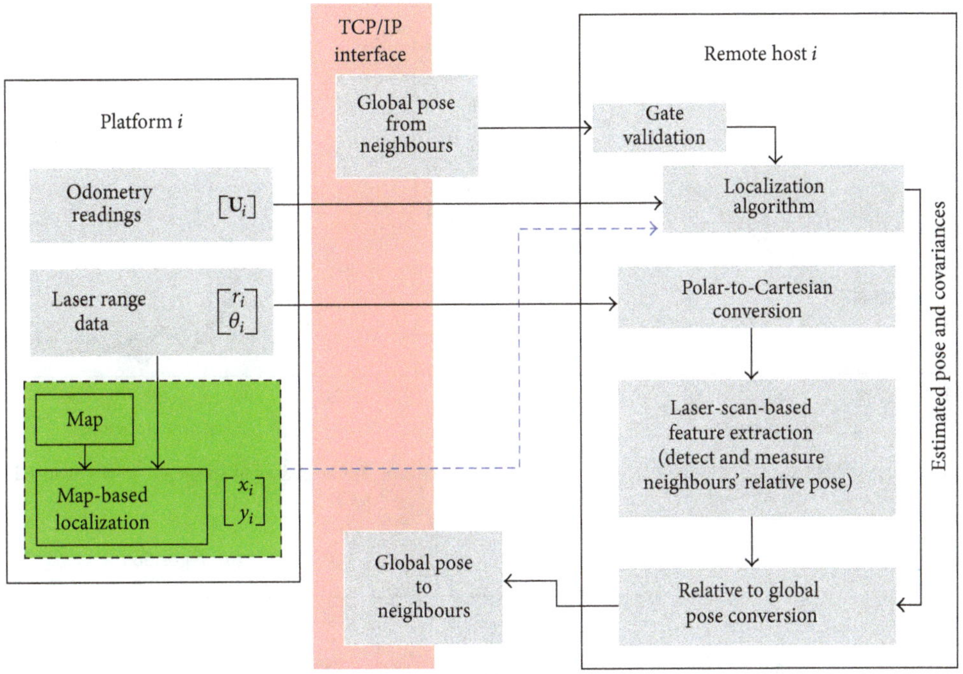

FIGURE 7: System architecture of the experiment setup to validate the proposed DCL scheme. Note that the map-based (scan-matching-based) localization information is available only for platform A.

Additionally, SICK laser scanners were attached to acquire range and bearing measurements for objects around the robot, periodically. Robots were navigated in indoor environment while maintaining triangular formation between them.

6.2. System Architecture. Figure 7 illustrates the system architecture of the experiment setup. Each robot acquires its odometry measurements and laser-scan readings periodically. The acquired measurements are transmitted to a host computer through a TCP/IP interface. Platform A was provided with the map of the navigation space and it performed scan-matching-based localization using this map. The position estimations of this scan-matching-based localization for platform A were considered as absolute position measurements for cooperative localization schemes and were transmitted to the host computer that executes the localization for platform A.

In the host computer, odometry readings were used for state propagation and the global pose measurements and the associated noise covariance from neighbours were used to correct the predicted pose. Note that the pose measurements from neighbours were first evaluated through ellipsoidal measurement validation gate in order to detect and discard outliers. Only platform A used scan-matching-based position calculation data at the sensor fusion. At each host processing unit, the received laser-scan data were first converted to the Cartesian coordinate frame from the polar coordinate system. This gives a set of points that represents the relative positions of objects around the corresponding robot. Laser-scan-based feature extraction algorithm was then exploited to detect and measure the relative pose of neighbouring robots. The data correspondence problem was addressed through the nearest neighbour data association technique. These relative pose measurements were then converted to global (reference) coordinate frame and then were communicated to the corresponding host.

6.3. Results. Figure 8 illustrates the comparison of pose estimates for platform B that were obtained from three different sensor fusion approaches: the centralized cooperative localization method, the proposed Split-CI based DCL algorithm, and the single-robot localization (dead-reckoning) method. The estimates obtained from the centralized cooperative localization approach serve as the benchmark for evaluating the proposed DCL algorithm. On the other hand, the estimates obtained from the single-robot localization represent the worst case pose estimates for each time step. These results suggest that the proposed Split-CI based DCL algorithm and the centralized cooperative localization algorithm generate approximately the same pose estimates for platform B. Although the two estimates are not identical to one another, the difference between the two estimates did not exceed the double-sided 3-σ error boundary, that is, the gray color region of Figure 8, of the proposed DCL algorithm. Pose estimates generated from dead-reckoning diverged from the true state (or the state obtained from the centralized approach) with the increase of experimental time period.

Figure 9 illustrates the comparison of pose uncertainty for three different sensor fusion approaches: the centralized cooperative localization method, the proposed split-CI based DCL algorithm, and single-robot localization method. These results verify that the cooperative localization

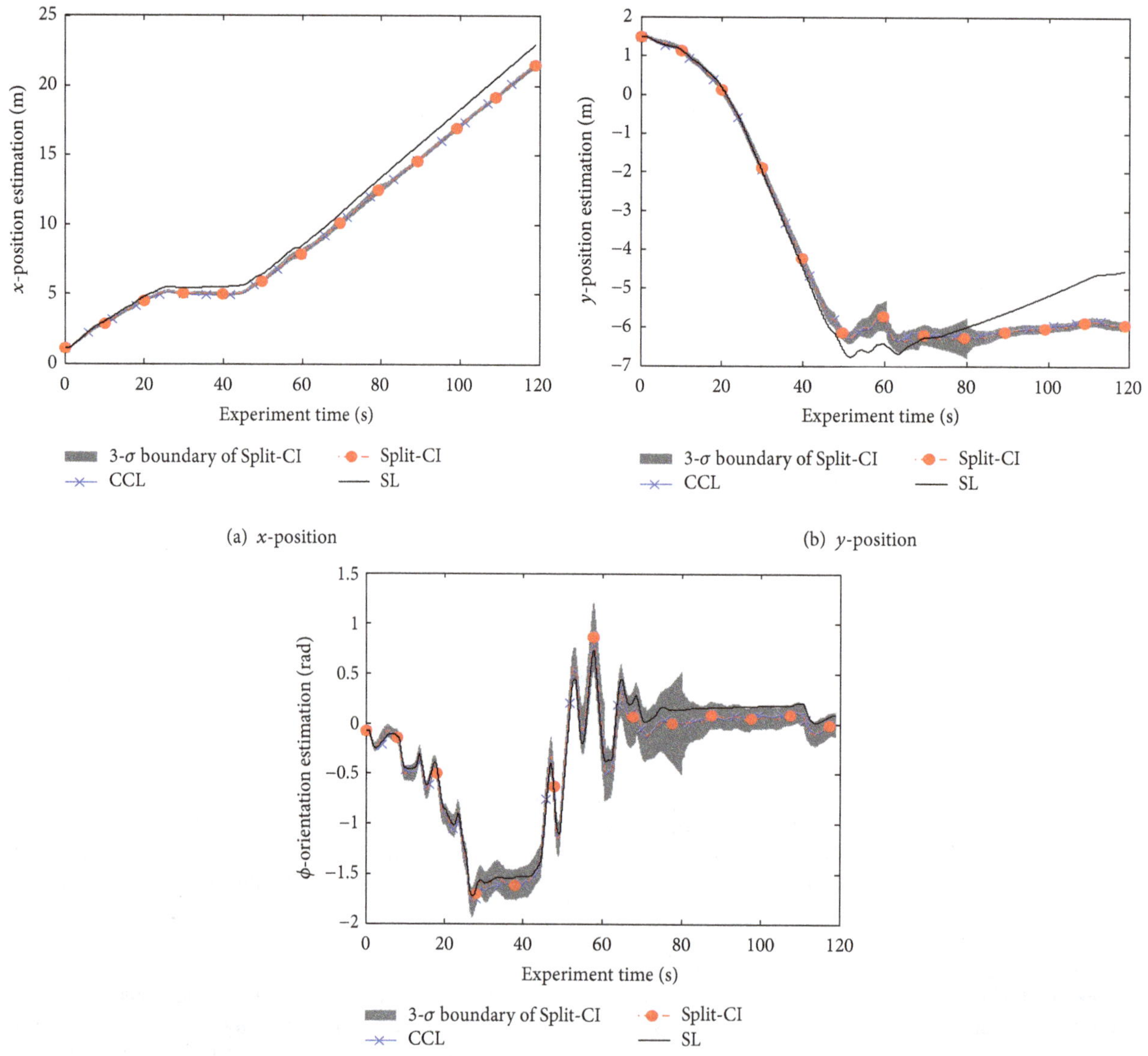

(a) x-position (b) y-position (c) ϕ-orientation

FIGURE 8: Pose estimation comparison of platform *B*.

approaches have bounded pose estimation uncertainty while the pose estimation uncertainty of the single-robot localization approach increases unboundedly. The lowest pose uncertainty is recorded in the centralized approach (see Figure 9(c)). The pose uncertainty found in the proposed Split-CI based DCL algorithm is slightly greater than that of the centralized approach. This is the expected result as the centralized estimator maintained the joint-state and associated dense covariance matrix in order to accurately represent the correlation between teammates' pose estimates.

7. Complexity

7.1. Computational Complexity. As the pose estimation of the proposed algorithm is decentralized, the computational complexity of the proposed DCL algorithm is increased linearly with the increase of number of neighbouring robots. In other words, the computational complexity of the proposed DCL algorithm is $\mathcal{O}(|\overline{N}_q|)$, where $|\overline{N}_q|$ is the number of neighbours, per robot per time step. This remains true for all the DCL algorithms while the computational complexity scales $\mathcal{O}(|N|^4)$ for the centralized cooperative localization where $|N|$ is the number of robots in the MRS.

7.2. Communicative Complexity. The proposed DCL algorithm does not require robot to communicate on-board high frequency proprioceptive sensory data with one another or with central processing unit. Only the interrobot measurements are required to exchange between neighbouring robots. These two properties considerably reduce the bandwidth requirement for communication network between robots. In general, communication complexity of the proposed algorithm remains $\mathcal{O}(|\overline{N}_q|)$ per robot, per interrobot observation event.

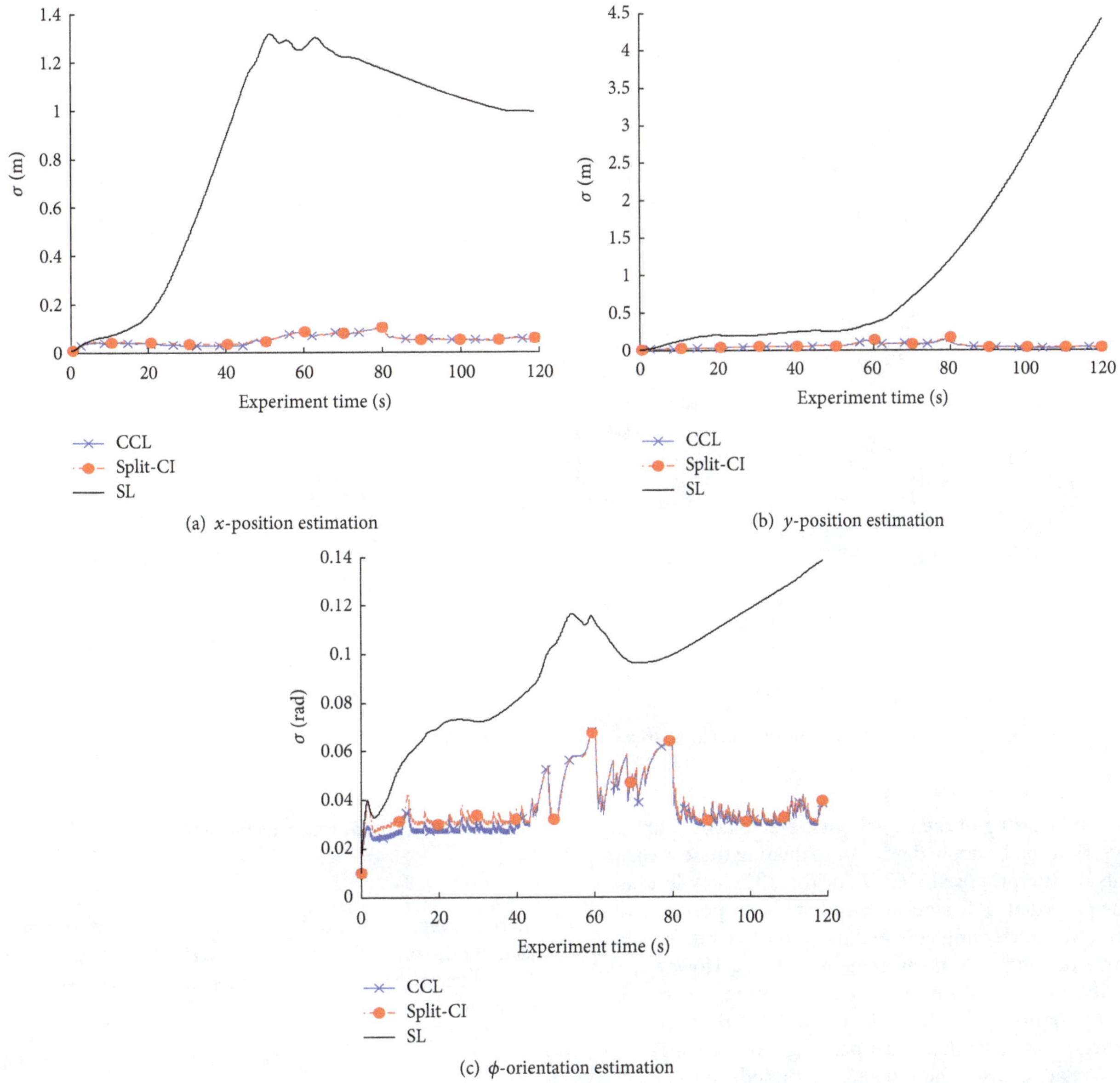

FIGURE 9: Comparison of pose estimation uncertainty (standard deviation) for platform B.

8. Conclusion

This study demonstrates the use of Split-CI algorithm and cubature Kalman filter for decentralized cooperative localization. Both the overall computational (processing) and communicative requirements of the proposed DCL algorithm remain $\mathcal{O}|\overline{N}_q|$ per robot per time step, where $|\overline{N}_q|$ is the number of neighbouring robots. This is a considerable reduction compared to state-of-the-art centralized cooperative localization approach in which computational cost scales $\mathcal{O}(|N|^4)$ and requires wider communication bandwidth to exchange high frequency ego-centric measurements between robots and/or central processing unit. Besides the reduced computational and communicative complexity, both the simulation and experiment results demonstrate that the estimation accuracy of the proposed method is comparable with the centralized cooperative localization. Therefore, the proposed DCL algorithm is more suitable for implementing cooperative localization for a MRS with large number of robots. Additionally, the simulation and experimental results verified that the estimation errors of the proposed DCL scheme are bounded and are not overconfidence. This can be attributed to the modification we added at line (19) in Algorithm 4. The results verified that the cooperative localization approaches outperform the single-robot localization method. Therefore, interrobot observation and flow of information between robots will be the most appropriate approach when implementing localization approach for heterogeneous MRS.

The proposed method can be directly applied to interrobot relative measurement systems that give either full relative pose or relative position of the neighbours. For the interrobot relative measurement system that measures

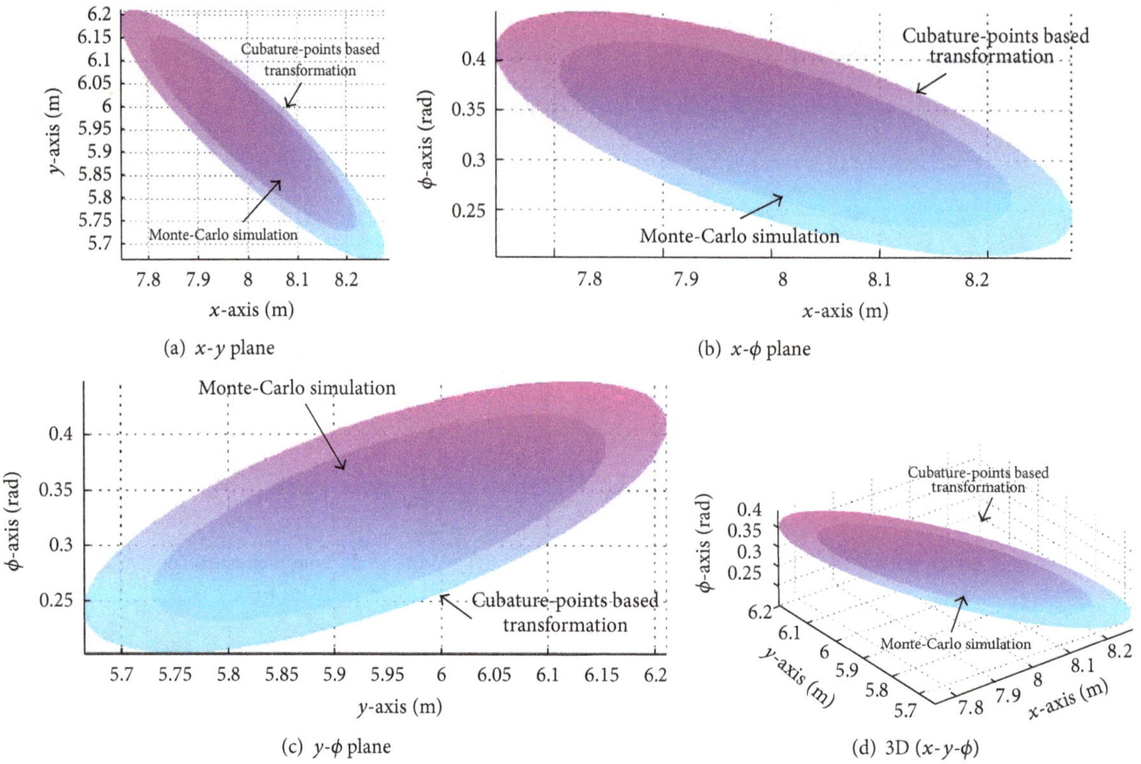

FIGURE 10: Comparison of estimated covariance matrixes.

range and bearing of the neighbours, the polar-to-Cartesian conversion can be applied prior to evaluating these measurements on the proposed DCL algorithm. The key limitation of the proposed DCL algorithm is that the proposed method cannot be directly implemented using relative range-only or relative bearing-only measurement systems. However, this limitation can be addressed by implementing a hierarchical filtering approach. In this hierarchical filtering approach, each robot runs a tracking filter per neighbour to track neighbours' relative pose. These tracks are periodically converted into the global coordinate frame and communicated to the corresponding neighbour. Once a robot receives pose measurement from a neighbour, the proposed DCL algorithm can be exploited for sensor fusion. Ongoing work attempts to implement this hierarchical filtering approach and evaluate its performance.

Appendix

Consistent and Debiased Method for Cartesian-to-Cartesian Conversion

Converting a relative pose measurement to a global pose measurement can be defined as a converting of uncertain information from one Cartesian coordinate frame to another Cartesian coordinate frame.

Assume $\mathbf{x}$ is a random variable with mean $\overline{\mathbf{x}}$ and covariance $\mathbf{P}_x$. Additionally, assume there is another random variable $\mathbf{y}$ which relates to $\mathbf{x}$ as follows:

$$\mathbf{y} = \mathbf{f}(\mathbf{x}), \tag{A.1}$$

where $\mathbf{f}(\cdot)$ represents nonlinear function. If the objective is to calculate the mean $\overline{\mathbf{y}}$ and covariance $\mathbf{P}_y$ of $\mathbf{y}$, given $\overline{\mathbf{x}}$, $\mathbf{P}_x$, and $\mathbf{f}(\cdot)$, the transformed statistics are said to be consistent if the inequality

$$\mathbf{P}_y - E\left[\{\mathbf{y}-\overline{\mathbf{y}}\}\{\mathbf{y}-\overline{\mathbf{y}}\}^T\right] \geq \mathbf{0} \tag{A.2}$$

holds [45]. Work presented in this study applies a cubature-point-based approach to perform Cartesian-to-Cartesian coordinate transformation (see Section 4.2 for more details). Here, we present a simulation study to verify that the Cartesian-to-Cartesian conversion algorithm presented in Algorithm 2 holds above inequality.

Consider a robot team with two robots $\mathscr{R}_1$, and $\mathscr{R}_2$. The global pose of $\mathscr{R}_1$ and $\mathscr{R}_2$ is $[5\ 3\ 0.6981]^T$ and $[8\ 6\ 0.3491]^T$, respectively (the format of the pose vector is $[x\ y\ \phi]^T$, where x and y coordinates are given in m while the orientation ϕ is given in rad). The objective is to find the global pose of $\mathscr{R}_2$ given the global pose of $\mathscr{R}_1$, relative pose of $\mathscr{R}_2$ with respect to $\mathscr{R}_1$, and the associated uncertainties. We compared the statistics we obtained from Algorithm 2 with those calculated by a Monte-Carlo simulation which used 10000 samples. Table 3 and Figure 10 illustrate the comparison of the statistics calculated from two methods. It can be seen that the mean values obtained from the proposed algorithm are approximately overlapped with those

TABLE 3: Comparison of mean global pose.

	True pose	Mean from Monte-Carlo simulation with 10000 samples	Mean from cubature-points based transformation
x (m)	8	8.0127	8.0112
y (m)	6	5.9410	5.9375
ϕ (rad)	0.3491	0.3240	0.3250

calculated by a Monte-Carlo simulation which used 10000 samples. Therefore, the conversion is unbiased. Further, it can be seen that the covariance ellipses of the cubature-point-based approach are always larger than that of the Monte-Carlo simulation. This implies that the proposed Cartesian-to-Cartesian transformation holds the inequality given in (A.2). Additionally, principal axis of covariance ellipse of the proposed approach is approximately overlapped with those of the Monte-Carlo localization. Therefore the proposed coordinate transformation algorithm is consistent.

Competing Interests

The authors declare that they have no competing interests.

Acknowledgments

The authors would like to thank the Natural Science and Engineering Research Council of Canada (NSERC) and Memorial University of Newfoundland for funding this research project.

References

[1] M. K. Habib and Y. Baudoin, "Robot-assisted risky intervention, search, rescue and environmental surveillance," *International Journal of Advanced Robotic Systems*, vol. 7, no. 1, pp. 1–8, 2010.

[2] N. Michael, S. Shen, K. Mohta et al., "Collaborative mapping of an earthquake-damaged building via ground and aerial robots," *Journal of Field Robotics*, vol. 29, no. 5, pp. 832–841, 2012.

[3] R. Kurazume, S. Nagata, and S. Hirose, "Cooperative positioning with multiple robots," in *Proceedings of the IEEE International Conference on Robotics and Automation (ICRA '94)*, vol. 2, pp. 1250–1257, May 1994.

[4] R. Kurazume, S. Hirose, S. Nagata, and N. Sashida, "Study on cooperative positioning system (basic principle and measurement experiment)," in *Proceedings of the 13th IEEE International Conference on Robotics and Automation (ICRA '96)*, vol. 2, pp. 1421–1426, Minneapolis, Minn, USA, April 1996.

[5] S. I. Roumeliotis and G. A. Bekey, "Distributed multirobot localization," *IEEE Transactions on Robotics and Automation*, vol. 18, no. 5, pp. 781–795, 2002.

[6] T. R. Wanasinghe, G. K. I. Mann, and R. G. Gosine, "Decentralized cooperative localization for heterogeneous multi-robot system using split covariance intersection filter," in *Proceedings of the Canadian Conference on Computer and Robot Vision (CRV '14)*, pp. 167–174, IEEE, Montreal, Canada, May 2014.

[7] K. Y. K. Leung, *Cooperative localization and mapping insparselycommunicating robot networks [Ph.D. thesis]*, Department of Aerospace Science and Engineering, University of Toronto, 2012.

[8] I. Rekleitis, G. Dudek, and E. Millios, "Probabilistic cooperative localization and mapping in practice," in *Proceedings of the IEEE International Conference on Robotics and Automation (ICRA '03)*, vol. 2, pp. 1907–1912, Taipei, Taiwan, September 2003.

[9] I. Rekleitis, G. Dudek, and E. Milios, "Experiments in free-space triangulation using cooperative localization," in *Proceedings of the IEEE/RSJ International Conference on Intelligent Robots and Systems (IROS '03)*, vol. 2, pp. 1777–1782, IEEE, October 2003.

[10] N. Trawny and T. Barfoot, "Optimized motion strategies for cooperative localization of mobile robots," in *Proceedings of the IEEE International Conference on Robotics and Automation (ICRA '04)*, vol. 1, pp. 1027–1032, May 2004.

[11] S. Tully, G. Kantor, and H. Choset, "Leap-frog path design for multi-robot cooperative localization," in *Field and Service Robotics*, A. Howard, K. Iagnemma, and A. Kelly, Eds., vol. 62 of *Springer Tracts in Advanced Robotics*, pp. 307–317, Springer, Berlin, Germany, 2010.

[12] E. D. Nerurkar and S. I. Roumeliotis, "Asynchronous multi-centralized cooperative localization," in *Proceedings of the IEEE/RSJ International Conference on Intelligent Robots and Systems (IROS '10)*, pp. 4352–4359, Taipei, Taiwan, October 2010.

[13] R. Sharma and C. Taylor, "Cooperative navigation of MAVs in GPS denied areas," in *Proceedings of the IEEE International Conference on Multisensor Fusion and Integration for Intelligent Systems (MFI '08)*, pp. 481–486, Seoul, Republic of Korea, August 2008.

[14] K. Y. K. Leung, T. D. Barfoot, and H. H. T. Liu, "Decentralized localization of sparsely-communicating robot networks: a centralized-equivalent approach," *IEEE Transactions on Robotics*, vol. 26, no. 1, pp. 62–77, 2010.

[15] E. D. Nerurkar, S. I. Roumeliotis, and A. Martinelli, "Distributed maximum a posteriori estimation for multi-robot cooperative localization," in *Proceedings of the IEEE International Conference on Robotics and Automation (ICRA '09)*, pp. 1402–1409, Kobe, Japan, May 2009.

[16] H. Durrant-Whyte, M. Stevens, and E. Nettleton, "Data fusion in decentralised sensing networks," in *Proceedings of the 4th International Conference on Information Fusion*, pp. 302–307, Montreal, Canada, August 2001.

[17] A. Howard, M. J. Mataric, and G. Sukhatme, "Putting the 'i' in 'team': an ego-centric approach to cooperative localization," in *Proceedings of the IEEE International Conference on Robotics and Automation (ICRA)*, vol. 1, pp. 868–874, Taipei, Taiwan, September 2003.

[18] S. Panzieri, F. Pascucci, and R. Setola, "Multirobot localisation using interlaced extended Kalman filter," in *Proceedings of the IEEE/RSJ International Conference on Intelligent Robots and Systems (IROS '06)*, pp. 2816–2821, IEEE, Beijing, China, October 2006.

[19] A. Bahr, M. R. Walter, and J. J. Leonard, "Consistent cooperative localization," in *Proceedings of the IEEE International Conference on Robotics and Automation (ICRA '09)*, pp. 3415–3422, Kobe, Japan, May 2009.

[20] T. Bailey, M. Bryson, H. Mu, J. Vial, L. McCalman, and H. Durrant-Whyte, "Decentralised cooperative localisation for heterogeneous teams of mobile robots," in *Proceedings of the IEEE International Conference on Robotics and Automation (ICRA '11)*, pp. 2859–2865, IEEE, Shanghai, China, May 2011.

[21] L. C. Carrillo-Arce, E. D. Nerurkar, J. L. Gordillo, and S. I. Roumeliotis, "Decentralized multi-robot cooperative localization using covariance intersection," in *Proceedings of the 26th IEEE/RSJ International Conference on Intelligent Robots and Systems (IROS '13)*, pp. 1412–1417, IEEE, Tokyo, Japan, November 2013.

[22] D. Fox, W. Burgard, H. Kruppa, and S. Thrun, "Probabilistic approach to collaborative multi-robot localization," *Autonomous Robots*, vol. 8, no. 3, pp. 325–344, 2000.

[23] N. E. Özkucur, B. Kurt, and H. L. Akın, "A collaborative multi-robot localization method without robot identification," in *RoboCup 2008: Robot Soccer World Cup XII*, vol. 5399 of *Lecture Notes in Computer Science*, pp. 189–199, Springer, Berlin, Germany, 2009.

[24] A. Prorok and A. Martinoli, "A reciprocal sampling algorithm for lightweight distributed multi-robot localization," in *Proceedings of the IEEE/RSJ International Conference on Intelligent Robots and Systems (IROS '11)*, pp. 3241–3247, IEEE, San Francisco, Calif, USA, September 2011.

[25] A. Prorok, A. Bahr, and A. Martinoli, "Low-cost collaborative localization for large-scale multi-robot systems," in *Proceedings of the IEEE International Conference on Robotics and Automation (ICRA '12)*, pp. 4236–4241, St. Paul, Minn, USA, May 2012.

[26] Y. Bar-Shalom, P. K. Willett, and X. Tian, *Tracking and Data Fusion: A Handbook of Algorithms*, YBS Publishing, 2011.

[27] D. Kurth, G. Kantor, and S. Singh, "Experimental results in rangeonly localization with radio," in *Proceedings of the IEEE/RSJ International Conference on Intelligent Robots and Systems*, vol. 1, pp. 974–979, October 2003.

[28] E. Olson, J. J. Leonard, and S. Teller, "Robust range-only beacon localization," *IEEE Journal of Oceanic Engineering*, vol. 31, no. 4, pp. 949–958, 2006.

[29] R. Sharma, S. Quebe, R. W. Beard, and C. N. Taylor, "Bearing-only cooperative localization," *Journal of Intelligent & Robotic Systems*, vol. 72, no. 3-4, pp. 429–440, 2013.

[30] K. E. Bekris, M. Glick, and L. E. Kavraki, "Evaluation of algorithms for bearing-only SLAM," in *Proceedings of the IEEE International Conference on Robotics and Automation (ICRA '06)*, pp. 1937–1943, IEEE, Orlando, Fla, USA, May 2006.

[31] O. De Silva, G. Mann, and R. Gosine, "Development of a relative localization scheme for ground-aerial multi-robot systems," in *Proceedings of the IEEE/RSJ International Conference on Intelligent Robots and Systems (IROS '12)*, pp. 870–875, Vilamoura-Algarve, Portugal, October 2012.

[32] T. R. Wanasinghe, G. K. I. Mann, and R. G. Gosine, "Pseudo-linear measurement approach for heterogeneous multi-robot relative localization," in *Proceedings of the 16th International Conference on Advanced Robotics (ICAR '13)*, pp. 1–6, IEEE, Montevideo, Uruguay, November 2013.

[33] M. W. Mehrez, G. K. I. Mann, and R. G. Gosine, "Nonlinear moving horizon state estimation for multi-robot relative localization," in *Proceedings of the IEEE 27th Canadian Conference on Electrical and Computer Engineering (CCECE '14)*, pp. 1–5, Toronto, Canada, May 2014.

[34] S. I. Roumeliotis and G. A. Bekey, "Collective localization: a distributed Kalman filter approach to localization of groups of mobile robots," in *Proceedings of the IEEE International Conference on Robotics and Automation (ICRA '00)*, vol. 3, pp. 2958–2965, April 2000.

[35] I. M. Rekleitis, G. Dudek, and E. E. Milios, "Multi-robot cooperative localization: a study of trade-offs between efficiency and accuracy," in *Proceedings of the IEEE/RSJ International Conference on Intelligent Robots and Systems ((IROS '02)*, vol. 3, pp. 2690–2695, IEEE, October 2002.

[36] I. Arasaratnam and S. Haykin, "Cubature Kalman filters," *IEEE Transactions on Automatic Control*, vol. 54, no. 6, pp. 1254–1269, 2009.

[37] K. P. B. Chandra, D.-W. Gu, and I. Postlethwaite, "Cubature Kalman filter based localization and mapping," in *Proceedings of the 18th International Federation of Automatic Control (IFAC '11) World Congress*, pp. 2121–2125, Milano, Italy, September 2011.

[38] Y. Song, Q. Li, Y. Kang, and Y. Song, "CFastSLAM: a new Jacobian free solution to SLAM problem," in *Proceedings of the IEEE International Conference on Robotics and Automation (ICRA '12)*, pp. 3063–3068, St. Paul, Minn, USA, May 2012.

[39] I. Arasaratnam, S. Haykin, and T. R. Hurd, "Cubature Kalman filtering for continuous-discrete systems: theory and simulations," *IEEE Transactions on Signal Processing*, vol. 58, no. 10, pp. 4977–4993, 2010.

[40] Y. Kosuge and T. Matsuzaki, "The optimum gate shape and threshold for target tracking," in *Proceedings of the SICE Annual Conference*, vol. 2, pp. 2152–2157, Fukui, Japan, August 2003.

[41] H. Li, F. Nashashibi, and M. Yang, "Split covariance intersection filter: theory and its application to vehicle localization," *IEEE Transactions on Intelligent Transportation Systems*, vol. 14, no. 4, pp. 1860–1871, 2013.

[42] S. Julier and J. K. Uhlmann, "General decentralized data fusion with covariance intersection (CI)," in *Handbook of Data Fusion*, chapter 12, CRC Press, Boca Raton, Fla, USA, 2001.

[43] K. Y. K. Leung, Y. Halpern, T. D. Barfoot, and H. H. T. Liu, "The UTIAS multi-robot cooperative localization and mapping dataset," *International Journal of Robotics Research*, vol. 30, no. 8, pp. 969–974, 2011.

[44] J. Sijs, M. Lazar, and P. P. J. V. D. Bosch, "State fusion with unknown correlation: ellipsoidal intersection," in *Proceedings of the American Control Conference (ACC '10)*, pp. 3992–3997, IEEE, Baltimore, Md, USA, June-July 2010.

[45] S. J. Julier and J. K. Uhlmann, "Consistent debiased method for converting between polar and cartesian coordinate systems," in *Acquisition, Tracking, and Pointing XI*, vol. 3086 of *Proceedings of SPIE*, pp. 110–121, International Society for Optics and Photonics, Orlando, Fla, USA, June 1997.

14

Theoretical Design and First Test in Laboratory of a Composite Visual Servo-Based Target Spray Robotic System

Dongjie Zhao,[1,2] Ying Zhao,[2] Xuelei Wang,[1] and Bin Zhang[1]

[1] *College of Engineering, China Agricultural University, Beijing 100083, China*
[2] *School of Mechanical and Automobile Engineering, Liaocheng University, Liaocheng 252059, China*

Correspondence should be addressed to Bin Zhang; zhangbin64@cau.edu.cn

Academic Editor: Gordon R. Pennock

In order to spray onto the canopy of interval planting crop, an approach of using a target spray robot with a composite vision servo system based on monocular scene vision and monocular eye-in-hand vision was proposed. Scene camera was used to roughly locate target crop, and then the image-processing methods for background segmentation, crop canopy centroid extraction, and 3D positioning were studied. Eye-in-hand camera was used to precisely determine spray position of each crop. Based on the center and area of 2D minimum-enclosing-circle (MEC) of crop canopy, a method to calculate spray position and spray time was determined. In addition, locating algorithm for the MEC center in nozzle reference frame and the hand-eye calibration matrix were studied. The processing of a mechanical arm guiding nozzle to spray was divided into three stages: reset, alignment, and hovering spray, and servo method of each stage was investigated. For preliminary verification of the theoretical studies on the approach, a simplified experimental prototype containing one spray mechanical arm was built and some performance tests were carried out under controlled environment in laboratory. The results showed that the prototype could achieve the effect of "spraying while moving and accurately spraying on target."

1. Introduction

With multiple advantages such as improving the efficiency of pesticide use and reducing environment pollution, target spray has become a hot topic in the field of precision agriculture. It automatically sprays on target after getting the target's information such as crop's location, morphology, diseases, and insect pests and combining with spray requirements and operation information of actuator such as sprayer's speed, mechanical arm motion control, and nozzle features. It has significance to increase the utilization ratio of pesticide, reduce pesticide residues on crops, and protect the environment and workers [1, 2].

Compared with robot control based on infrared sensor, ultrasonic sensor, and laser radar, visual servo control exhibits high flexibility and intelligence and is more suitable for agricultural environment [3–5]. In recent years, many scholars in America, Europe, Japan, China, and South Korea have carried out research on the target spray technology based on visual servo and have made notable progress. Giles and Slaughter developed a visual servo spray system, which could make spray nozzle move along the center line of crop row [6]. Precision herbicide application system, reported by Tian, can separately control each nozzle to do variable spray in each control zone according to weed infestation conditions detected by vision [7, 8]. Steward and Tang designed one set of target spraying systems based on machine vision and finite state machine controller, used for accurate spraying of herbicide to remove weeds in the field [9]. Li et al. studied a spray system to detect diseases and insect pests of plant in greenhouse and spray automatically [10]. Zhang et al. developed a mobile spray robotic system in greenhouse, which achieved spray control based on regional differences in terms of plant diseases and insect pests detected by machine vision [11, 12]. Yufeng et al. established indoor smart pesticide spray system based on machine vision and accomplished in-depth study on some problems such as image processing, application decision-making, and data exchange [13]. Based

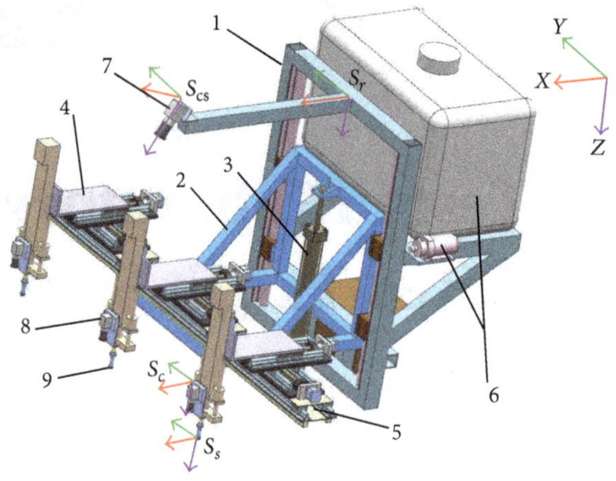

(1) Rigid frame
(2) Movable frame
(3) Lifting linear actuator
(4) Spray mechanical arm
(5) Translation stage
(6) Liquid box and dc pump
(7) Scene camera
(8) Eye-in-hand camera
(9) Nozzle
S_r: robot reference frame
S_{cs}: scene camera reference frame
S_c: eye-in-hand camera reference frame
S_s: nozzle reference frame

FIGURE 1: Robot body.

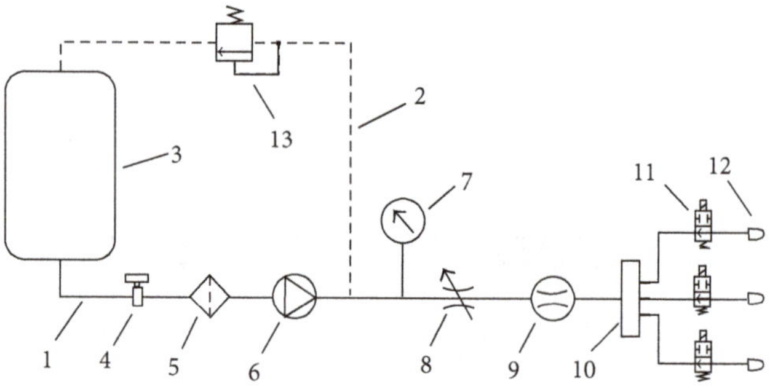

(1) Main line
(2) Return line
(3) Liquid box
(4) Manual switch
(5) Filter
(6) dc pump
(7) Pressure sensor
(8) Throttle
(9) Flow sensor
(10) Liquid dispenser
(11) Solenoid valve
(12) Nozzle
(13) Overflow valve

FIGURE 2: Spray system.

on existing knowledge, however, accurate spraying onto the canopy of interval planting crops directly has not been implemented.

To better solve the problem of accurate spraying onto the canopy of interval planting crops, this paper proposed a schematic design of target spray robot system based on composite visual servo and built a simplified experimental prototype to verify partial performances of the design in laboratory. Section 2 of the paper describes the construction and working method of the system. Section 3 details the solutions to four key issues involved in the realization of system function. Section 4 builds a simplified prototype containing one spray mechanical arm to test the accuracy of spray position and spray time and other functions of the design.

2. Material and Methods of the Target Spray Robotic System

2.1. System Structure. The target spray robot system based on composite visual servo is mainly composed of robot body, spray system, visual processing, and servo control system.

The robot body is shown in Figure 1. Rigid frame, the mounting base of other parts, is fixed on the front of motion platform (not shown). Movable frame, mounted on the rigid frame, can move up and down (Z direction) along guide rail of the rigid frame under the action of lifting linear actuator. Translation stage is fixed on the front end of the movable frame. Spray mechanical arms, mounted on the translation stage and arranged for one or more as desired (three in Figure 1), can move horizontally (Y direction) along guide rail of the translation stage. The mechanical spray arms are a three-degree-of-freedom (DOF) serial mechanism and can achieve nozzle movement in X, Y, and Z directions.

The composition of spray system is shown in Figure 2. Manual switch is used to manually control on/off of the liquid box. dc pump provides power for spray process. Filter is used to filter the impurities in the liquid, to protect pipeline and nozzle. Pressure sensor and flow sensor acquire relevant information of main line, which is provided to the control system for decision-making. Throttle in main line and overflow valve in return line are used to adjust flow and pressure of the main line. Each nozzle is equipped with an independent solenoid valve to control the start-stop of spray process.

Visual processing and servo control system is composed of visual unit, sensor unit, and control unit. Among them, the vision unit includes one scene camera and several eye-in-hand cameras. The sensor unit mainly consists of one speed sensor to detect the speed of the mobile platform and several position sensors to detect the positions of all moving parts of the spray mechanical arms. The control unit consists of one PC (host computer) and several DSP (slave computer). DSP acquires and processes the information of eye-in-hand camera, position sensor, speed sensor, flow sensor, and pressure sensor and combines with PC instructions to control the spray mechanical arm and solenoid valve to do corresponding actions and uploads relevant information to PC. PC analyses the scene camera images and DSP feedback information to generate control instructions, which will be transmitted to the corresponding DSP through the switch. The working principle of the visual processing and servo control system is shown in Figure 3.

2.2. Working Method. The basic working process of the system can be expressed as follows: before spraying, control the movement of the lifting linear actuator to adjust initial height of the spray mechanical arm, and make the nozzles basically consistent with average spray height of the crops.

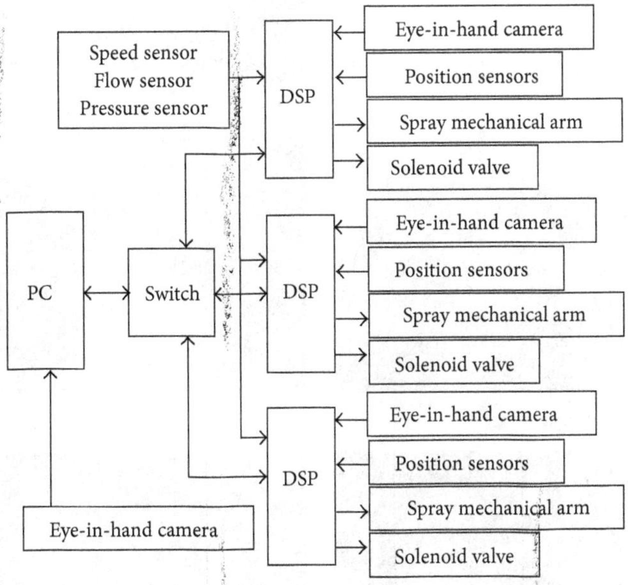

FIGURE 3: Visual processing and servo control system.

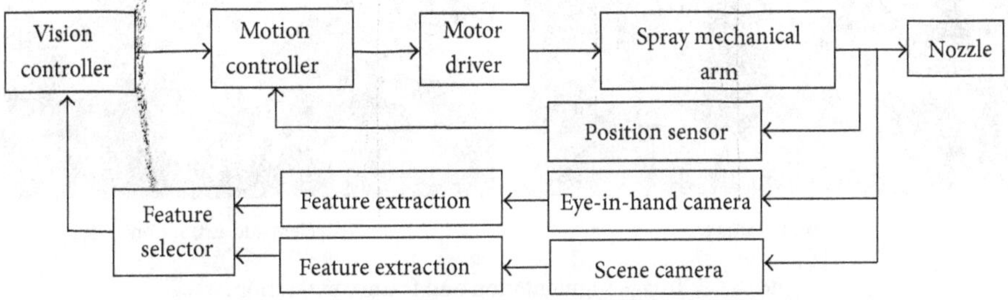

FIGURE 4: Block diagram of visual servo system.

And then, adjust the position of each spray mechanical arm on the translation stage to make the distance between them basically consistent with the row spacing and each nozzle corresponding to a row of crops. In the process of spraying, the scene camera roughly positions crops and transmits the information to the processing system which will control the spray mechanical arm move in advance. While the crops come into the field-of-view (FOV) of the eye-in-hand camera, it will pinpoint the crop and guide nozzle to move to appropriate location above the centroid of crop canopy through controlling the spray mechanical arm to complete target spraying.

2.3. Composite Visual Servo System Structure. Monocular vision system can indirectly obtain depth information through motion compensation method. Despite complex algorithm, it costs smaller computing resource and has better real-time performance than binocular system. Compared with scene vision system, eye-in-hand vision system shows higher accuracy of target positioning, but lower accuracy in working space [14, 15]. Considering spray robot has the features of big working space and real-time and high target positioning accuracy, a kind of composite vision servo method has been designed based on monocular scene vision and monocular eye-in-hand vision. The scene camera, mounted at the front of the robot body, had a fixed angle with the ground and could acquire crops' image information in real time during the process of marching with the robot in the field. Centroids' coordinates of crops canopy in the field, acquired through recognizing crops' image information, could be used to roughly position the crops and to direct the spray mechanical arm moving in advance to make the crops smoothly enter the FOV of eye-in-hand camera. The image information acquired by eye-in-hand cameras which were mounted at the end of spray mechanical arms with nozzles was recognized to get accurate spray position and time of target crop and guided nozzle to complete spraying.

The composite visual servo system used a double closed-loop structure (as shown in Figure 4). The inner ring adopted location information given by position sensors of the spray mechanical arms, while the outer ring adopted location information of crop acquired by the visual image. The sampling period of the outer ring depended on real-time processing speed of the image detection system, and its sampling frequency was far lower than that of the inner ring. Such control structure could help to improve the stability and dynamic performance of the system. Feature selector controlled the source of feedback information. When target

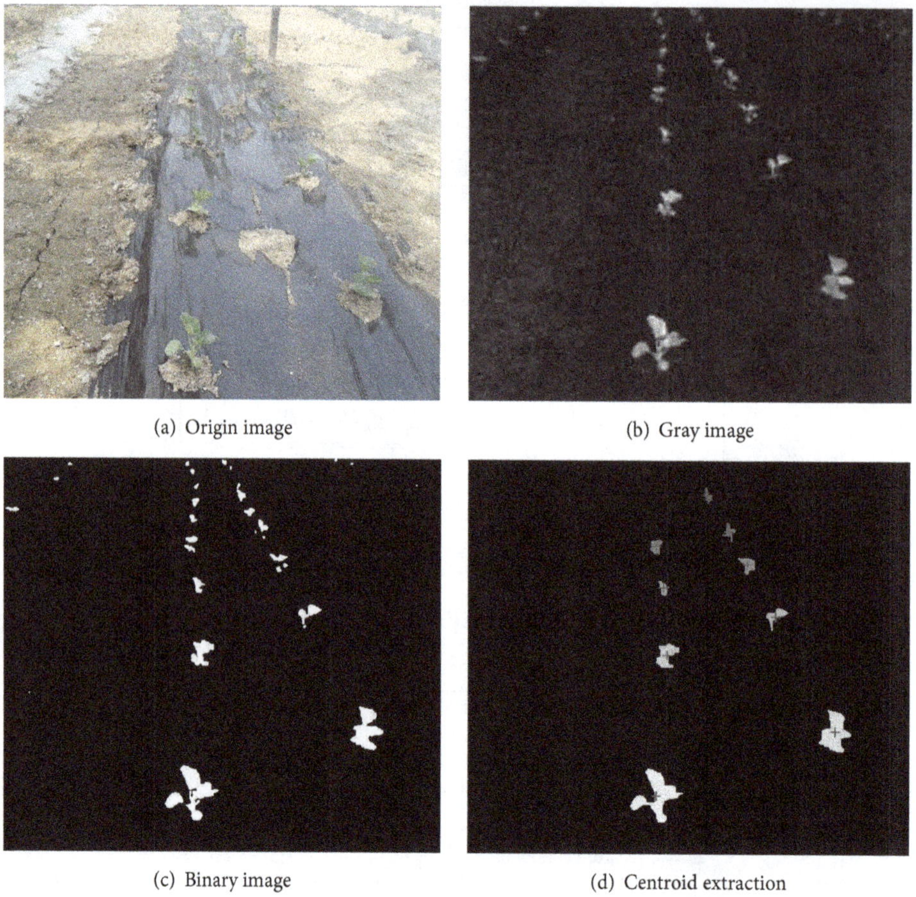

FIGURE 5: Image segmentation and feature extraction.

crop entered the FOV of eye-in-hand camera, the decision information of vision controller would use features detected by eye-in-hand vision and would not switch to scene vision features until spraying is completed.

3. Principle and Method of Implementation

To realize visual servo of the spray robot, two aspects are needed to be solved. One is image interpretation, that is, how to quickly and accurately extract effective features from images. And the other is servo control, that is, how to map visual features to control space of the spray mechanical arm. The solution involves four key issues such as background segmentation, features extraction and target location, camera calibration, and visual servo method. This study uses spraying operation of field gourd seedlings as an example.

3.1. Background Segmentation. Generally, gourd seedlings use spacing planting, with space of 0.4~0.6 m. When 4–6 true leaves sprout from crop, its foliage needs to have ethephon sprayed 2-3 times to promote the formation of female flowers and increase production. The information of crop image acquired under natural light environment mainly includes gourd seedlings, soil, dead leaves, small weeds, and black (or white) plastic film (as shown in Figure 5(a)). Gourd seedlings and weeds are green and the others are nongreen while weeds are smaller than gourd seedlings.

CIVE operator (*0.441R-0.811G + 0.385B + 18.78745*) [16] and OTSU method [17] are selected to carry out the background segmentation of acquired RGB images (R: red, G: green, and B: blue) through experiments. Figure 5(b) is the gray image after graying origin image by CIVE operator, where crop and noncrop (background) can be easily distinguished.

When the target occupies a proper proportion in the image, adaptive threshold segmentation based on OTSU method can effectively solve the impact of environment change on image quality [18]. Figure 5(c) is the binary image after segmentation by the method, which has less noise points and ideal segmentation effect.

3.2. Feature Extraction and Target Positioning. Target positioning is to acquire the location of target crop in robot reference frame, by controlling the spray mechanical arm move to make the nozzle arrive at the operation location to complete spraying. Target crop moves backward relative to the spray robot during work. Provisions of robot reference frame can be determined by right-hand rule. Positive X-axis is the horizontal forward direction. Positive Z-axis is the vertical downward direction (as shown in Figure 1).

(a) Actual spray effect

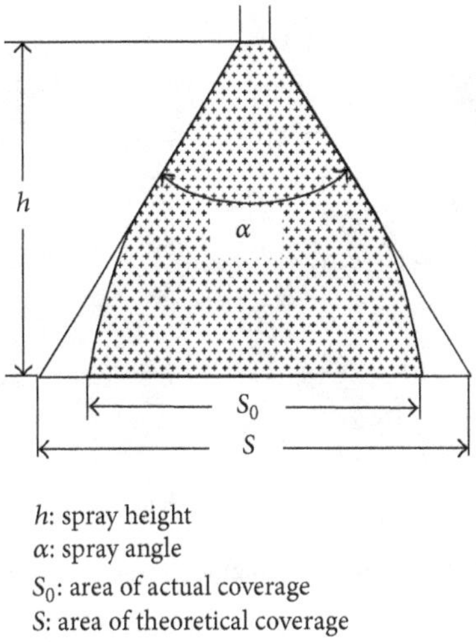

(b) Spray sketch map

FIGURE 6: Spray with full cone nozzle.

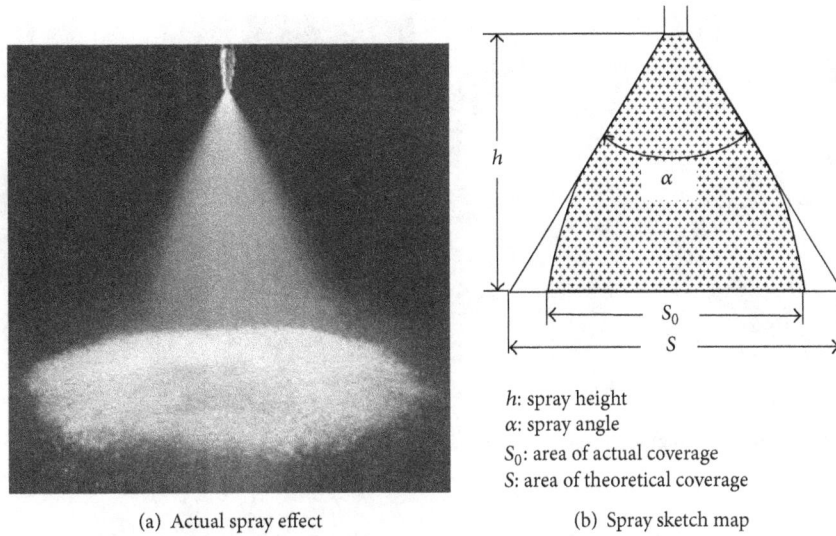

3.2.1. Scene Camera Image. Scene camera image is used for coarse positioning of target crop to guide the spray mechanical arm move in advance. The centroid of crop canopy is selected as the reference feature for positioning. It is possible to extract image coordinates of the centroid first and then calculate its 3D coordinates in robot reference frame according to transformation relationship between the image reference frame and robot reference frame, which is shown in Section 3.3.1. The extraction method of centroid is as follows.

Generally, the binary image after threshold segmentation will include some defects, such as weed image and random noise and individual crop's leaf separation. A processing method can be as follows: Firstly, apply morphological closing operation to improve image connectivity. Secondly, label connected regions with dual scanning method, calculate pixel area of each region, and remove small separation regions such as weeds, random noise, and far crops with area filtering method. And then calculate the average values of pixel coordinates of each connected region, and make them as the centroid's coordinates of target crops. The results are shown in Figure 5(d). The centroids' position of the crops was cross marked.

3.2.2. Eye-in-Hand Camera Image. Eye-in-hand camera image provides feature parameters used for determining spray position and spray time of target crops.

For vertical downward spraying with full cone nozzle (such as TeeJet D5 and TeeJet DVP-4, Spraying System Co., IL, USA), droplet coverage is approximately circular (as shown in Figure 6), and its area can be approximated as

$$S_0 = \xi \cdot \pi \cdot \left(h \cdot \tan\left(\frac{\alpha}{2}\right)\right)^2, \quad (1)$$

where ξ denotes proportion coefficient of actual coverage of droplet relative to its theoretical coverage, which is related to liquid viscosity, liquid temperature, liquid surface tension and spray height, and so forth. Let Q denote flow through nozzle per unit time and let q denote amount of liquid needed for crop per unit area. Then, the spray time for target crop will be

$$t = \frac{q \cdot S_0}{Q}. \quad (2)$$

By formulas (1) and (2), we know that if ξ, α, Q, and q keep constant during spraying, then area of actual coverage (denoted by S_0) is proportional to spray height (denoted by h), and t is proportional to S_0.

In order to meet the requirement of precision spray, we can build a 2D minimum-enclosing-circle (MEC) of crop canopy and make actual coverage of droplet match it during spraying. Then, the center of MEC can be used to determine nozzle's X, Y position, the area (or radius) to determine relative height between nozzle and crop canopy h (namely, nozzle's Z position) by formula (1) and t by formula (2).

Figure 7(a) shows original image in eye-in-hand camera of crop and Figure 7(b) shows the binary image processed with the above segmentation and filtering method, in which MEC of crop canopy and circular mark in the center are made at the same time.

The Z coordinate of MEC center in the eye-in-hand camera reference frame can be determined according to corresponding relationship between target's image size and camera position. Target's image size, according to pinhole imaging model (Figure 8), is related to the movement of camera along the optical axis (Z direction), but basically not with the movement of camera along X, Y directions.

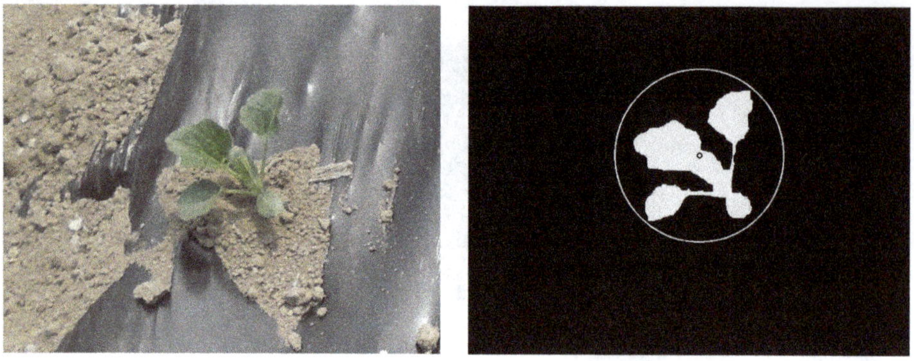

FIGURE 7: Segmentation and feature extraction of eye-in-hand camera image.

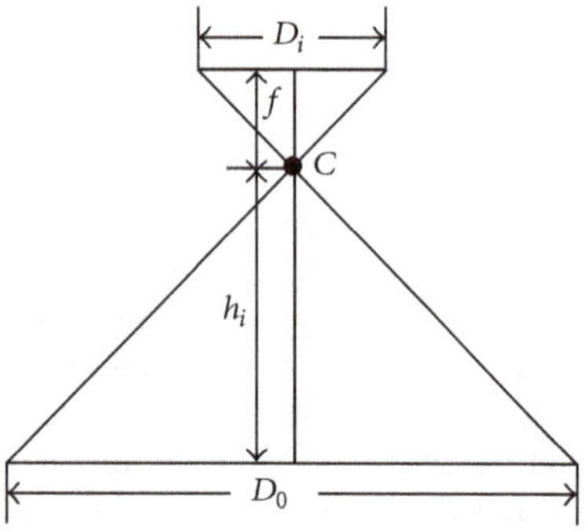

C: camera projection center
f: focal length
D_0: actual diameter of MEC
h_i: distance between crop canopy and camera
D_i: image diameter of MEC

FIGURE 8: Imaging geometry of target.

Suppose that the camera moves from starting position h_1 to end position h_2 along Z-axis; the distance moved is d; based on geometrical relationship it can be seen that

$$h_1 = f \cdot \frac{D_0}{D_1},$$
$$h_2 = f \cdot \frac{D_0}{D_2}, \quad (3)$$
$$d = h_1 - h_2.$$

Combine the above and solve; the following can be obtained:

$$h_2 = d \cdot \frac{D_1}{(D_2 - D_1)}$$
$$D_0 = d \cdot D_1 \cdot \frac{D_2}{(f \cdot (D_2 - D_1))}. \quad (4)$$

The distance in Z direction between target crop canopy and current camera is h_2, which is Z coordinate of MEC center in current camera reference frame. It is possible to obtain 3D coordinates of MEC center in the camera reference frame in accordance with h_2, image projection coordinates of MEC center, and intrinsic parameter matrix of the camera. Besides, combining with eye-in-hand calibration matrix provided in Section 3.3.2, the 3D coordinates of MEC center in nozzle reference frame can be obtained.

3.3. Camera Calibration. In order to accurately acquire crop's location information from images, intrinsic and extrinsic parameters of camera need to be calibrated. The calibration of intrinsic parameters can use classic Zhang calibration algorithm [19, 20]. Extrinsic parameters to be calibrated mainly include transformation matrix between scene camera reference frame and robot reference frame, transformation matrix between eye-in-hand camera reference frame and nozzle reference frame. Intrinsic and extrinsic parameters, supposed to be fixed in normal use, can be calibrated offline.

3.3.1. Extrinsic Parameter Calibration of Scene Camera and Target Location Estimation. While calibrating camera intrinsic parameters by Zhang calibration algorithm, we can obtain transformation matrix between each calibration target reference frame and camera reference frame (extrinsic parameters) at the same time. If calibration target is placed at a setting position of robot reference frame and we make the two reference frames in the same direction, there is only translation transformation between the two reference frames. Let ${}^{b}T_r$ denote the translation vector and let $[{}^{cs}R_b \ {}^{cs}T_b]$ denote the transformation matrix between calibration plate reference frame and camera reference frame; then the transformation matrix M_2 between robot reference frame and camera reference frame will be

$$M_2 = \begin{bmatrix} {}^{cs}R_b & {}^{cs}T_b \\ 0 & 1 \end{bmatrix} \begin{bmatrix} E & {}^{b}T_r \\ 0 & 1 \end{bmatrix} = \begin{bmatrix} {}^{cs}R_b & {}^{cs}R_b{}^{b}T_r + {}^{cs}T_b \\ 0 & 1 \end{bmatrix}. \quad (5)$$

Assume that the height of crops is equal, and the canopies are within $Z = H_0$ plane of robot reference frame. Let ${}^{cs}p$

denote the image projection point of the centroid (denoted by ${}^{r}P$) of a crop canopy and let $[u\ \ v]^T$ denote the image coordinates of ${}^{cs}p$; then the coordinates of ${}^{cs}p$ in camera reference frame will be

$$\begin{bmatrix} x_p \\ y_p \\ f \end{bmatrix} = M_1^{-1} f \begin{bmatrix} u \\ v \\ 1 \end{bmatrix}, \tag{6}$$

where M_1 denotes intrinsic parametric matrix of the scene camera and f denotes its focal length. The homogeneous coordinates of ${}^{cs}p$ may be represented as $[x_p\ \ y_p\ \ f\ \ 1]^T$. Let ${}^{r}p$ denote the mapping point of ${}^{cs}p$ in robot reference frame; then

$$ {}^{r}p = M_2^{-1} \cdot {}^{cs}p. \tag{7}$$

Similarly, homogeneous coordinates of the mapping point (denoted by ${}^{r}O_{cs}$) in robot reference frame of the origin of the scene camera reference frame will be $M_2^{-1}\,[0\ \ 0\ \ 0\ \ 1]^T$. From geometrical relationship it can be seen that ${}^{r}P$ means the intersection point between $Z = H_0$ plane and extension line of ${}^{r}O_{cs}$ and ${}^{r}p$. Therefore, the coordinates of ${}^{r}P$ in robot reference frame can be approximately expressed as

$$\begin{bmatrix} P_x \\ P_y \\ P_z \end{bmatrix} = k \begin{bmatrix} p_x - O_{cs-x} \\ p_y - O_{cs-y} \\ 0 \end{bmatrix} + \begin{bmatrix} O_{cs-x} \\ O_{cs-y} \\ H_0 \end{bmatrix}, \tag{8}$$

where $k = (H_0 - O_{cs-z})/(p_z - O_{cs-z})$. $[P_x\ \ P_y\ \ P_z]^T$, $[p_x\ \ p_y\ \ p_z]^T$, and $[O_{cs-x}\ \ O_{cs-y}\ \ O_{cs-z}]^T$ denote the 3D coordinates of ${}^{r}P$, ${}^{r}p$, and ${}^{r}O_{cs}$, respectively.

3.3.2. Eye-in-Hand Calibration. Eye-in-hand calibration means estimating the rigid transformation matrix between the eye-in-hand camera reference frame and nozzle reference frame.

Let ${}^{c}K$ denote one point in eye-in-hand camera reference frame and let ${}^{s}K$ denote the mapping point of ${}^{c}K$ in nozzle reference frame; then

$$ {}^{s}K = \begin{bmatrix} {}^{s}R_c & {}^{s}T_c \\ 0 & 1 \end{bmatrix} {}^{c}K, \tag{9}$$

where ${}^{s}R_c$ and ${}^{s}T_c$ denote rotation matrix and translation vector of the transformation from eye-in-hand camera reference frame to nozzle reference frame, respectively. And they will remain unchanged because the relative position between eye-in-hand camera and nozzle is fixed. Let the optical axis of eye-in-hand camera be arranged vertically downward and parallel to the axis of nozzle; then the two reference frames will have no relative rotation and ${}^{s}R_c$ is an identity matrix. Therefore, we only need to calculate ${}^{s}T_c$.

Similarly, let ${}^{r}K$ denote the mapping point of ${}^{s}K$ in robot reference frame; then

$$ {}^{r}K = \begin{bmatrix} {}^{r}R_s & {}^{r}T_s \\ 0 & 1 \end{bmatrix} {}^{s}K, \tag{10}$$

where ${}^{r}R_s$ and ${}^{r}T_s$ denote, respectively, rotation matrix and translation vector of the transformation from nozzle reference frame to robot reference frame. They are determined by the position and orientation of nozzle. The matrix ${}^{r}R_s$ is also an identity matrix because the spray mechanical arm only can make nozzle translation motion along X, Y, and Z directions.

Combining formula (9) and formula (10) gives

$$ {}^{r}K = \begin{bmatrix} {}^{r}R_s & {}^{r}T_s \\ 0 & 1 \end{bmatrix} \begin{bmatrix} {}^{s}R_c & {}^{s}T_c \\ 0 & 1 \end{bmatrix} {}^{c}K = \begin{bmatrix} E & {}^{r}T_s + {}^{s}T_c \\ 0 & 1 \end{bmatrix} {}^{c}K. \tag{11}$$

Then

$$\begin{bmatrix} {}^{r}K_{xyz} \\ 1 \end{bmatrix} = \begin{bmatrix} E & {}^{r}T_s + {}^{s}T_c \\ 0 & 1 \end{bmatrix} \begin{bmatrix} {}^{c}K_{xyz} \\ 1 \end{bmatrix}, \tag{12}$$

where ${}^{r}K_{xyz}$ and ${}^{c}K_{xyz}$ denote 3D coordinates of K in the two reference frames, respectively.

Using formula (12) the following can be obtained:

$$ {}^{s}T_c = {}^{r}K_{xyz} - {}^{c}K_{xyz} - {}^{r}T_s. \tag{13}$$

In that, ${}^{r}T_s$ can be directly acquired from the state of spray mechanical arm. K can be some special point on a regular object (such as center point and centroid point of square and midpoint of edge), while ${}^{r}K_{xyz}$ can be obtained by measuring and ${}^{c}K_{xyz}$ can be calculated according to the method provided in Section 3.2.2. This study selected several K points and used least-square method to improve computational accuracy when practically calculating ${}^{s}T_c$.

3.4. Visual Servo Method. The processing of the spray mechanical arm guiding nozzle to complete spraying can be divided into reset stage, alignment stage, and hovering spray stage while the movable platform is moving forward. Reset stage begins at the end of spraying former crop and stops when the distance between the robot and target crop is equal to a set value which should ensure target crop completely into FOV of eye-in-hand camera. During the stage, X linear actuator returns to positive limit position, Y linear actuator returns to middle position, and Z linear actuator returns to upper limit position. Alignment stage begins at the end of reset stage and stops when nozzle moves to spraying position above the canopy of target crop. During the stage, X linear actuator shall be fixed, Z linear actuator moves downward, and Y linear actuator moves to target position. During hovering spray stage, X linear actuator goes back, and Y and Z linear actuators track target position, to make the nozzle hover over the spraying position until the end of spray process.

During reset stage, the system acquires crop location information in real time from scene camera images to control the spray mechanical arm. At the end of reset stage, the target crop has completely entered FOV of eye-in-hand camera, and control information will be switched and provided by eye-in-hand camera images. That is to say, eye-in-hand camera images will be used for accurate and real-time control of nozzle during alignment and hovering spray stages.

3.5. Speed Control of Three Electric Cylinders at Each Stage

3.5.1. Reset Stage. Let V denote the forward speed of the movable platform. d_x, d_y, and d_z denote the reset distances of three electric cylinders detected by the position sensor, respectively. Also let l denote X direction space between the centroid of target crop canopy and the movable platform detected by visual system, and D_s denotes the set space between them at the end of reset stage. Then, the reset speeds of three linear actuators are, respectively,

$$v_{cx} \geq d_x \cdot \frac{V}{(l - D_s)},$$
$$v_{cy} \geq d_y \cdot \frac{V}{(l - D_s)}, \quad (14)$$
$$v_{cz} \geq d_z \cdot \frac{V}{(l - D_s)}.$$

3.5.2. Alignment Stage. X linear actuator is fixed, so

$$v_{cx} = 0. \quad (15)$$

The nozzle moves to target position with the movable platform and its X direction movement speed is

$$v_{sx} = \frac{(X_{T-1} - X_T)}{T_c}. \quad (16)$$

The time spent for nozzle arriving at target position is $t_{sx} = X_T / v_{sx} = X_T \cdot T_c / (X_{T-1} - X_T)$.

In that, T_c denotes sampling period and X_{T-1} and X_T denote X coordinate of spray target position in nozzle reference frame at $T - 1$ and T moment, respectively.

When the movable platform moves for t_{sx} again after $t_{sx} \leq T_c$, it can be judged that the nozzle has arrived at target position in X direction and alignment stage is over. The Y and Z linear actuators should make the nozzle arrive at target position in Y and Z direction before the end of alignment stage, respectively. The speeds are

$$v_{cy} = \begin{cases} 0 & \text{if } \Delta_{yT} \in [-\delta_y \ \ \delta_y] \\ k \cdot \dfrac{\Delta_{yT}}{t_{sx}} = k \cdot \Delta_{yT} \cdot \dfrac{(X_{T-1} - X_T)}{(X_T \cdot T_c)} & \text{else,} \end{cases}$$
$$v_{cz} = \begin{cases} 0 & \text{if } \Delta_{zT} \in [-\delta_z \ \ \delta_z] \\ k \cdot \dfrac{\Delta_{zT}}{t_{sx}} = k \cdot \Delta_{zT} \cdot \dfrac{(X_{T-1} - X_T)}{(X_T \cdot T_c)} & \text{else.} \end{cases} \quad (17)$$

In the above, k is a scale factor, in order to make nozzle's Y and Z directions arrive at target position in advance of its X direction. Take $k \geq 1.1$; Δ_{yT} and Δ_{zT} denote Y and Z direction deviation of nozzle's position relative to target position at T moment, and $[-\delta_y \ \ \delta_y]$ and $[-\delta_z \ \ \delta_z]$ are allowable variation range of Y and Z direction deviation, respectively.

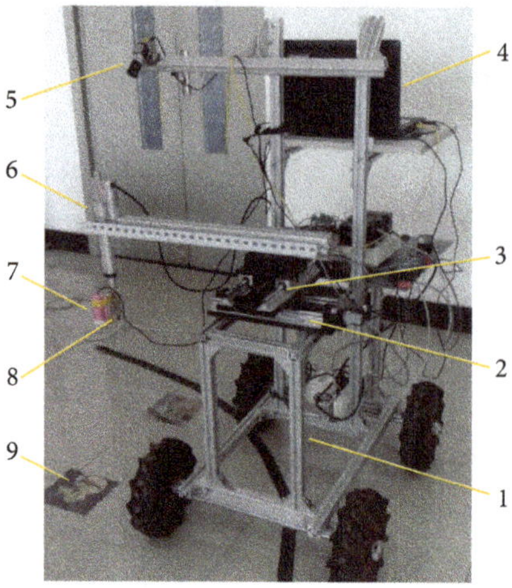

(1) Frame
(2) Y linear actuator
(3) X linear actuator
(4) Portable computer
(5) Scene camera
(6) Z linear actuator
(7) Eye-in-hand camera
(8) Laser light
(9) Crop picture

FIGURE 9: Experimental prototype.

3.5.3. Hovering Spray Stage. To ensure nozzle hover over the target position during the process of spraying, the astern speed of X linear actuator should be matched with movable platform's forward speed; Δ_{yT} and Δ_{zT} should be restricted within corresponding allowable ranges $[-\delta_y \ \ \delta_y]$ and $[-\delta_z \ \ \delta_z]$ by controlling Y and Z linear actuators. The speeds of three linear actuators are, respectively,

$$v_{cx} = \frac{(X_{T-1} - X_T)}{T_c},$$
$$v_{cy} = \begin{cases} 0 & \text{if } \Delta_{yT} \in [-\delta_y \ \ \delta_y] \\ k_1 \cdot \dfrac{(Y_{T-1} - Y_T)}{T_c} & \text{else,} \end{cases} \quad (18)$$
$$v_{cz} = \begin{cases} 0 & \text{if } \Delta_{zT} \in [-\delta_z \ \ \delta_z] \\ k_1 \cdot \dfrac{(Z_{T-1} - Z_T)}{T_c} & \text{else,} \end{cases}$$

where k_1 is another scale factor, and take $k_1 \geq 1.5$. Y_{T-1}, Z_{T-1}, Y_T, and Z_T denote Y and Z coordinates of spray target position in nozzle reference frame at $T - 1$ and T moment, respectively.

4. Experiment Verification

A simplified prototype containing one spray mechanical arm (as shown in Figure 9) was built to demonstrate the feasibility of the design. The stroke of X and Y linear actuator (01TS, Jike Instrument Co., Beijing, China) of the spray mechanical arm was 200 mm, and that of Z linear actuator (XC800, Xunchi Electric Co., Zhejiang, China) was 150 mm. The scene camera (acA1300, Basler AG, Ahrensburg,

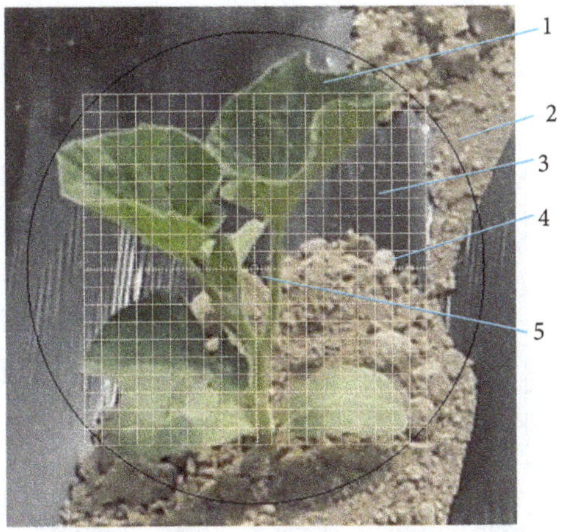

(1) Crop (4) Ruler
(2) MEC (5) Center of the MEC
(3) Grids

FIGURE 10: Printed crop picture.

Germany), which had a resolution of 1296 × 966 pixels and a frame rate of 30 fps, was mounted on the top right of the frame with its optical axis declining 46 degrees. The eye-in-hand camera (a1200, Lanseyaoji Co., Jiangsu, China), which had a resolution of 640 × 480 pixels and a frame rate of 30 fps, was vertically mounted on the extending end of Z linear actuator. A quad processing (Core i5) portable computer was used as the main image-processing computer. A low-cost microcontroller (STC89C52RC) was used as a controller to control the three linear actuators and all sensors. The link between PC and controller was a RS-232 serial communication line. To measure spray position more easily, the nozzle was replaced with a laser light (SU6505, Sulei Laser Technology Co., Guangdong, China), and the crop was replaced with printed crop pictures, in which there were marked MEC, ruler, and grids (as shown in Figure 10). The ruler's origin was the center of the MEC and its index value was 1 mm.

For spray process, spray position and spray time are important indicators, the accuracy of which was tested through prototype experiments under controlled environment in laboratory. In the experiments, expected spray position of nozzle in X and Y directions coincides with the center of MEC in crop picture, and actual spray position in X and Y directions can be obtained by measuring the spot of laser light dropped on crop picture. Expected spray position of nozzle in Z direction is calculated by $h = 5.4r$ (origin: according to formula (1), let $S_0 = \pi \cdot r^2$, suppose $\xi = 0.8$, $\alpha = 45^\circ$, and get $h = 2.7r$, due to the limitations of prototype structure, which need to be multiplied by a factor of 2; then we get $h = 5.4r$), and actual spray position in Z direction is provided by the position sensor of Z linear actuator. Expected spray time of nozzle is calculated by $t = 75\pi \cdot r^2$ (origin: according to formula (2), let $q/Q = 75\,\text{s/m}^2$), and actual spray time is obtained by software system through the difference between starting and stopping of spraying.

Five crop pictures of spacing placement were as experimental objects, with space about 400~600 mm in X direction and 30~60 mm in Y direction. The experiments, measuring spray position and spray time of each crop picture, had been repeated for 20 times in the same circumstances. During the experiments, actual spray position in X and Y directions had some fluctuation, and the center of fluctuation range was treated as actual spray position. Because their desired value can be regarded as 0, the deviation (actual value − expected value) is equal to actual value. The deviation of spray position in X, Y, and Z directions is shown in Figure 11(a), and we can think that the positioning accuracies of prototype in X, Y, and Z directions are about ±6 mm, ±3.8 mm, and ±1.9 mm, respectively. Figure 11(b) shows the fluctuation quality of actual spray position in X and Y directions. The fluctuation in X direction is obviously larger than that in Y direction, which was mainly related to velocity fluctuation of the motion platform. Figures 11(c) and 11(d) compare average spray position with expected position in X, Y directions and in Z direction, respectively. And we can see that the deviation of average spray position relative to expected position in X, Y, and Z directions is about (−2.5 mm~1 mm), (−0.5 mm~1 mm), and (−1 mm~1 mm), respectively. Figure 11(e) shows the deviation of actual spray time relative to expected spray time, and we can think that the spray time accuracy is about ±0.17 s. Figure 11(f) compares average spray time with expected time, and the deviation of average spray time relative to expected time is about (−0.05 s~0.07 s). Moreover, the switching process between scene visual servo and eye-in-hand visual servo is fluent and real-time performance of the system is good during the whole experiments.

5. Conclusions

(1) A schematic design of target spraying robotic system used for spraying onto the canopy of interval planting crop was proposed. And composite vision servo method based on monocular scene vision and monocular eye-in-hand vision and double closed-loop control structure for the system were studied.

(2) On the basis of image segmentation and feature extraction, methods of roughly positioning target based on single scene camera and positioning spray location based on single eye-in-hand camera were studied. A kind of mechanical arm servo control method which divided the process of spray into reset stage, alignment stage, and hovering spray stage was proposed.

(3) The experimental results showed that the system exhibits higher control accuracy and acceptable real-time performance. The switch between scene visual servo and eye-in-hand visual servo was smooth. Using this system, combined spraying and moving and accurately spraying on target can be achieved.

Competing Interests

The authors declare that there is no conflict of interests regarding the publication of this paper.

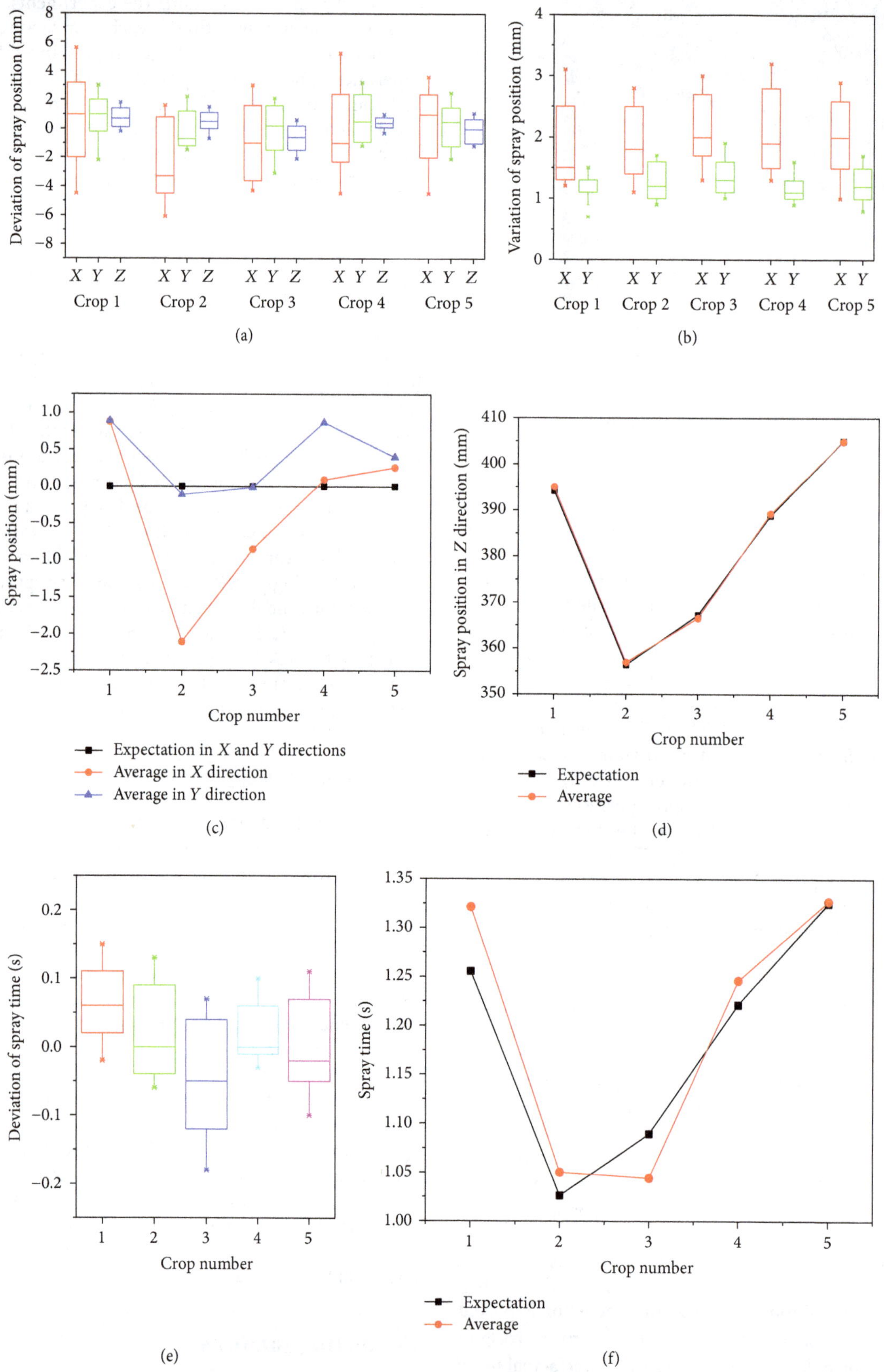

FIGURE 11: Experimental results.

Acknowledgments

This work was supported by the Specialized Research Fund for the Doctoral Program of Higher Education, China (Grant no. 20120008110046), and the Natural Science Foundation of Shandong Province, China (Grant no. ZR2012CQ026), and the Scientific Research Fund of Liaocheng University, China (Grant no. 318011519).

References

[1] B. Qiu, R. Yan, J. Ma, X. Guan, and M. Ou, "Research progress analysis of variable rate sprayer technology," *Transactions of the Chinese Society for Agricultural Machinery*, vol. 46, no. 3, pp. 59–72, 2015.

[2] X. He, "Improving severe draggling actuality of plant protection machinery and its application techniques," *Transactions of the Chinese Society of Agricultural Engineering*, vol. 20, no. 1, pp. 13–15, 2004.

[3] M. Kise, Q. Zhang, and F. Rovira Más, "A stereovision-based crop row detection method for tractor-automated guidance," *Biosystems Engineering*, vol. 90, no. 4, pp. 357–367, 2005.

[4] W. Deng, X.-K. He, L.-D. Zhang, A.-J. Zeng, J.-L. Song, and J.-J. Zou, "Target infrared detection in target spray," *Spectroscopy and Spectral Analysis*, vol. 28, no. 10, pp. 2285–2289, 2008.

[5] J. R. Rosell, J. Llorens, R. Sanz et al., "Obtaining the three-dimensional structure of tree orchards from remote 2D terrestrial LIDAR scanning," *Agricultural and Forest Meteorology*, vol. 149, no. 9, pp. 1505–1515, 2009.

[6] D. K. Giles and D. C. Slaughter, "Precision band spraying with machine-vision guidance and adjustable yaw nozzles," *Transactions of the ASAE*, vol. 40, no. 1, pp. 29–36, 1997.

[7] L. Tian, J. F. Reid, and J. W. Hummel, "Development of a precision sprayer for site-specific weed management," *Transactions of the American Society of Agricultural Engineers*, vol. 42, no. 4, pp. 893–900, 1999.

[8] L. Tian, "Development of a sensor-based precision herbicide application system," *Computers and Electronics in Agriculture*, vol. 36, no. 2-3, pp. 133–149, 2002.

[9] B. L. Steward and L. F. Tang, "Distance-based control system for machine vision-based selective spraying," *Transactions of the ASAE*, vol. 45, no. 5, pp. 1255–1262, 2002.

[10] Y. Li, C. Xia, and J. Lee, "Vision-based pest detection and automatic spray of greenhouse plant," in *Proceedings of the IEEE International Symposium on Industrial Electronics*, pp. 909–914, Seoul, South Korea, July 2009.

[11] J. Zhang, Z. Cao, C. Geng, and W. Li, "Research on precision target spray robot in greenhouse," *Transactions of the Chinese Society of Agricultural Engineering*, vol. 25, supplement 2, pp. 70–73, 2009.

[12] Z. Cao, J. Zhang, C. Geng, and W. Li, "Control system of target spraying robot in greenhouse," *Transactions of the Chinese Society of Agricultural Engineering*, vol. 26, no. 2, pp. 228–233, 2010.

[13] G. Yufeng, Z. Hongping, Z. Jiaqiang et al., "Indoor pesticide smart spraying system based on machine vision," *Transactions of the Chinese Society of Agricultural Machinery*, vol. 36, no. 3, pp. 86–89, 2005.

[14] Y. Fang, "A survey of robot visual servoing," *CAAI Transactions on Intelligent Systems*, vol. 3, no. 2, pp. 109–113, 2008.

[15] L. Wang, D. Xu, and M. Tan, "Survey of research on robotic visual servoing," *Robot*, vol. 26, no. 3, pp. 277–282, 2004.

[16] T. Kataoka, T. Kaneko, H. Okamoto, and S. Hata, "Crop growth estimation system using machine vision," in *Proceedings of the IEEE/ASME International Conference on Advanced Intelligent Mechatronics (AIM '03)*, vol. 2, pp. b1079–b1083, IEEE, Kobe, Japan, July 2003.

[17] N. Otsu, "A threshold selection method from gray-level histogram," *IEEE Transactions on Systems, Man, and Cybernetics*, vol. 9, no. 1, pp. 62–66, 1979.

[18] J. Romeo, G. Pajares, M. Montalvo, J. M. Guerrero, M. Guijarro, and J. M. de la Cruz, "A new Expert System for greenness identification in agricultural images," *Expert Systems with Applications*, vol. 40, no. 6, pp. 2275–2286, 2013.

[19] Z. Y. Zhang, "Flexible camera calibration by viewing a plane from unknown orientations," in *Proceedings of the 7th IEEE International Conference on Computer Vision (ICCV '99)*, pp. 666–673, Kerkyra, Greece, September 1999.

[20] Z. Y. Zhang, "Camera calibration with one-dimensional objects," *IEEE Transactions on Pattern Analysis and Machine Intelligence*, vol. 26, no. 7, pp. 892–899, 2004.

15

Dynamic Surface Adaptive Robust Control of Unmanned Marine Vehicles with Disturbance Observer

Pengchao Zhang

Key Laboratory of Industrial Automation of Shaanxi Province, Shaanxi University of Technology, Hanzhong, Shaanxi 723000, China

Correspondence should be addressed to Pengchao Zhang; snutzpc@126.com

Academic Editor: Keigo Watanabe

This paper presents a dynamic surface adaptive robust control method with disturbance observer for unmanned marine vehicles (UMV). It uses adaptive law to estimate and compensate the disturbance observer error. Dynamic surface is introduced to solve the "differential explosion" caused by the virtual control derivation in traditional backstepping method. The final controlled system is proved to be globally uniformly bounded based on Lyapunov stability theory. Simulation results illustrate the effectiveness of the proposed controller, which can realize the three-dimensional trajectory tracking for UMV with the systematic uncertainty and time-varying disturbances.

1. Introduction

Unmanned marine vehicle (UMV) has attracted a number of researchers from all over the world. For the high nonlinearity and the characteristic of being easy to be disturbed by external environment, the control of the UMV is challenged especially for trajectory. With the continuous development of the military and marine economy, UMV needs to complete more complicated tasks accurately. To explore the new nonlinear control strategy for UMV position and trajectory is of great theoretical and practical significance [1–5].

Due to the nonlinear characteristics of the UMV, the backstepping and Lyapunov theory are combined to solve the problem of trajectory tracking control. In [1], considering the Coriolis force and damping force, the backstepping control law is proposed and proved the global exponential stability by the Lyapunov theory. The method of adaptive backstepping is given to design the trajectory tracking controller in view of the slow disturbance from external environment in [2].

Sliding mode control characterized by rapid response, simplicity, and high robustness is used widely for the trajectory control of marine vehicles [3, 4]. But the performance is declined by the high-frequency vibration for the switch mode in basic sliding mode. The equipment can be destroyed in the serious case. So, the saturation function and dead zone correction are presented to eliminate the chattering [5, 6]. In [7], an adaptive fuzzy backstepping is studied to guarantee the semiglobal congruent eventually boundedness of the close-loop system.

UMV trajectory is inevitably influenced by the external environment in the navigation. The disturbance observer can estimate the external disturbance of the system and observe its characteristics. A nonlinear disturbance observer is adopted for the forward compensation to decline the switch gain of the backstepping controller in [8]. The boundary layer adaptive sliding mode controller based on disturbance observer is also effective in eliminating the chattering for the uncertainty and disturbance in [9]. Aschemann [10] proposed two variable gain feedback nonlinear control methods based on extended linearization technology. Yang et al. [11] give the ship trajectory tracking controller which can resist time-varying environmental disturbance based on disturbance observation, backstepping, and Lyapunov theory. For the "number explosion" in traditional backstepping, Swaroop et al. [12] put forward the dynamic surface control.

This paper presents a composite dynamic surface adaptive robust control method for UMV with disturbance observer to design the position, attitude, and time-varying velocity controllers. The dynamic surface adaptive robust controller is designed for the UMV with disturbance observer. The disturbance observer is for estimating external unknown disturbance and the forward control is for compensation to

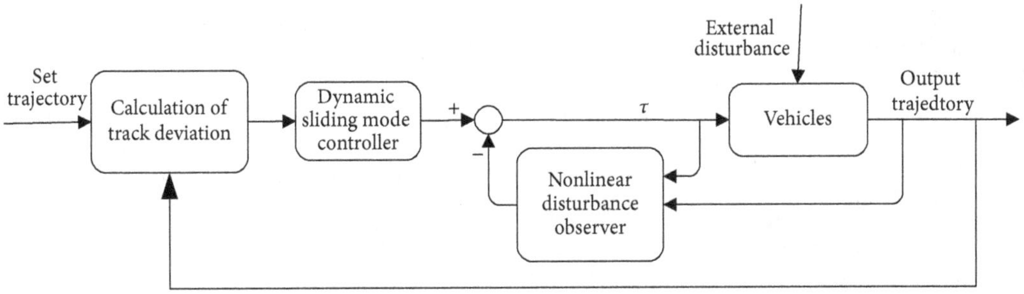

FIGURE 1: Schematic of the control system for UMV.

weaken buffeting. The limit of the disturbance observer error is estimated by the adaptive law. The final controlled system is proved to be globally uniformly bounded based on Lyapunov stability theory.

The remaining parts of this paper are organized as follows. In Section 2, the model of UMV is proposed. In Section 3, the controller design and analysis are shown in detail. At last, the simulation results show the effectiveness of the method in Section 4 and the conclusion is drawn in Section 5.

2. Model of the UMV

For a type of UMV, the mathematical model can be expressed as

$$\begin{aligned}\dot{\eta} &= J(\psi)\nu \\ M\dot{\nu} &= \tau + d - C(\nu)\nu - D(\nu)\nu,\end{aligned} \quad (1)$$

where $\eta = [x, y, \psi]^T$ is the vector of position (x, y) and yaw angle ψ in the ground coordinate, $\nu = [u, v, r]^T$ is the vector of advance velocity u, drifting velocity v, and yaw angular velocity r in hull coordinate. M is the inertial matrix including the added mass. $J(\psi)$ is the transformation matrix and $\mathbf{J}^{-1}(\psi) = \mathbf{J}^T(\psi)$, $\|\mathbf{J}(\psi)\| = 1$:

$$\mathbf{J}(\psi) = \begin{bmatrix} \cos\psi & -\sin\psi & 0 \\ \sin\psi & \cos\psi & 0 \\ 0 & 0 & 1 \end{bmatrix}. \quad (2)$$

$C(\nu)$ is the matrix of centripetal force and the Coriolis force. $D(\nu)$ is the hydrodynamic resistance and the lifting force moment. $\tau = [\tau_1, \tau_2, \tau_3]^T$ is the control input vector and τ_1 is advance force, τ_2 is drifting force, and τ_3 is yaw force. $d = [d_1, d_2, d_3]^T$ is the external disturbance force and d_1 is lateral disturbance, d_2 is vertical disturbance, and d_3 is yaw disturbance.

Assumptions of the UMV are as follows:

(a) The reference trajectory $\boldsymbol{\eta}_d$ is smooth, derivable, and bounded.

(b) The disturbance is bounded and $\|\dot{\mathbf{d}}(t)\| \le C_d < \infty$.

3. Design of the UMV Tracking Controller

3.1. Design of the Nonlinear Disturbance Observer. The schematic of the UMV tracking control system is shown in Figure 1. The UMV trajectory is subjected to the environmental disturbances which influence the position and velocity from the expected values. The composite dynamic sliding mode control system can calculate the perfect forces and moment.

Considering the nonlinear kinematic and dynamic equations of the UMV, the nonlinear disturbance observer is designed as

$$\hat{\mathbf{d}} = \boldsymbol{\beta} + \mathbf{K}_0\mathbf{M}\boldsymbol{\nu} \quad (3)$$

$$\dot{\boldsymbol{\beta}} = -\mathbf{K}_0\boldsymbol{\beta} - \mathbf{K}_0\left[-\mathbf{C}(\boldsymbol{\nu})\boldsymbol{\nu} - \mathbf{D}\boldsymbol{\nu} + \boldsymbol{\tau} + \mathbf{K}_0\mathbf{M}\boldsymbol{\nu}\right], \quad (4)$$

where $\hat{\mathbf{d}} = [\hat{d}_1, \hat{d}_2, \hat{d}_3]^T \in \mathbf{R}^3$ is a vector of the disturbance estimate of the observer output and $\mathbf{K}_0 \in \mathbf{R}^{3\times3}$ is a positive definite diagonal matrix, and $\boldsymbol{\beta} \in \mathbf{R}^3$ is the intermediate vector for the design.

Define observation error vector $\tilde{\mathbf{d}} = [\tilde{d}_1, \tilde{d}_2, \tilde{d}_3]^T$:

$$\tilde{\mathbf{d}} = \mathbf{d} - \hat{\mathbf{d}}. \quad (5)$$

Through the derivation of (3) combined with (1), (4), and (5), there is

$$\dot{\hat{\mathbf{d}}} = \dot{\boldsymbol{\beta}} + \mathbf{K}_0\mathbf{M}\dot{\boldsymbol{\nu}} = \mathbf{K}_0\left[\mathbf{d} - (\boldsymbol{\beta} + \mathbf{K}_0\mathbf{M}\boldsymbol{\nu})\right] = \mathbf{K}_0\tilde{\mathbf{d}} \quad (6)$$

$$\dot{\tilde{\mathbf{d}}} = \dot{\mathbf{d}} - \dot{\hat{\mathbf{d}}} = \dot{\mathbf{d}} - \mathbf{K}_0\tilde{\mathbf{d}}. \quad (7)$$

3.2. Design of Dynamic Surface Adaptive Robust Controller. The control method is based on dynamic surface combined with sliding backstepping. There are two steps in the process.

Step 1. Define the position error vector $\mathbf{z}_1 \in \mathbf{R}^3$:

$$\mathbf{z}_1 = \boldsymbol{\eta} - \boldsymbol{\eta}_d, \quad (8)$$

where $\boldsymbol{\eta}_d = [x_d, y_d, \psi_d]^T$ is the vector of the expected position values and yaw angles.

Through the derivation of (8), there is

$$\dot{\mathbf{z}}_1 = \mathbf{J}(\psi)\boldsymbol{\nu} - \dot{\boldsymbol{\eta}}_d. \quad (9)$$

Define virtual vector $\boldsymbol{\alpha}_1 \in \mathbf{R}^3$ and

$$\boldsymbol{\alpha}_1 = \mathbf{J}^{-1}(\psi)(-\mathbf{K}_1\mathbf{z}_1 + \dot{\boldsymbol{\eta}}_d), \quad (10)$$

where $\mathbf{K}_1 \in \mathbf{R}^{3\times3}$ is the positive definite diagonal matrix.

To avoid the "differential explosion" by virtual control derivation in traditional backstepping control, the first-order filter is introduced according to dynamic surface control. $\boldsymbol{\nu}_d \in \mathbf{R}^3$ is the output of the first-order filter of virtual control vector $\boldsymbol{\alpha}_1$. That is,

$$T\dot{\boldsymbol{\nu}}_d + \boldsymbol{\nu}_d = \boldsymbol{\alpha}_1, \qquad \boldsymbol{\nu}_d(0) = \boldsymbol{\alpha}_1(0), \tag{11}$$

where T is the time constant of the filter.

Step 2. Define the velocity error vector $\mathbf{z}_2 \in \mathbf{R}^3$ as

$$\mathbf{z}_2 = \boldsymbol{\nu} - \boldsymbol{\nu}_d. \tag{12}$$

According to (1) and (12), there is

$$\mathbf{M}\dot{\mathbf{z}}_2 = -\mathbf{C}(\boldsymbol{\nu})\boldsymbol{\nu} - \mathbf{D}\boldsymbol{\nu} + \boldsymbol{\tau} + \mathbf{d} - \mathbf{M}\dot{\boldsymbol{\nu}}_d. \tag{13}$$

Define the filter error vector $\mathbf{y} \in \mathbf{R}^3$ as

$$\mathbf{y} = \boldsymbol{\nu}_d - \boldsymbol{\alpha}_1. \tag{14}$$

Design the state feedback control law as

$$\boldsymbol{\tau} = \mathbf{C}(\boldsymbol{\nu})\boldsymbol{\nu} + \mathbf{D}\boldsymbol{\nu} + \mathbf{M}\dot{\boldsymbol{\nu}}_d - \mathbf{K}_2\mathbf{z}_2 - \hat{\mathbf{d}} - \Xi\hat{\boldsymbol{\delta}}, \tag{15}$$

where $\mathbf{K}_2 \in \mathbf{R}^{3\times3}$ is the designed positive definite diagonal matrix and $\Xi = \text{diag}\{\tanh(z_{21}/\varepsilon_1), \tanh(z_{22}/\varepsilon_2), \tanh(z_{23}/\varepsilon_3)\} \in \mathbf{R}^3$. ε_i is the positive constant. z_{2i} $(i = 1,2,3)$ is the weight of $\mathbf{z}_2$. $\hat{\boldsymbol{\delta}} = [\hat{\delta}_1, \hat{\delta}_2, \hat{\delta}_3]^{\mathrm{T}} \in \mathbf{R}^3$ is the vector of the upper limit of uncertain item $\boldsymbol{\delta}$.

Design the adaptive law of $\hat{\boldsymbol{\delta}}$ as

$$\dot{\hat{\boldsymbol{\delta}}} = \boldsymbol{\Gamma}\left[\Xi\mathbf{z}_2 - \boldsymbol{\Lambda}\left(\hat{\boldsymbol{\delta}} - \hat{\boldsymbol{\delta}}^0\right)\right], \tag{16}$$

where $\boldsymbol{\Gamma} = \text{diag}\{\gamma_1, \gamma_2, \gamma_3\} \in \mathbf{R}^{3\times3}$ and $\boldsymbol{\Lambda} = \text{diag}\{\sigma_1, \sigma_2, \sigma_3\} \in \mathbf{R}^{3\times3}$ are the positive definite diagonal matrix. $\hat{\boldsymbol{\delta}}^0 = [\hat{\delta}_1^{\ 0}, \hat{\delta}_2^{\ 0}, \hat{\delta}_3^{\ 0}]^{\mathrm{T}} \in \mathbf{R}^3$ is the a priori estimate of $\hat{\boldsymbol{\delta}}_i$ $(i = 1,2,3)$.

3.3. Analysis of Stability. Choose the Lyapunov function as

$$V = \frac{1}{2}\mathbf{z}_1^{\mathrm{T}}\mathbf{z}_1 + \frac{1}{2}\mathbf{z}_2^{\mathrm{T}}\mathbf{M}\mathbf{z}_2 + \frac{1}{2}\mathbf{y}^{\mathrm{T}}\mathbf{y} + \frac{1}{2}\tilde{\boldsymbol{\delta}}^{\mathrm{T}}\boldsymbol{\Gamma}^{-1}\tilde{\boldsymbol{\delta}} + \frac{1}{2}\tilde{\mathbf{d}}^{\mathrm{T}}\tilde{\mathbf{d}}, \tag{17}$$

where $\tilde{\boldsymbol{\delta}} = \hat{\boldsymbol{\delta}} - \boldsymbol{\delta}$ is the defined error estimation vector. Through the derivation of (17), there is

$$\dot{V} = \mathbf{z}_1^{\mathrm{T}}\dot{\mathbf{z}}_1 + \mathbf{z}_2^{\mathrm{T}}\mathbf{M}\dot{\mathbf{z}}_2 + \mathbf{y}^{\mathrm{T}}\dot{\mathbf{y}} + \tilde{\boldsymbol{\delta}}^{\mathrm{T}}\boldsymbol{\Gamma}^{-1}\dot{\hat{\boldsymbol{\delta}}} + \tilde{\mathbf{d}}^{\mathrm{T}}\dot{\tilde{\mathbf{d}}}. \tag{18}$$

For the first item $\mathbf{z}_1^{\mathrm{T}}\dot{\mathbf{z}}_1$ of (18), considering (9), (10), (12), (13), (14), (15), and $\|\mathbf{J}(\psi)\| = 1$, we have

$$\begin{aligned}\mathbf{z}_1^{\mathrm{T}}\dot{\mathbf{z}}_1 &= \mathbf{z}_1^{\mathrm{T}}\left[\mathbf{J}(\psi)(\mathbf{z}_2 + \mathbf{y} + \boldsymbol{\alpha}_1) - \dot{\boldsymbol{\eta}}_d\right] \\ &= -\mathbf{z}_1^{\mathrm{T}}\mathbf{K}_1\mathbf{z}_1 + \mathbf{z}_1^{\mathrm{T}}\mathbf{J}(\psi)\mathbf{z}_2 + \mathbf{z}_1^{\mathrm{T}}\mathbf{J}(\psi)\mathbf{y} \\ &\le -\mathbf{z}_1^{\mathrm{T}}\mathbf{K}_1\mathbf{z}_1 + a_1\mathbf{z}_1^{\mathrm{T}}\mathbf{z}_1 + \frac{\mathbf{z}_2^{\mathrm{T}}\mathbf{z}_2}{4a_1} + a_2\mathbf{z}_1^{\mathrm{T}}\mathbf{z}_1 + \frac{\mathbf{y}^{\mathrm{T}}\mathbf{y}}{4a_2},\end{aligned} \tag{19}$$

where a_1 and a_2 are positive constants.

For the second item $\mathbf{z}_2^{\mathrm{T}}\mathbf{M}\dot{\mathbf{z}}_2$ in (18), considering (13) and (15), there is

$$\begin{aligned}\mathbf{z}_2^{\mathrm{T}}\mathbf{M}\dot{\mathbf{z}}_2 &= \mathbf{z}_2^{\mathrm{T}}\left[-\mathbf{C}(\boldsymbol{\nu})\boldsymbol{\nu} - \mathbf{D}\boldsymbol{\nu} + \mathbf{C}(\boldsymbol{\nu})\boldsymbol{\nu} + \mathbf{D}\boldsymbol{\nu} + \mathbf{M}\dot{\boldsymbol{\nu}}_d \right. \\ &\quad \left. - \mathbf{K}_2\mathbf{z}_2 - \hat{\mathbf{d}} - \Xi\hat{\boldsymbol{\delta}} + \mathbf{d} - \mathbf{M}\dot{\boldsymbol{\nu}}_d\right] = -\mathbf{z}_2^{\mathrm{T}}\mathbf{K}_2\mathbf{z}_2 - \mathbf{z}_2^{\mathrm{T}}\Xi\tilde{\boldsymbol{\delta}} \\ &\quad + \mathbf{z}_2^{\mathrm{T}}\tilde{\mathbf{d}}.\end{aligned} \tag{20}$$

For the third item $\mathbf{y}^{\mathrm{T}}\dot{\mathbf{y}}$ in (18), there is

$$\begin{aligned}\dot{\mathbf{y}} &= \dot{\mathbf{v}}_d - \dot{\boldsymbol{\alpha}}_1 \\ &= -\frac{\mathbf{y}}{T} + \dot{\mathbf{J}}^{-1}(\psi)\mathbf{K}_1\mathbf{z}_1 + \mathbf{J}^{-1}(\psi)\mathbf{K}_1\dot{\mathbf{z}}_1 - \dot{\mathbf{J}}^{-1}(\psi)\dot{\boldsymbol{\eta}}_d \\ &\quad - \mathbf{J}^{-1}(\psi)\ddot{\boldsymbol{\eta}}_d.\end{aligned} \tag{21}$$

Consider the compact set

$$\begin{aligned}\Omega_1 &= \left\{\left[\mathbf{z}_1^{\mathrm{T}}, \mathbf{z}_2^{\mathrm{T}}, \mathbf{y}^{\mathrm{T}}\right]^{\mathrm{T}} : V \le \varpi_0\right\}, \\ \Omega_d &= \left\{\left[\boldsymbol{\eta}_d^{\ \mathrm{T}}, \dot{\boldsymbol{\eta}}_d^{\ \mathrm{T}}, \ddot{\boldsymbol{\eta}}_d^{\ \mathrm{T}}\right]^{\mathrm{T}} : \|\boldsymbol{\eta}_d\|^2 + \|\dot{\boldsymbol{\eta}}_d\|^2 + \|\ddot{\boldsymbol{\eta}}_d\|^2 \le B_0\right\},\end{aligned} \tag{22}$$

where B_0 and ϖ_0 are given positive constants. For $\Omega_1 \times \Omega_d$ is also compact set, there are nonnegative continuous functions $\beta(\cdot)$ and the maximum of $\beta(\cdot)$ is N.

$$\begin{aligned}&\left\|\dot{\mathbf{y}} + \frac{\mathbf{y}}{T}\right\| \le \beta(\mathbf{z}_1, \mathbf{z}_2, \mathbf{y}, \boldsymbol{\eta}_d, \dot{\boldsymbol{\eta}}_d, \ddot{\boldsymbol{\eta}}_d) \\ &\mathbf{y}^{\mathrm{T}}\dot{\mathbf{y}} = -\frac{\mathbf{y}^{\mathrm{T}}\mathbf{y}}{T} + \frac{\mathbf{y}^{\mathrm{T}}\mathbf{y}}{T} + \mathbf{y}^{\mathrm{T}}\dot{\mathbf{y}} = -\frac{\mathbf{y}^{\mathrm{T}}\mathbf{y}}{T} + \mathbf{y}^{\mathrm{T}}\left(\frac{\mathbf{y}}{T} + \dot{\mathbf{y}}\right) \\ &\quad \le -\frac{\mathbf{y}^{\mathrm{T}}\mathbf{y}}{T} + \alpha_3\mathbf{y}^{\mathrm{T}}\mathbf{y} + \frac{N^2}{4\alpha_3},\end{aligned} \tag{23}$$

where α_3 is positive constant.

For the fourth item $\tilde{\boldsymbol{\delta}}^{\mathrm{T}}\boldsymbol{\Gamma}^{-1}\dot{\hat{\boldsymbol{\delta}}}$ of (18) and considering (16), there is

$$\begin{aligned}\tilde{\boldsymbol{\delta}}^{\mathrm{T}}\boldsymbol{\Gamma}^{-1}\dot{\hat{\boldsymbol{\delta}}} &= \tilde{\boldsymbol{\delta}}^{\mathrm{T}}\boldsymbol{\Gamma}^{-1}\boldsymbol{\Gamma}\left[\Xi\mathbf{z}_2 - \boldsymbol{\Lambda}\left(\hat{\boldsymbol{\delta}} - \hat{\boldsymbol{\delta}}^0\right)\right] \\ &= \mathbf{z}_2^{\mathrm{T}}\Xi\tilde{\boldsymbol{\delta}} - \left(\hat{\boldsymbol{\delta}} - \boldsymbol{\delta}\right)^{\mathrm{T}}\boldsymbol{\Lambda}\left(\hat{\boldsymbol{\delta}} - \hat{\boldsymbol{\delta}}^0\right) \\ &\le \mathbf{z}_2^{\mathrm{T}}\Xi\tilde{\boldsymbol{\delta}} - \frac{1}{2}\tilde{\boldsymbol{\delta}}^{\mathrm{T}}\boldsymbol{\Lambda}\tilde{\boldsymbol{\delta}} + \frac{1}{2}\left(\boldsymbol{\delta} - \hat{\boldsymbol{\delta}}^0\right)^{\mathrm{T}}\boldsymbol{\Lambda}\left(\boldsymbol{\delta} - \hat{\boldsymbol{\delta}}^0\right).\end{aligned} \tag{24}$$

For the fifth item $\tilde{\mathbf{d}}^{\mathrm{T}}\dot{\tilde{\mathbf{d}}}$ of (18), considering (7), there is

$$\begin{aligned}\tilde{\mathbf{d}}^{\mathrm{T}}\dot{\tilde{\mathbf{d}}} &= \tilde{\mathbf{d}}^{\mathrm{T}}\left(\dot{\mathbf{d}} - \mathbf{K}_0\tilde{\mathbf{d}}\right) = \tilde{\mathbf{d}}^{\mathrm{T}}\dot{\mathbf{d}} - \tilde{\mathbf{d}}^{\mathrm{T}}\mathbf{K}_0\tilde{\mathbf{d}} \\ &\le a_4\tilde{\mathbf{d}}^{\mathrm{T}}\tilde{\mathbf{d}} + \frac{C_d^{\ 2}}{4a_4} - \tilde{\mathbf{d}}^{\mathrm{T}}\mathbf{K}_0\tilde{\mathbf{d}},\end{aligned} \tag{25}$$

where a_4 is positive constant.

According to (19), (20), (23), (24), and (25), there is

$$\dot{V} \le -\mathbf{z}_1^{\mathrm{T}}\mathbf{K}_1\mathbf{z}_1 + a_1\mathbf{z}_1^{\mathrm{T}}\mathbf{z}_1 + \frac{\mathbf{z}_2^{\mathrm{T}}\mathbf{z}_2}{4a_1} + a_2\mathbf{z}_1^{\mathrm{T}}\mathbf{z}_1 + \frac{\mathbf{y}^{\mathrm{T}}\mathbf{y}}{4a_2} - \mathbf{z}_2^{\mathrm{T}}\mathbf{K}_2\mathbf{z}_2 + \mathbf{z}_2^{\mathrm{T}}\tilde{\mathbf{d}} - \frac{\mathbf{y}^{\mathrm{T}}\mathbf{y}}{T} + a_3\mathbf{y}^{\mathrm{T}}\mathbf{y} + \frac{N^2}{4a_3} - \mathbf{z}_2^{\mathrm{T}}\Xi\boldsymbol{\delta} - \frac{1}{2}\tilde{\boldsymbol{\delta}}^{\mathrm{T}}\Lambda\tilde{\boldsymbol{\delta}} + \frac{1}{2}\left(\boldsymbol{\delta} - \hat{\boldsymbol{\delta}}^0\right)^{\mathrm{T}}\Lambda\left(\boldsymbol{\delta} - \hat{\boldsymbol{\delta}}^0\right) + a_4\tilde{\mathbf{d}}^{\mathrm{T}}\tilde{\mathbf{d}} + \frac{C_d^{\,2}}{4a_4} - \tilde{\mathbf{d}}^{\mathrm{T}}\mathbf{K}_0\tilde{\mathbf{d}}. \tag{26}$$

Considering $0 \le |a| - a\tanh(a/\varepsilon) \le 0.2785\varepsilon$ (for $\varepsilon > 0$, $a \in R$) and $z_{2i}\tilde{d}_i \le |z_{2i}| \cdot |\tilde{d}_i| \le |z_{2i}|\delta_i$, we can have

$$\dot{V} \le -\left[\lambda_{\min}(\mathbf{K}_1) - a_1 - a_2\right]\mathbf{z}_1^{\mathrm{T}}\mathbf{z}_1 - \left[\lambda_{\min}(\mathbf{K}_2) - \frac{1}{4a_1}\right]\mathbf{z}_2^{\mathrm{T}}\mathbf{z}_2 - \left[\frac{1}{T} - \frac{1}{4a_2} - a_3\right]\mathbf{y}^{\mathrm{T}}\mathbf{y} - \left[\lambda_{\min}(\mathbf{K}_0) - a_4\right]\tilde{\mathbf{d}}^{\mathrm{T}}\tilde{\mathbf{d}} - \frac{1}{2}\lambda_{\min}(\Lambda\Gamma)\frac{1}{2}\tilde{\boldsymbol{\delta}}^{\mathrm{T}}\Gamma^{-1}\tilde{\boldsymbol{\delta}} + 0.2785\mathbf{E}^{\mathrm{T}}\boldsymbol{\delta} + \frac{N^2}{4a_3} + \frac{1}{2}\left(\boldsymbol{\delta} - \hat{\boldsymbol{\delta}}^0\right)^{\mathrm{T}}\Lambda\left(\boldsymbol{\delta} - \hat{\boldsymbol{\delta}}^0\right) + \frac{C_d^{\,2}}{4a_4} \le -\mu V + C, \tag{27}$$

where

$$\mathbf{E} = \left[\varepsilon_1, \varepsilon_2, \varepsilon_3\right]^{\mathrm{T}} \tag{28}$$

$$\mu = \min\left\{2\left[\lambda_{\min}(\mathbf{K}_1) - a_1 - a_2\right], 2\left[\frac{1}{T} - \frac{1}{4a_2} - a_3\right], \frac{2\left[\lambda_{\min}(\mathbf{K}_2) - 1/4a_1\right]}{\lambda_{\max}(\mathbf{M})}, \lambda_{\min}(\Lambda\Gamma), 2\left[\lambda_{\min}(\mathbf{K}_0) - a_4\right]\right\} \tag{29}$$

$$C = 0.2785\mathbf{E}^{\mathrm{T}}\boldsymbol{\delta} + \frac{N^2}{4a_3} + \frac{1}{2}\left(\boldsymbol{\delta} - \hat{\boldsymbol{\delta}}^0\right)^{\mathrm{T}}\Lambda\left(\boldsymbol{\delta} - \hat{\boldsymbol{\delta}}^0\right) + \frac{C_d^{\,2}}{4a_4} \tag{30}$$

$$\lambda_{\min}(\mathbf{K}_1) - a_1 - a_2 > 0 \tag{31}$$

$$\lambda_{\min}(\mathbf{K}_2) - \frac{1}{4a_1} > 0 \tag{32}$$

$$\frac{1}{T} - \frac{1}{4a_2} - a_3 > 0 \tag{33}$$

$$\lambda_{\min}(\Lambda\Gamma) > 0 \tag{34}$$

$$\lambda_{\min}(\mathbf{K}_0) - a_4 > 0, \tag{35}$$

where $\lambda_{\min}(\cdot)$ is the minimal eigenvalue of the matrix and $\lambda_{\max}(\cdot)$ is the maximal eigenvalue of the matrix.

According to the above analysis, the theorem is drawn as follows.

Theorem 1. *Considering the kinematic dynamic model of the trajectory tracking ship with three degrees of freedom satisfies the assumptions, there are adaptive disturbance observers as (3) and (4), first-order filter as (11), and control laws as (15) and (16) which can guarantee the global uniform ultimate boundedness of the tracking system. Choose the proper* $\mathbf{K}_0$, $\mathbf{K}_1$, $\mathbf{K}_2$, Λ, Γ, ε_i, $\hat{\boldsymbol{\delta}}_i^{\,0}$, *and* T *to satisfy (31)–(35). The UMV can be tracked with high precision.*

Proof. To solve (27), there is

$$0 \le V(t) \le \frac{C}{\mu} + \left[V(0) - \frac{C}{\mu}\right]e^{-\mu t}. \tag{36}$$

So, $V(t)$ is uniform ultimate boundedness. Combining (17) and (36), there is

$$0 \le \|\mathbf{z}_1\| \le \sqrt{\frac{2C}{\mu} + 2\left[V(0) - \frac{C}{\mu}\right]e^{-\mu t}}. \tag{37}$$

For arbitrary $\mu_{\mathbf{z}_1} \ge \sqrt{2C/\mu}$ and existing constant $T_{\mathbf{z}_1}$, there is $\|\mathbf{z}_1\| \le \mu_{\mathbf{z}_1}$ for all $t > T_{\mathbf{z}_1}$. That is, the position error vector $\mathbf{z}_1$ is convergent to the compact set $\Omega_{\mathbf{z}_1} = \{\mathbf{z}_1 \in R^3 \mid \|\mathbf{z}_1\| \le \mu_{\mathbf{z}_1}\}$. The proper adjustment of $\mathbf{K}_0$, $\mathbf{K}_1$, $\mathbf{K}_2$, Λ, Γ, ε_i, $\hat{\boldsymbol{\delta}}_i^{\,0}$, and T can make $\Omega_{\mathbf{z}_1}$ arbitrarily small. That is, we can get the high precision tacking. □

4. Simulation Example

In this paper, a supply vessel is taken as the simulation example in [13]. The length of the vessel is 76.2 m and the mass is 4.591×10^6 kg. The parameter matrixes are as follows:

$$\mathbf{M} = 10^6 \times \begin{bmatrix} 5.3122 & 0 & 0 \\ 0 & 8.2831 & 0 \\ 0 & 0 & 3745.4 \end{bmatrix}$$

$$\mathbf{C}(\boldsymbol{\nu}) = 10^6 \times \begin{bmatrix} 0 & 0 & 8.2831v \\ 0 & 0 & 5.3122u \\ 8.2831v & -5.3122u & 0 \end{bmatrix} \tag{38}$$

$$\mathbf{D} = 10^4 \times \begin{bmatrix} 5.0242 & 0 & 0 \\ 0 & 27.299 & -439.33 \\ 0 & -439.33 & 41894 \end{bmatrix}.$$

The disturbance from external environment is described by a first-order Gauss-Markov process as

$$\mathbf{d} = \mathbf{J}^{\mathrm{T}}(\psi)\mathbf{b}, \qquad \dot{\mathbf{b}} = -\mathbf{T}_c\mathbf{b} + \boldsymbol{\rho}\mathbf{n}, \tag{39}$$

where $\mathbf{b} \in \mathbf{R}^3$ is the external disturbance in the earth reference coordinate.

$\mathbf{T}_c = \text{diag}\{10^3, 10^3, 10^3\}$ is the designed time constant diagonal matrix. $\mathbf{n} \in \mathbf{R}^3$ is the zero mean Gaussian white noise vector. $\boldsymbol{\rho} = 10^4 \times \text{diag}\{5, 5, 50\}$ is amplitude matrix.

Assume that the expected trajectory is

$$x_d = 500 \sin(0.02t),$$
$$y_d = 500[1 - \cos(0.02t)], \quad (40)$$
$$\psi_d = 0.01t.$$

The initial position and velocity vector are

$$[x(0), y(0), \psi(0), u(0), v(0), r(0)]^{\mathrm{T}}$$
$$= \left[100\,\text{m}, 200\,\text{m}, \frac{\pi}{4}, 0\,\text{m/s}, 0\,\text{m/s}, 0\,\text{rad/s}\right]^{\mathrm{T}}. \quad (41)$$

The initial state of observer is $\mathbf{b}(0) = [0, 0, 0]^{\mathrm{T}}$ and control parameters matrixes, respectively, are

$$K_0 = \text{diag}\{2, 2, 2\},$$
$$\mathbf{K}_1 = 0.08 \times \text{diag}\{1, 1, 1\},$$
$$\mathbf{K}_2 = 10^6 \times \text{diag}\{1, 1, 10^3\},$$
$$\boldsymbol{\Lambda} = 10^{-8} \times \text{diag}\{1, 1, 10^{-3}\}, \quad (42)$$
$$\boldsymbol{\Gamma} = 10^6 \times \text{diag}\{1, 1, 10^3\},$$
$$\varepsilon_1 = \varepsilon_2 = 0.005,$$
$$\varepsilon_3 = 0.0001,$$
$$\hat{\delta}_1^0 = \hat{\delta}_2^0 = \hat{\delta}_3^0 = 0.1.$$

The filter time constant is $T = 0.3$. So, in the conditions of $a_1 > 2.5 \times 10^{-7}$, $0 < a_2 < 0.2 - \alpha_1$, $0 < a_3 < 10/3 - 1/(4a_2)$, and $0 < a_4 < 2$, (37)–(39) are satisfied.

The simulation results are shown in Figures 2–4.

Figure 2 shows the expected and actual position and yaw angle curves. From the results we can see that the UMV can reach the expected trajectory at 15 seconds despite the external disturbance. Figure 3 shows the external disturbance $\mathbf{d}$ and its estimation $\hat{\mathbf{d}}$. So, the disturbance observer can estimate the external unknown disturbance. From Figures 3 and 4 we can see that the estimation $\hat{\mathbf{d}}$ can reach actual $\mathbf{d}$ using the nonlinear disturbance observer and the controller switch gain $\hat{\boldsymbol{\delta}}$ can be very small. This can weaken the drifting of the controller. So, the controller has the high robust and good performance.

5. Conclusions

During the actual ocean voyage, the UMV suffers external environmental disturbance. This paper assumes that external

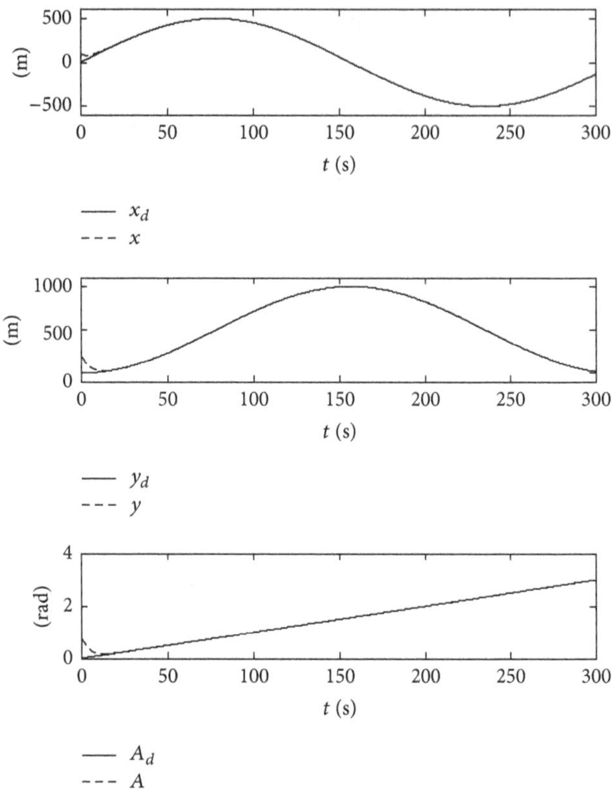

FIGURE 2: Curves of expected trajectory $\boldsymbol{\eta}_d$ and actual trajectory $\boldsymbol{\eta}$.

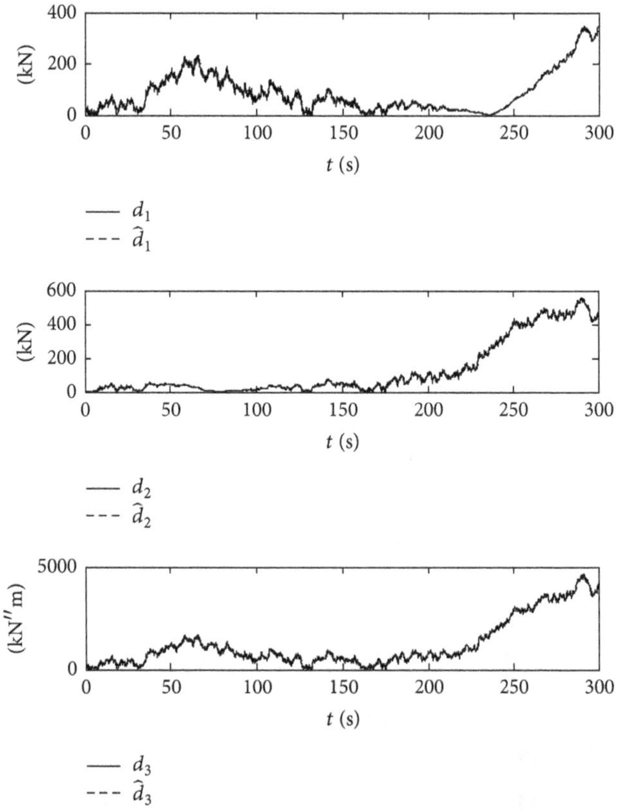

FIGURE 3: Curves of external environment disturbances $\mathbf{d}$ and estimations $\hat{\mathbf{d}}$.

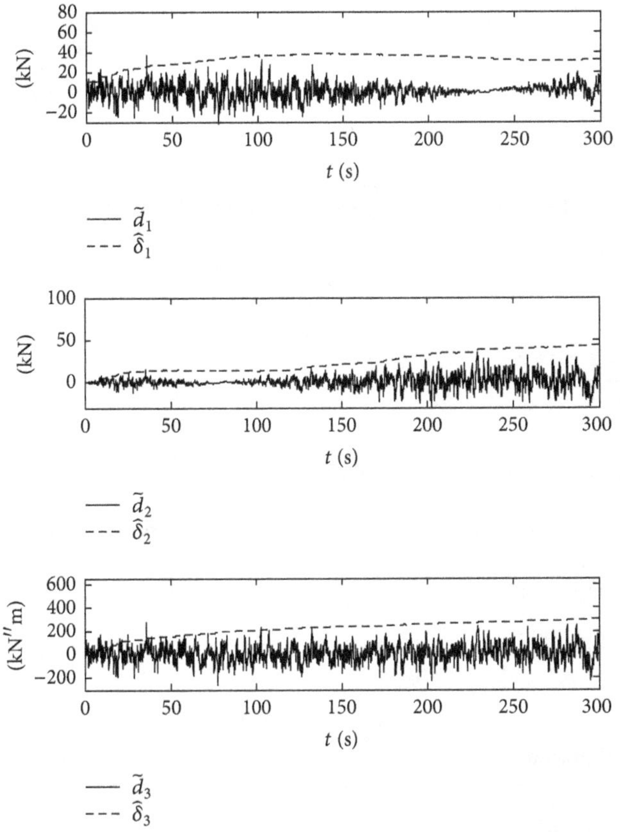

FIGURE 4: Curves of disturbances observer errors $\tilde{\mathbf{d}}$ and their upper bound estimations $\hat{\boldsymbol{\delta}}$.

disturbance and its boundedness are unknown. Combined with nonlinear disturbance observer, dynamic surface control, and adaptive robust backstepping, the dynamic surface adaptive robust controller is designed for the UMV with disturbance observer. The disturbance observer is for estimation of external unknown disturbance and the forward control is for compensation to weaken buffeting. The limit of the disturbance observer error is estimated by the adaptive law. The dynamic surface control is introduced to solve the "differential explosion." At last, the simulation of an UMV shows the high precision of trajectory tacking. This is meaningful in engineering practice.

Conflicts of Interest

The author declares that there are no conflicts of interest regarding the publication of this paper.

Acknowledgments

This work was supported by Key Project of Shaanxi Provincial Education Department under Grant no. 16JS017 and Shaanxi Provincial Science and Technology Department Industrial Research Project under Grant no. 2016GY-070.

References

[1] A. P. Mouritz, E. Gellert, P. Burchill, and K. Challis, "Review of advanced composite structures for naval ships and submarines," *Composite Structures*, vol. 53, no. 1, pp. 21–41, 2001.

[2] J. M. Godhavn, T. I. Fossen, and S. P. Berge, "Non-linear and adaptive backstepping designs for tracking control of ships," *International Journal of Adaptive Control and Signal Processing*, vol. 12, no. 8, pp. 649–670, 1998.

[3] X. Yang, T. Wang, J. Liang, G. Yao, and M. Liu, "Survey on the novel hybrid aquatic-aerial amphibious aircraft: Aquatic unmanned aerial vehicle (AquaUAV)," *Progress in Aerospace Sciences*, vol. 74, pp. 131–151, 2014.

[4] M. N. Azzeri, F. A. Adnan, and M. Z. M. Zain, "Review of course keeping control system for unmanned surface vehicle," *Jurnal Teknologi*, vol. 74, no. 5, pp. 11–20, 2015.

[5] T. I. Fossen and S. P. Berge, "Nonlinear vectorial backstepping design for global exponential tracking of marine vessels in the presence of actuator dynamics," in *Proceedings of the IEEE Conference on Decission and Control*, pp. 4237–4242, IEEE, San Diego, SD, USA, 1997.

[6] C. Wen, J. Zhou, Z. Liu et al., "Robust adaptive control of uncertain nonlinear systems in the presence of input saturation and external disturbance," *System Identification*, vol. 56, no. 7, pp. 1672–1678, 2011.

[7] Y. M. Li, S. C. Tong, and T. S. Li, "Direct adaptive fuzzy backstepping control of uncertain nonlinear systems in the presence of input saturation," *Neural Computing and Applications*, vol. 23, no. 5, pp. 1207–1216, 2013.

[8] M. Chen and W.-H. Chen, "Sliding mode control for a class of uncertain nonlinear system based on disturbance observer," *International Journal of Adaptive Control and Signal Processing*, vol. 24, no. 1, pp. 51–64, 2010.

[9] A. A. Pereira, J. Binney, G. A. Hollinger, and G. S. Sukhatme, "Risk-aware path planning for autonomous underwater vehicles using predictive ocean models," *Journal of Field Robotics*, vol. 30, no. 5, pp. 741–762, 2013.

[10] H. Aschemann, "Nonlinear control and disturbance compensation for underactuated ships using extended linearisation techniques," *Control Applications in Marine Systems*, pp. 167–172, 2010.

[11] Y. Yang, J. Du, H. Liu, C. Guo, and A. Abraham, "A trajectory tracking robust controller of surface vessels with disturbance uncertainties," *IEEE Transactions on Control Systems Technology*, vol. 22, no. 4, pp. 1511–1518, 2013.

[12] D. Swaroop, J. Hedrick K, P. Yip et al., "Dynamic surface control for a class of nonlinear systems," *IEEE Transactions on Automatic Control*, vol. 45, no. 10, pp. 1893–1899, 2000.

[13] T. I. Fossen and S. I. Sagatun, "Identification of dynamically positioned ships," *Modeling Identification & Control*, vol. 17, no. 2, pp. 369–376, 1996.

16

Long Short-Term Memory Projection Recurrent Neural Network Architectures for Piano's Continuous Note Recognition

YuKang Jia,[1] Zhicheng Wu,[1] Yanyan Xu,[1] Dengfeng Ke,[2] and Kaile Su[3]

[1] *School of Information Science and Technology, Beijing Forestry University, No. 35 Qinghuadong Road, Haidian District, Beijing 100083, China*
[2] *National Laboratory of Pattern Recognition, Institute of Automation, Chinese Academy of Sciences, No. 95 Zhongguancundong Road, Haidian District, Beijing 100190, China*
[3] *College of Information Science and Technology, Jinan University, No. 601, West Huangpu Avenue, Guangzhou, Guangdong 510632, China*

Correspondence should be addressed to Yanyan Xu; xuyanyan@bjfu.edu.cn

Academic Editor: Keigo Watanabe

Long Short-Term Memory (LSTM) is a kind of Recurrent Neural Networks (RNN) relating to time series, which has achieved good performance in speech recogniton and image recognition. Long Short-Term Memory Projection (LSTMP) is a variant of LSTM to further optimize speed and performance of LSTM by adding a projection layer. As LSTM and LSTMP have performed well in pattern recognition, in this paper, we combine them with Connectionist Temporal Classification (CTC) to study piano's continuous note recognition for robotics. Based on the Beijing Forestry University music library, we conduct experiments to show recognition rates and numbers of iterations of LSTM with a single layer, LSTMP with a single layer, and Deep LSTM (DLSTM, LSTM with multilayers). As a result, the single layer LSTMP proves performing much better than the single layer LSTM in both time and the recognition rate; that is, LSTMP has fewer parameters and therefore reduces the training time, and, moreover, benefiting from the projection layer, LSTMP has better performance, too. The best recognition rate of LSTMP is 99.8%. As for DLSTM, the recognition rate can reach 100% because of the effectiveness of the deep structure, but compared with the single layer LSTMP, DLSTM needs more training time.

1. Introduction

Piano's continuous note recognition is important for a robot, whether it is a bionic robot, a dance robot, or a music robot. There have been companies researching on music robots. For example, Vadi produced by Vatti is able to identify a voiceprint.

Most of the existing piano's note recognition techniques use Hidden Markov Model (HMM) and Radial Basis Function (RBF) to recognize musical notes with one musical note at a time and therefore are not suitable for continuous note recognition. Fortunately, in the field of pattern recognition, Deep Neural Networks (DNNs) have shown great advantages. DNNs are used to recognize features extracted from a large number of hidden nodes [1] and seek reverse partial guidance through the chain rule and at the same time make the neural network weight matrix convergence through training data iteratively and then achieve recognition [2]. RNN adds a time series based on DNN [3], which makes features have time continuity [4, 5]. However, in experiments, we find that RNN's time characteristics will disappear completely after four iterations [6], and a music note is generally longer than a frame [7], so RNN is not suitable for piano's continuous note recognition [8]. Fortunately, a variant of RNN, named LSTM, is proposed [9–12], in which an input gate, an output gate, and a forgotten gate are added to memorize a long-term cell state to maintain long-term memory [8, 9, 13–16]. Furthermore, LSTMP adds a projection layer to LSTM to increase its efficiency and effectiveness.

This paper studies LSTM and LSTMP for piano's continuous note recognition, and in order to solve the temporal classification problem, we combine LSTM and LSTMP with

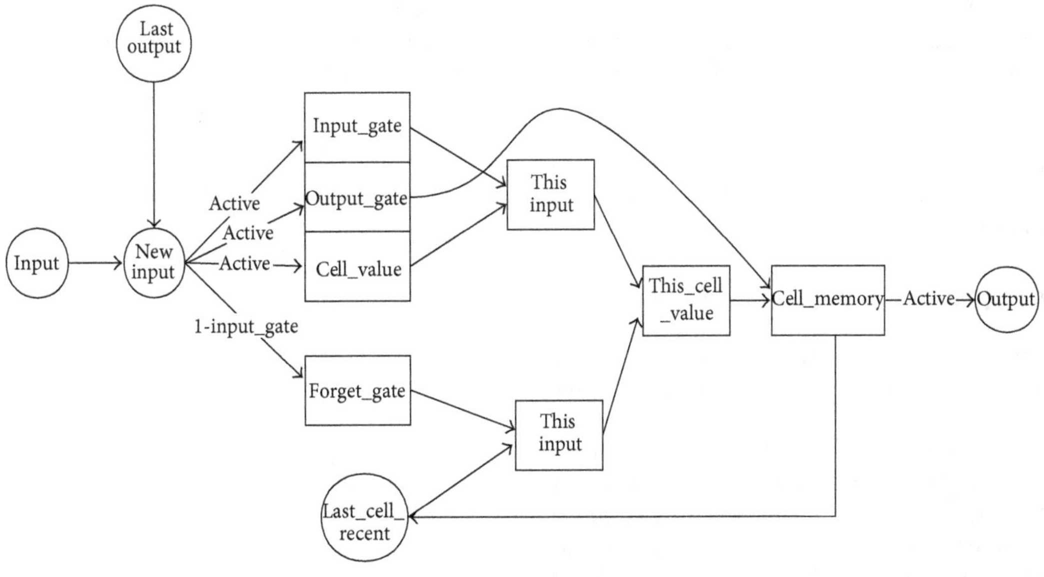

FIGURE 1: The LSTM network structure.

a method named CTC [17]. In experiments, we test the performance of a single layer LSTM, Deep LSTM, and a single layer LSTMP with different parameters. Compared with the traditional piano's note recognition methods, LSTM and LSTMP can recognize continuous notes, that is, some simple piano music. The experimental results show that a single layer LSTMP can attain a recognition rate of 99.8% and Deep LSTM can reach 100%, which proves that our methods are quite effective.

The rest of this paper is organised as follows. In Section 2, we first introduce the LSTM network architecture, and then Deep LSTM. LSTMP is illustrated in Section 3. In Section 4, we discuss CTC. The experimental results are presented in Section 5, and finally, in Section 6, we draw conclusions and give our future work.

2. LSTM

2.1. The LSTM Network Architecture. LSTM is a kind of RNN which succeeds to keep memory for a period of time by adding a "memory cell." The memory cell is mainly controlled by "the input gate," "the forgetting gate," and "the output gate." The input gate activates the input of information to the memory cell, and the forgetting gate selectively obliterates some information in the memory cell and activates the storage to the next input [18]. Finally, the output gate decides what information will be outputted by the memory cell [19].

The LSTM network structure is illustrated in Figure 1. Each box represents different data, and the lines with arrows mean data flow among these data. From Figure 1, we can understand how LSTM stores memory for a long period of time.

The recognition procedure of LSTM begins with a set of input sequences $x = (x_1, x_2, \ldots, x_t)$ (x_i is a vector) and finally outputs a set of $y = (y_1, y_2, \ldots, y_t)$ (y_i is also a vector), which is calculated according to the following equations:

$$i_t = \sigma\left(W_{ix}x_t + W_{im}m_{t-1} + b_i\right) \quad (1)$$

$$o_t = \sigma\left(W_{ox}x_t + W_{om}m_{t-1} + b_o\right) \quad (2)$$

$$f_t = 1 - i_t \quad (3)$$

$$tc_t = g\left(W_{cx}x_t + W_{cm}m_{t-1} + b_c\right) \quad (4)$$

$$c_t = f_t \odot c_{t-1} + i_t \odot tc_t \quad (5)$$

$$m_t = o_t \odot hc_t \quad (6)$$

$$y_t = \phi\left(W_{ym}m_t + b_y\right). \quad (7)$$

In these equations, i means the input gate, and o and f are the output gate and the forget gate, respectively. tc is the information input to the memory cell, and c includes cell activation vectors, and m is the information the memory cell outputs. W represents weight matrices (e.g., W_{ix} represents the weight matrix from input x to the input gate i). b is the bias (b_i is the input gate bias vector), and g and h are the activation function of cell inputting and cell outputting, respectively, regarded as *tanh* and *linear* in most of the models and also in this paper. $\odot$ is the point multiplication in a matrix. ϕ is the activation function of the neural network output, and we use *softmax* in this paper.

After conducting some experiments, we find that, compared with the f_t standard equation, (3) is more simple and easier to converge. Not only does the training time become less but also the number of iterations becomes smaller. Therefore, in the neural networks in this paper, we use (3) to calculate f_t instead of the f_t standard equation.

2.2. Deep LSTM. In piano's continuous note recognition, we also build a multilayer neural network to further increase the recognition rate. Deep LSTM adds an LSTM after another and so on [10]. The added LSTMs have the same structure as the original one. Each layer regards the output from the last layer as the input of the next layer. We hope that the neural networks in different LSTM layers will learn different characteristics, so as to learn the various features of musical notes from different aspects and therefore improve the recognition rate.

3. LSTMP-LSTM with a Projection Layer

In LSTM, there are a large number of calculations in the various gates, calculating the number of parameters N in the neural network. The weight matrix dimension input by the input gate, the output gate, and the cell state at this time is $ni * nc$, and the weight matrix dimension at the last time is $nc * nc$, and the output matrix dimension connected to the output of the neural network is $no * nc$, where ni and no are the dimensions of the input and the output, respectively, and nc is the number of memory cells. We can easily get the following formula:

$$N_{\text{LSTM}} = 3 * ni * nc + 3 * nc * nc + no * nc; \quad (8)$$

that is,

$$N_{\text{LSTM}} = 3 * ni * nc + nc * (3 * nc + no) . \quad (9)$$

As we increase nc, N_{LSTM} grows in a square pattern. Therefore, increasing the number of memory cells to increase the amount of memory costs a lot, but a smaller cell number will bring a lower recognition rate, so we propose an architecture named LSTMP, which can not only improve the accuracy, but also effectively reduce the computations.

In the output layer of the neural network, LSTM outputs a matrix of $nc^2 * m_t$. Then, m_t is sent into the output matrix to be outputted and also serves as the input to the neural network at the next time. We add a *projection* layer to the LSTM architecture, and after passing this layer, m_t becomes an $nc * nr$ matrix called r_t, which replaces m_t as the input of the next neural network. When the memory cell number of the neural network increases, the number of parameters in the neural network is

$$N_{\text{LSTMP}} = 3 * ni * nc + 3 * nc * nr + no * nr + nc * nr; \quad (10)$$

that is,

$$N_{\text{LSTMP}} = 3 * ni * nc + nr * (4 * nc + no) . \quad (11)$$

Calculating $N_{\text{LSTM}} - N_{\text{LSTMP}}$, we have

$$N_{\text{LSTM}} - N_{\text{LSTMP}} = nc * (3 * nc + no) - nr * (4 * nc + no) . \quad (12)$$

Therefore, in LSTMP, the factor that affects the total number of parameters changes from $nc * nc$ to $nc * nr$. We can change the value of nc/nr to reduce the computational complexity. When $3 * nc > 4 * nr$, LSTMP can speed up the training model. Moreover, with the projection layer, LSTMP can converge faster to ensure the convergence of the model. The mathematical formulae of LSTMP are as follows:

$$\begin{aligned} i_t &= \sigma\left(W_{ix}x_t + W_{ir}r_{t-1} + b_i\right) \\ o_t &= \sigma\left(W_{ox}x_t + W_{or}r_{t-1} + b_o\right) \\ f_t &= 1 - i_t \\ tc_t &= g\left(W_{cx}x_t + W_{cr}r_{t-1} + b_c\right) \\ c_t &= f_t \odot c_{t-1} + i_t \odot tc_t \\ m_t &= o_t \odot hc_t \\ r_t &= W_{rm}m_t \\ y_t &= \phi\left(W_{yr}r_t + b_y\right). \end{aligned} \quad (13)$$

In these formulae, r_t represents the *projection* layer, and the other equations are the same as LSTM.

Figure 2 is the structure of LSTMP, and the part marked with red dashed lines is the projection. By comparing Figure 1 with Figure 2, we can see that LSTMP is LSTM with a projection layer.

Algorithm 1 is the pseudocode of LSTMP. w is the input weight matrix, and u is the weight matrix of the last result. b is bias and r is the projection matrix. We put the extracted musical notes features into the neural network and the algorithm executes until we get an acceptable recognition rate.

4. CTC

The output layer of our LSTM and LSTMP is called CTC [20]. We use CTC because it does not need presegmented training data or external postprocessing to extract label sequences from the network outputs.

To be the same as many latest neural networks, CTC has forward and backward algorithms. When it comes to the forward algorithm, the key point is to estimate the distribution through probabilities. Given the length T, the input sequence x, and the training set S at time t, the activation $y^t{}_k$ of the output unit k at time t is interpreted as the probability of observing label k (*or blank if* $k = |L| + 1$):

$$p(\pi \mid x, S) = \prod_{t=1}^{T} y^t_{\pi t}. \quad (14)$$

We refer to the elements $\pi \in L'^T$ as paths, where L'^T is the set of the length T sequences over the alphabet $L' = L \cup blank$. Then we define a many-to-one map ϕ to remove first the repeated labels and then the blanks from the paths. With glance at the paths, will find they are mutually exclusive. According to the characteristic, the conditional probability of some labelling $l \in L^{\leqslant T}$ can be calculated by summing the probabilities of all the paths mapped onto it by ϕ:

$$p(l \mid x) = \sum_{\pi \in \phi^{-1}(l)} p(\pi \mid x). \quad (15)$$

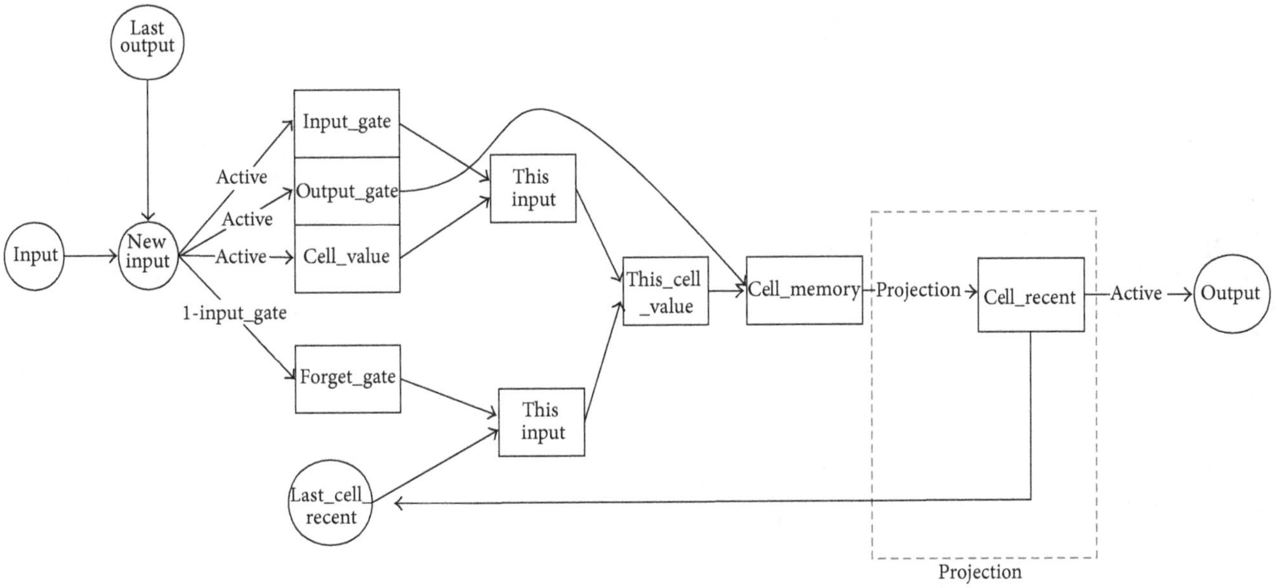

FIGURE 2: The LSTMP structure.

Require: $input\ Result_{last}\ Cell_{last}\ w\ u\ b\ r$
Ensure: $Cell_{value}\ Result$
 while *When having input* **do**
  $temp_{input} = (Result_{last} * u_{input}) + (input * w_{input}) + b$
  $temp_{output} = (Result_{last} * u_{output}) + (input * w_{output}) + b$
  $temp_{cell} = (Result_{last} * u_{cell}) + (input * w_{cell}) + b$
  $input_{gate} = sigmoid(temp_{input})$
  $output_{gate} = sigmoid(temp_{output})$
  $forget_{gate} = 1 - input_{gate}$
  $Cell_{value} = activate(temp_{cell}) * input_{gate} + Cell_{last} * forget_{gate}$
  $Result = output_{gate} * activate(cell_{value})$
  $Cell_{value} = (Result, r)$
 end while

ALGORITHM 1: LSTM-projection.

After all these procedures, CTC will complete its classification task.

5. Experiments

We conduct all our experiments on a server with 4 Intel Xeon E5-2620 CPUs and 512 GB memories. A NVIDIA Tesla M2070-Q graphics card is used to train all the models. The programming language we use is python 3.5.

We choose the piano as our instrument. We record 445 note sequences as our dataset and the length of each sequence is around 8 seconds.

In the extraction of features, we carry out Hamming window processing and then take Fast Fourier Transform (FFT) for the real part and the imaginary part of each window. Then we let the FFT result to be orthogonal by adding the square of the real part and that of the imaginary part together. Apart from that, we gain the log of the quadratic sum. Finally, the normalization of the input data is performed.

In the experiments, the number of kinds of notes is 8, and the number of input nodes is 9. We try different numbers of cell units in our models, from 20 to 320. The initial value of the neural network is set as a random value within [−0.2, 0.2], and the learning rate is 0.001. In terms of the structures, all the neural networks are connected to a single layer CTC. As for the dataset, we choose 80% of the samples as the development set and 20% as the test set.

5.1. Experimental Results. Table 1 shows the recognition rates and how many times LSTM, DLSTM, and LSTMP with different parameters need to iterate until their recognition rates are stable, and the best results are in bold.

In Table 1, "LSTMP-80 to 20" means the LSTMP model projecting 80 cell states to 20 cell states. From Table 1, we see

TABLE 1: The number of iterations and the recognition rates of different models.

Model	Number of iterations	Recognition rate
LSTM-20	300	77.2%
LSTM-40	300	87.1%
LSTM-80	300	**90.7%**
LSTM-160	400	82.5%
DLSTM-two layers	61	85.7%
DLSTM-three layers	71	99.5%
DLSTM-four layers	109	**100.0%**
DLSTM-five layers	76	97.5%
DLSTM-six layers	103	**100.0%**
LSTMP-10 to 20	43	56%
LSTMP-30 to 20	35	94.2%
LSTMP-40 to 20	45	96.0%
LSTMP-40 to 30	30	94.0%
LSTMP-60 to 30	**22**	98.0%
LSTMP-80 to 20	58	**99.8%**
LSTMP-80 to 30	27	97.3%
LSTMP-80 to 40	29	97.3%
LSTMP-160 to 40	39	99.0%
LSTMP-160 to 80	50	95.0%
LSTMP-320 to 160	93	93.4%

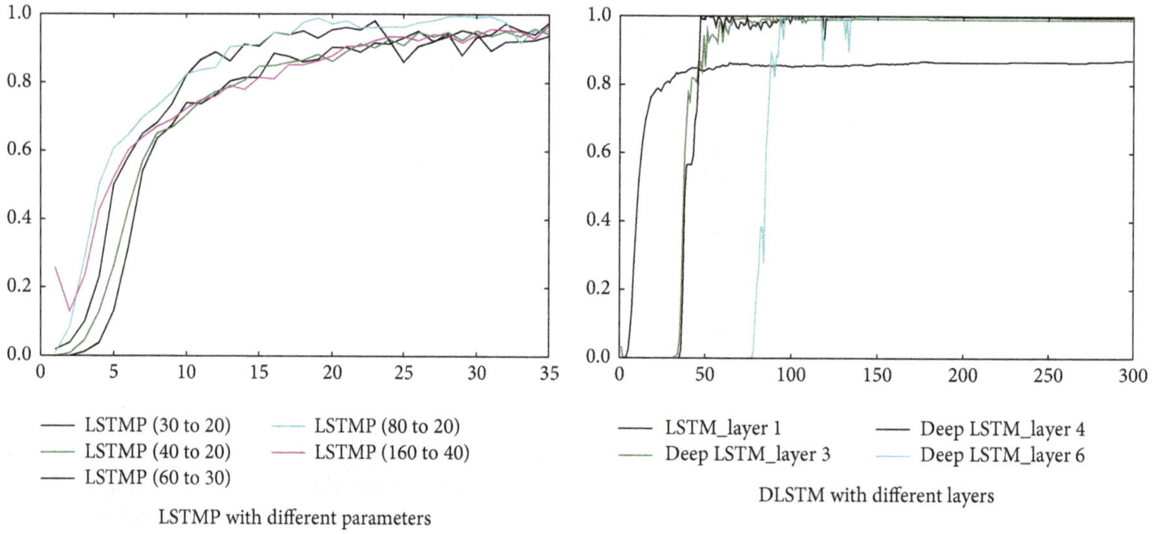

FIGURE 3: The recognition rates of LSTMP with different parameters and DLSTM with different layers.

that DLSTM and LSTMP perform much better than LSTM, and their best recognition rates are almost the same, which are 100% and 99.8%, respectively. As for the numbers of iterations, LSTMP needs much less iterations than LSTM and DLSTM, which makes LSTMP more suitable for piano's continuous note recognition for robotics considering the efficiency.

5.2. LSTMP and DLSTM with Different Parameters. Figure 3 illustrates LSTMP with different parameters and DLSTM with different layers. The x axis means the number of iterations and the y axis means the recognition rate. We see that for LSTMP the model projecting 80 cell states to 20 cell states has the best result, but all LSTMP results are very close. As for DLSTM, we see clearly that Deep LSTM is much better than LSTM with only one layer.

5.3. Comparisons of LSTM, LSTMP, and DLSTM. We compare LSTM, LSTMP, and DLSTM in Figure 4. Given the same parameters, LSTMP performs much better than LSTM. As for LSTMP and DLSTM, we find that when the number of

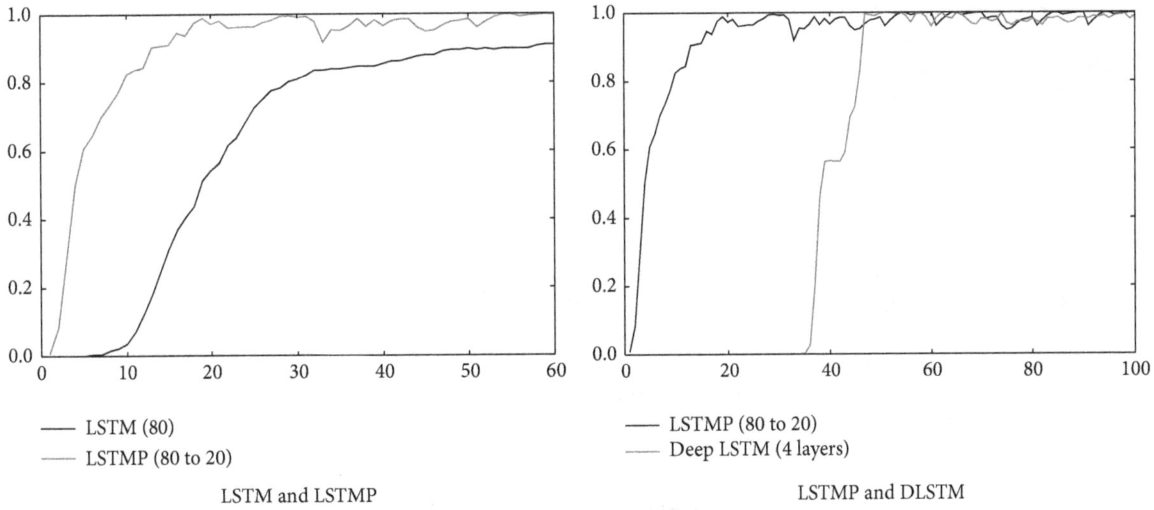

FIGURE 4: Comparisons of LSTM, LSTMP, and DLSTM.

iterations is small, LSTMP has great advantages, but as the number of iterations increases, DLSTM becomes better.

6. Conclusions and Future Work

In this paper, we have used neural network structures called LSTM with CTC to recognize continuous musical notes. On the basis of LSTM, we have also tried LSTMP and DLSTM. Among them, LSTMP worked best when projecting 80 cell states to 20 cell states, which needed much less iterations than LSTM and DLSTM, making it most suitable for piano's continuous note recognition.

In the future, we will use LSTM, LSTMP, and DLSTM to recognize more complex continuous chord music, such as piano music, violin pieces, or even symphony, which will greatly improve the development of music robots.

Conflicts of Interest

The authors declare that they have no conflicts of interest.

Acknowledgments

Thanks are due to Yanlin Yang for collecting the recording materials. This work is supported by the Fundamental Research Funds for the Central Universities (no. 2016JX06) and the National Natural Science Foundation of China (no. 61472369).

References

[1] A. K. Jain, J. Mao, and K. M. Mohiuddin, "Artificial neural networks: a tutorial," *IEEE Computational Science & Engineering*, vol. 29, no. 3, pp. 31–44, 1996.

[2] J. R. Zhang, T. M. Lok, and M. R. Lyu, "A hybrid particle swarm optimization-back-propagation algorithm for feedforward neural network training," *Applied Mathematics and Computation*, vol. 185, no. 2, pp. 1026–1037, 2007.

[3] L. Liang, Y. Tang, L. Bin, and W. Xiaohua, Artificial neural network based universal time series, May 11 2004. US Patent 6,735,580.

[4] R. J. Williams and D. Zipser, "A learning algorithm for continually running fully recurrent neural networks," *Neural Computation*, vol. 1, no. 2, pp. 270–280, 1989.

[5] L. R. Medsker and L. C. Jain, "Recurrent neural networks," *Design and Applications*, vol. 5, 2001.

[6] G. Shalini, O. Vinyals, B. Strope, R. Scott, T. Dean, and L. Heck, "Contextual lstm (clstm) models for large scale nlp tasks," https://arxiv.org/abs/1602.06291.

[7] A. Karpathy, The unreasonable effectiveness of recurrent neural networks. http://karpathy.github.io/2015/05/21/rnn-effectiveness/, 2015.

[8] D. Turnbull and C. Elkan, "Fast recognition of musical genres using RBF networks," *IEEE Transactions on Knowledge and Data Engineering*, vol. 17, no. 4, pp. 580–584, 2005.

[9] H. Sak, A. Senior, and F. Beaufays, "Long short-term memory recurrent neural network architectures for large scale acoustic modeling," in *Proceedings of the 15th Annual Conference of the International Speech Communication Association: Celebrating the Diversity of Spoken Languages, INTERSPEECH 2014*, pp. 338-342, September 2014.

[10] A. Graves, A.-R. Mohamed, and G. Hinton, "Speech recognition with deep recurrent neural networks," in *Proceedings of the 38th IEEE International Conference on Acoustics, Speech, and Signal Processing (ICASSP '13)*, pp. 6645–6649, May 2013.

[11] J. Schmidhuber, "Deep learning in neural networks: an overview," *Neural Networks*, vol. 61, pp. 85–117, 2015.

[12] K. M. Hermann, T. Kočiský, E. Grefenstette et al., "Teaching machines to read and comprehend," in *Proceedings of the 29th Annual Conference on Neural Information Processing Systems, NIPS 2015*, pp. 1693–1701, December 2015.

[13] B. Bakker, "Reinforcement learning with long short-term memory," in *In Advances in Neural Information Processing Systems*, pp. 1475–1482.

[14] S. Hochreiter and J. Schmidhuber, "Long short-term memory," *Neural Computation*, vol. 9, no. 8, pp. 1735–1780, 1997.

[15] P. Goelet, V. F. Castellucci, S. Schacher, and E. R. Kandel, "The long and the short of long-term memory - A molecular framework," *Nature*, vol. 322, no. 6078, pp. 419–422, 1986.

[16] Q. Lyu, Z. Wu, and J. Zhu, "Polyphonic music modelling with LSTM-RTRBM," in *Proceedings of the 23rd ACM International Conference on Multimedia, MM 2015*, pp. 991–994, aus, October 2015.

[17] S. Haim, A. Senior, R. Kanishka, and F. Beaufays, "Fast and accurate recurrent neural network acoustic models for speech recognition," https://arxiv.org/abs/1507.06947.

[18] F. A. Gers, J. Schmidhuber, and F. Cummins, "Learning to forget: Continual prediction with LSTM," *Neural Computation*, vol. 12, no. 10, pp. 2451–2471, 2000.

[19] F. A. Gers, N. N. Schraudolph, and J. Schmidhuber, "Learning precise timing with LSTM recurrent networks," *Journal of Machine Learning Research (JMLR)*, vol. 3, no. 1, pp. 115–143, 2003.

[20] A. Graves, S. Fernández, F. Gomez, and J. Schmidhuber, "Connectionist temporal classification: Labelling unsegmented sequence data with recurrent neural networks," in *Proceedings of the 23rd International Conference on Machine Learning, ICML 2006*, pp. 369–376, Pittsburgh, Pa, USA, June 2006.

17

Modular Self-Reconfigurable Robotic Systems: A Survey on Hardware Architectures

S. Sankhar Reddy Chennareddy, Anita Agrawal, and Anupama Karuppiah

EEE Department, NH-17B, BITS Pilani KK Birla Goa Campus, Goa 403726, India

Correspondence should be addressed to S. Sankhar Reddy Chennareddy; sankhar@goa.bits-pilani.ac.in

Academic Editor: Shahram Payandeh

Modular self-reconfigurable robots present wide and unique solutions for growing demands in the domains of space exploration, automation, consumer products, and so forth. The higher utilization factor and self-healing capabilities are most demanded traits in robotics for real world applications and modular robotics offer better solutions in these perspectives in relation to traditional robotics. The researchers in robotics domain identified various applications and prototyped numerous robotic models while addressing constraints such as homogeneity, reconfigurability, form factor, and power consumption. The diversified nature of various modular robotic solutions proposed for real world applications and utilization of different sensor and actuator interfacing techniques along with physical model optimizations presents implicit challenges to researchers while identifying and visualizing the merits/demerits of various approaches to a solution. This paper attempts to simplify the comparison of various hardware prototypes by providing a brief study on hardware architectures of modular robots capable of self-healing and reconfiguration along with design techniques adopted in modeling robots, interfacing technologies, and so forth over the past 25 years.

1. Introduction

Modular robotics provides a unique advantage over traditional robotic technologies in terms of reconfigurability, reusability, and ease in manufacturing. The traditional robotic designs such as robotic arms and hexapods. provide unique solutions to each real world application and the generated prototypes are generally inflexible for the rest of the applications. Most of the traditional robotic solutions are operated in a controlled environment and any changes in environments often make these traditional solutions inflexible due to lack of their adaptive nature. The repair and maintenance of such conventional designs generally require separate trained personnel for each model and hence increasing the average resource consumption in industries. The next phase of robotic designs is developed in the perspective of assembly of modular units for increasing ease of repairing, replacing, control, and so forth. The researchers in later phases of development introduced the concept of automation, self-healing, reconfiguration, and so forth creating a modular self-reconfigurable robots (MSRR). Many applications such as management of large facilities [1], space exploration [2], surveillance in military zones, disaster management, and prosthetics for physically disabled often require adaptable and self-healing abilities and MSRR is often considered as a viable solution for the same. The major difference of MSRR designs over modular robotic designs can be visualized as the abilities of designs to attach/detach in/from a formation as per the requirement of application with minimal human intervention.

The growing demand for reusable, space constrained, and multipurpose solutions for real world applications is a great motivator for research in the field of MSRR. The researchers in domain of MSRR provided numerous solutions via various prototype designs, communication algorithms, coordination, and dispersion techniques using selected test scenarios. The development of novel prototypes for MSRR is an analytical process that often has deep roots in intuition and derives better fruits from experience on basic locomotion and laws of physics. Majority of the robotic modules developed so far utilized limited resources available at time of development and have restricted capabilities due to slow technological advancements in the areas of sensors, hardware prototyping, actuators, communication interfaces, and so forth. The different approaches adopted by researchers to validate the designs and prototypes make relative comparison between the robotic modules a fairly difficult process and present challenges in quantifying and understanding merits/demerits of various designs.

A detailed survey on wide range of solutions for modular designs of outer structures, physical interfaces between modules, communication protocols, sensor technologies for docking and alignment, coordinate movement algorithms, environment characteristics, and so forth provides better understanding to novel researchers about merits and demerits of previous designs so that better solutions can be provided with the utilization of latest technologies. This paper scope is limited to study and summarize the hardware architectures along with sensor and interfacing technologies of various MSRR.

2. Modular Robots, Hardware Architectures

The hardware architectures in MSRR are evolving along with the technologies and so does the paradigm used for categorizing the robots. The first prototype developed in MSRR research is CEBOT consisting of heterogeneous separate units capable of binding together and since then the research was directed to development of systems capable for forming different structures mimicking biological organisms. Yim et al. [2, 3] suggested two classifications of modular robotic systems: classification based on structures formed by MSRR and classification based on reconfiguration strategies. Gilpin and Rus [4] added few more subclassifications under structures category by including research from microelectronic mechanical systems (MEMS) and other latest developments in MSRR by the time of publication. Moubarak and Ben-Tzvi [5] categorized MSRR systems based on locomotion of the individual modules and coordinated structures along with form factors. The classifications proposed so far are categorized as per the state-of-the-art research in recent MSRR technologies, prototypes, and so forth available till the data of publication. The recent research in MSRR is generating solutions that are falling in the middle of earlier classifications and the identification of a category and subcategory for MSRR robots is becoming difficult due to recent sophisticated designs and features of robots.

The classification of MSRR based on various categories and subcategories such as physical characteristics, abilities, and so forth is provided in Figure 1. The widely accepted classification is in the perspective of possible structural formations when the independent MSRR are brought together and five subcategories are recognized under structures as per the current MSRR research: *lattice*, *chain*, *hybrid*, *truss*, and *free-form*.

The MSRR designed for lattice structures are inspired from atomic structures like cubic centered lattice, tetrahedron, and so forth and are equipped with actuators to form similar structures. The individual robotic units occupy discrete positions in space and lack capabilities to reach random positions/orientations if necessary due to limitation in actuator assemblies. The lattice architectures provide easy control mechanisms and do not often require closed loop control due to their defined actuator positions in 2D and 3D space. The robotic units under chain category are generally serially connected robotic units and are capable of forming complex structures like snakes, centipedes, and so forth. The actuators of these robots are assembled to provide end effectors random positions in space. The control of chained systems is more complex and often requires feedback to confirm the position of modules in space for reconfiguring structures. The hybrid designs provide more advantages compared to lattice and chain robotic structures due to their capabilities in easy adaptation to surroundings by forming both lattice, chained and mixture of both. The MSRR with truss based designs support formation of random structures due to the employment of telescopic links and heterogeneous units for forming structures but require complex algorithms for handling assembly and formation of structures. The free-form category MSRR are generally more flexible in the perspective of attaching and detaching from the system. They can form arbitrary structures and normally maintain weak bonds between the neighbors. The chain and hybrid differ from the free-form structures generally in terms of rigidity in bonding.

The sophisticated locomotion capabilities of MSRR are resultant of coordinated actions of many individual units aggregated in various structures. The capabilities (autonomous/semiautonomous/manual) of aggregation in MSRR for facilitating complex movements by reconfiguration rely on the actuator-sensor assemblies embedded in individual robotic designs. The MSRR designs equipped with wheels are capable of forming lattice or chain structures depending on the design and hence can be placed in mobile subcategory under locomotion in Figure 1. The majority of lattice and chain systems are designed without wheels on individual units and hence mobility is realized only by employment of coordination of robots. The aggregation of individual units requires human intervention to a certain degree for nonwheeled systems. These MSRR designs can be placed in coordinated subcategory under locomotion category shown in Figure 1. The external subcategory under locomotion in Figure 1 refers to the MSRR designs that rely on environmental stimulus/disturbances for locomotion as well as reconfiguration.

The recent contributions to MSRR attempted the concept of employing disturbances and vibrations in environment for assembly of robots introducing along with addressing uncertainty in reconfiguration structures and hence creating two subcategories based on reconfiguration: *deterministic* and *stochastic*. The deterministic reconfiguration type of MSRR has precise control over the structures, assembly, and reconfiguration by employing either closed loop control or advanced actuator assemblies. The stochastic type of MSRR mostly does not have control over the assembly of units but generally retains the ability for disassembly. Hence the reconfiguration after completing a particular structure assembly and the time required for the same has major contribution from environmental factors.

Many researchers developed designs in Micro to Macro sizes for addressing various scenarios in MSRR. The form

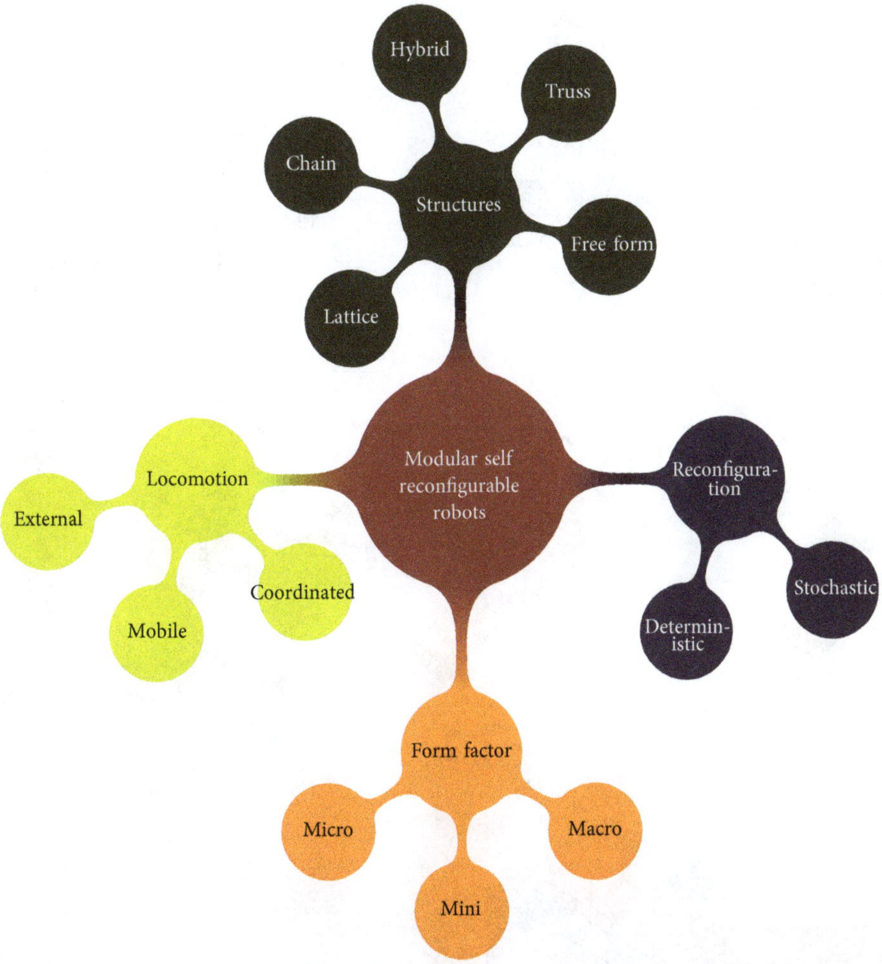

FIGURE 1: Classification of MSRR designs based on hardware characteristics.

factor scaling is completed at a trade-off with capabilities and also increased dependency on events happening in the surrounding environment. Henceforth in this paper, the MSRR robots occupying volume equal to and more than a cube of 5 cm side are referred to as macro structures, models occupying less than volume of macro designs but visible to naked eye are referred to as mini structures, and designs not easily visible to naked eye are referred to as micro structures. In this paper we adopted the widely accepted classification of MSRR-classification based on structures for broadly summarizing the research so far. The other categories are related implicitly while providing the details of locomotion, dimension, and mobility.

2.1. Lattice Structured Systems. The metamorphic robotic system [6–8] is the first lattice structure category robotic design capable of changing structures in 2D environment. The authors explored the idea of hexagonal and square lattice structures using the metamorphic robotics systems. A hexagonal skeleton was developed for mimicking the robot outer structure with 6 servo motors at each corner and male and female connectors on alternate sides for docking as shown in Figure 2(a). After successful docking between the cells, each cell can revolve around the periphery of neighboring cell by gradually changing their structure. The square structured prototypes for lattice structures employ sliding mechanism using gender based connectivity for movement along the lattice structures.

Murata et al. developed a 2D lattice category MSRR called Fracta [9]. The individual robot in Fracta consists of a top, a bottom module with permanent magnets, and a middle module equipped with electromagnets. The assembly is shown in Figure 2(b). The docking process begins with insertion of middle layer into the empty space between the top and bottom layer of neighboring modules by activating electromagnets. The operating principle was tested using modules equipped with castors on frictional less surface.

Molecule [10] is a 3D structure supporting design developed by Rus D. and each unit consists of two atoms and a right angle rigid bond binding them. The connectors equipped with electromagnets are present on side faces of each atom. The bonded two-atom system is referred to as "Molecule" and each atom has two degrees of freedom (DOF) with one provided by motor at a connector on face and another due to

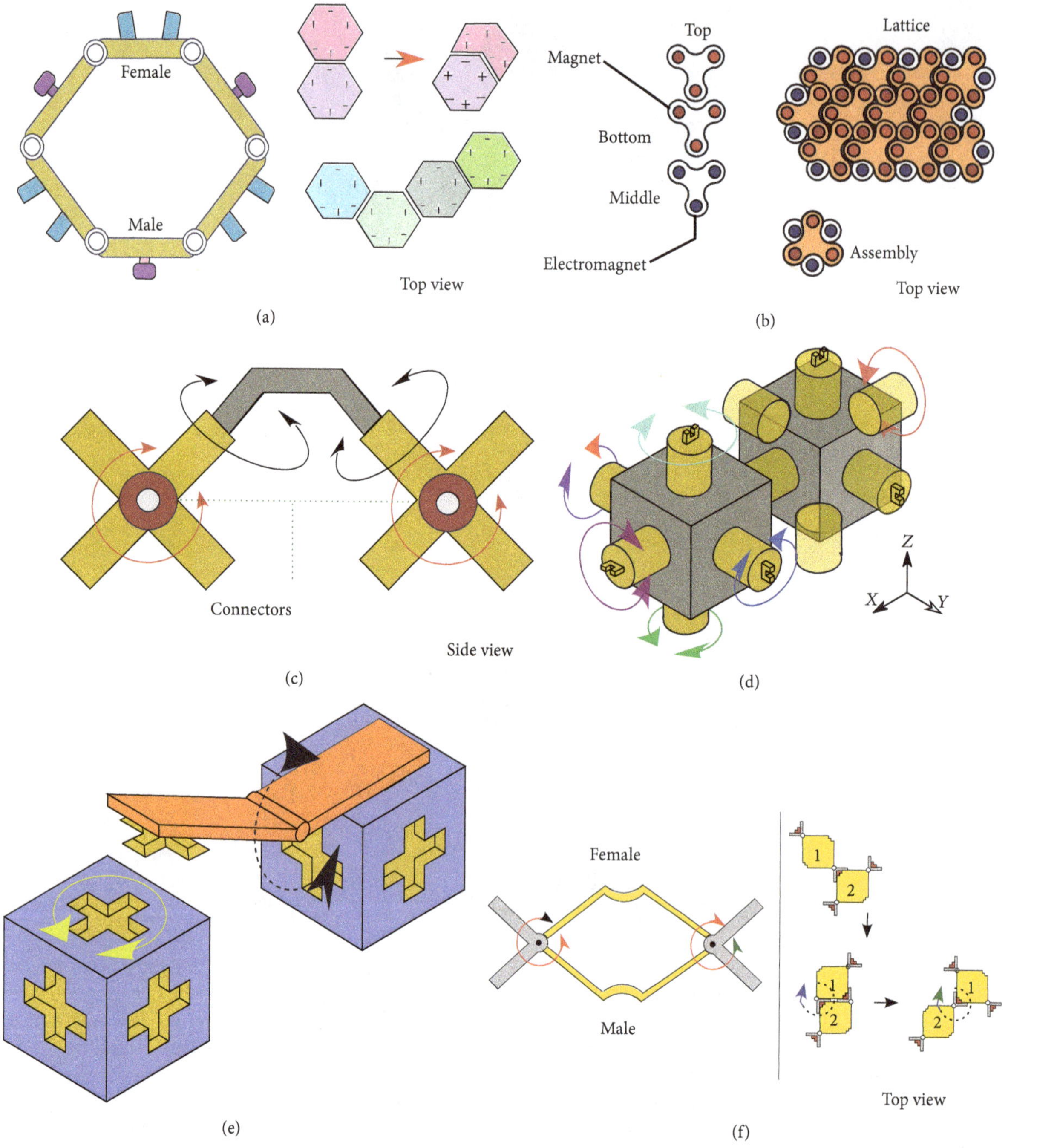

FIGURE 2: Lattice MSRR hardware models: (a) metamorphic, (b) Fracta, (c) molecule, (d) 3D unit, (e) I-cubes, and (f) Microunit 1.

motor at the bond as shown in Figure 2(c). The Molecule as a whole provides 4 DOF and can be used for creating arbitrary structures like walls.

Kurokawa et al. prototyped a 3D unit [11, 12] in cubical structure with connectors on all faces. Each connector can rotate independently along their axis providing the 3D unit 6 DOF as shown in Figure 2(d). The connectors on all faces are connected to a single 7 W motor using worm gear mechanism controlled by independent solenoid driven switching technique for each arm. The arms are connected using connection cuff capable of moving back and forth along the axis of arm. The connection hand mounted on cuff closes at one extreme of sliding displacement and opens at the other.

The I-Cubes proposed by Ünsal et al. in [13, 14] is another cubical structure robotic design with two units: cubes and links. The faces of cubes have female connectors to mount the links using lock and key mechanism. A cube at a given time can have zero to six links connected to its faces. The links are independently controlled multijoint unit shared and transferred between cubes. The horizontal beam of link constitutes a joint at the center of two horizontal beams and can be rotated as shown in Figure 2(e). The cubes can rotate with respect to link after successful latching and hence providing locomotion to the cubes present in the system.

A mini form factor design referred to as Microunit [15, 16] further is developed by Yoshida et al. Microunit was prototyped in two different models and each module in the system has square skeleton structure with two static female connecting parts at two ends of a diagonal and rotating male connecting parts at the end of other diagonal as shown in Figure 2(f). The first prototype designed can form structures in 2D with docking controlled by torsion springs made from shape memory alloys (SMA). The design employs torsion springs and stoppers coupled with SMA for generating rotation mechanism. The authors also attempted further miniaturization of modules by removing the control unit present in earlier prototype and designed second model providing capabilities for forming structures in 3D.

The vertical robot published in [17] is a cubical structure of 90 mm side independent units. Each cube is equipped with two hands each lying on parallel side faces similar to human hands and the rest of the faces are equipped with magnetic sheets. The cells are capable of extending hands and rotation of the same only along the axis normal to surface they are mounted on. The design facilitates movement of robots only along vertical plane and hence stacking is the only method supported for navigation. The hands of two robots can be docked for lifting and the docking technique is facilitated by a genderless lock and key passive connector. The extension of hands is controlled using sliding mechanism.

Crystalline [18] is a cuboid structured robot with expansion and retraction capabilities on side faces. The expansion and retraction of faces are performed on all sides simultaneously using rack and pinion mechanism. The active connection mechanism is present on two neighboring side faces and passive connector mechanism is present on others. Since the system is not designed to docking on top and bottom faces, the crystalline MSRR structures are limited to 2D scenarios. The telecubes module [19] developed by Suh et al. is an improvisation to crystalline design with support for 3D structures. The six faces on each module can expand and contract in the direction normal to face similar to crystalline. Unlike crystalline, telecubes can move in vertical axis and hence has capabilities of forming 3D structures. Each face on telecubes module is divided into four quadrants with magnet pole pieces in odd and magnetic metal in even quadrants with chamfered borders for passive docking. The modules couple when they are close to each other since the connection plates on them are mirror images and the SMA springs present in the system pull magnetic pole pieces inside for undocking. The cubic structured module, EM-cubes published in [20], also employed magnets on four faces for docking and locomotion. The permanent magnets installed provide firm bonding and electromagnets facilitate locomotion. The electromagnets are activated alternatively to create attractive and repulsive forces simultaneously generating couple force at two ends of cube for locomotion.

M-blocks developed by Romanishin et al. are cubical MSRR prototyped in two versions: M-blocks [21] and 3D M-blocks [22]. The M-blocks and 3D M-blocks are equipped with an inertial actuator at center of body for applying

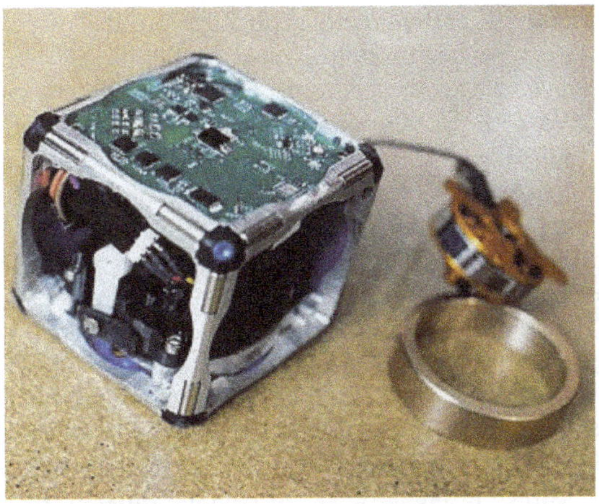

FIGURE 3: M-block MSRR (picture courtesy of M. Scott Brauer, *source:* http://news.mit.edu/2013/simple-scheme-for-self-assembling-robots-1004).

controlled torque at center of mass of the module and hence rotating the M-block MSRR in clockwise and anticlockwise directions. The M-block cells have capabilities of individual movement for docking and coordinate movement. The faces and edges of both models are embedded with permanent magnets as shown in Figure 3. The rapidly accelerating and decelerating internal rotation mechanism sources the locomotion and edge magnets control locomotion of modules around other robots using pivot action. The face magnets support alignment between the modules after locomotion. The M-blocks provide actuation in single direction and 3D M-blocks can actuate in six directions by changing inertial actuator orientation to any of three orthogonal axes for 3D movements. A mini form factor MSRR, MICHE [23], is designed for forming lattice structures in 3D with the aid of environment. Three faces of cubic structured MICHE are equipped with switchable magnets and the rest of the faces are covered with steel plates. The magnets are placed away from geometric center of plates for avoiding repulsion forces between magnets of two robots during docking. The magnet switching is controlled by internal microcontrollers communicating via IR transceivers and hence providing capabilities for retaining structures to MICHE MSRR. The MICHE MSRR falls under stochastic category for its dependence on environment for aggregation and locomotion. Pebbles [24] is another stochastic category cubic structure designed to form lattice structures in 2D. The four side faces on the robot can act as a connection plates due to their internal contact with four custom designed electropermanent magnets.

White et al. proposed stochastic robotic modules prototyped in two models with one supporting only 2D structures [25] and another supporting 3D structures [26]. The 2D structure modules are designed in both triangular and square base structures. The sides of a module's base are equipped with electromagnet for coupling. The docking and undocking are controlled by actuation and deactuation of electromagnets enabled via H-bridges. The stochastic 3D version modules are cubic structures of 10 cm side with permanent magnets placed radially from center and electromagnets at the center of each face. The latching/unlatching is controlled by polarity of the electromagnet. The Programmable Parts [27] MSRR is another stochastic category robot with a triangular chassis equipped with latching mechanism on all sides. Each side is equipped with a fixed magnet and a rotating magnet controlled by DC motor placed adjacent to each other. During latching process, the fixed magnets of a module face rotating magnets of other robotic modules. Hence the modules can undock by retracting magnets by rotation in the self- and neighboring module. The IR sensors inserted into sides handle communication between the modules. The X-BOT [28, 29] MSRR consists of "X" shaped cuboid modules capable of forming 2D structures stochastically. Each leg in "X" shape is equipped with a pair of compliant arms with magnets at their tips as shown in Figure 4. The arms bond different modules together and the coupling/decoupling process is controlled by push-pull process regulated by SMA wires wounded around the frame and arms.

The ATRON module proposed in [30, 31] is a lattice structured design along with minimal flexibility for forming chain structures in 3D as shown in Figure 5. The modules are composed of two hemispheres mounted upon each other on flat side and each hemisphere is capable of rotating 180° independently. The two hooks (active male) and two passive female connectors placed equidistantly around periphery of each hemisphere in alternate positions facilitate docking. The hooks are driven by worm gears and female connectors are two rigid bars firmly connected to chassis of the module. The rotation of a hemisphere with respect to another provides locomotion in the structures.

The tetrapod structured PetRo MSRR [32] developed by Salem et al. is a self-mobile lattice category design proposed for forming 3D structures. The central hub and four legs together constitute a single unit in tetrapod shape. Each free end of legs is connected to wheels providing one DOF along leg axis and another DOF is added at the central hub perpendicular to leg axis with a rotation of ±45°. The wheels are also proposed to play a role in connection plate between various PetRo modules forming complex structures similar to pets. The IR sensors present on the connector faces aids in alignment for docking. The grooved pins and chamfered holes on connection surfaces come opposite to each other after alignment and rotation and along with support from magnets the docking is completed successfully.

Figure 4: Model diagram of X-Bot MSRR.

Figure 5: ATRON MSRR (http://modular.tek.sdu.dk/index.php?page=robots).

Table 1: CEBOT, cell physical characteristics.

	Series I	Series II
Dimensions (mm)	190 * 90 * 50	176 * 126 * 90
Weight (Kg)	1.2 (mobile cell) 1.2 (target cell)	2.7 (mobile cell) 1.0 (target cell)
Connectivity surface	Flat	Tapered
Coupler actuator	SMA	DC motor

2.2. Chain Structured Systems. CEBOT [33, 34] MSRR belongs to mobile category comprising heterogeneous modules and has two hardware prototypes referred to as Series I and Series II. The design facilitates 3D structure formation and comprises three types of cells:

(a) Wheel mobile cell.

(b) Rotation joint cell.

(c) Bending joint cell.

The cells are fitted with castors at bottom for frictional less movement and are equipped with male and female connectors for docking. The wheel mobile cell shown in Figure 6 having mobile capabilities initiates docking with the necessary cell. The cells are equipped with SMA couplers for active latching of the male connector during docking and position sensors mounted on cells provide time to feedback on the docking process. The cells in series I require precise control and alignment for docking. The cells in series II are replaced with tapered female socket with worm gear for the active latch mechanism instead of SMA while maintaining the same docking process. The physical characteristics of cells in CEBOT are listed in Table 1.

Endo et al. developed ACM MSRR for mimicking snake alike chain structures in 2D. The ACM MSRR have three different versions: ACM [35], ACM-R2 [36], and ACM-R3 [37]. Each unit in ACM is a wheeled square chassis robot without any actuator present for controlling individual

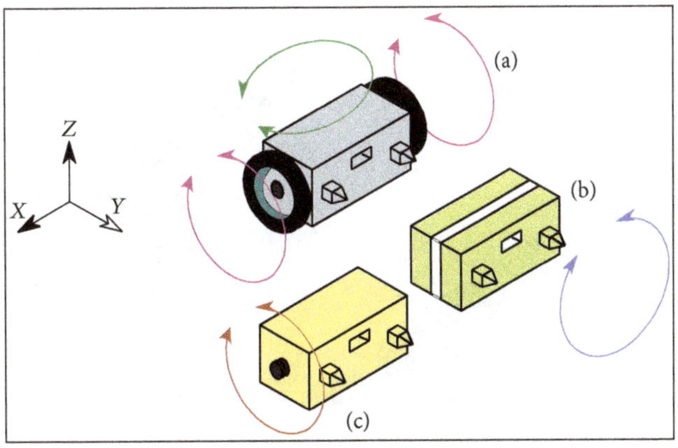

FIGURE 6: Structure of cells in CEBOT MSRR: (a) wheel mobile cell, (b) rotation joint cell, and (c) bending joint cell.

mobility. A servo motor is equipped behind every unit to rotate the robot at joint axis. The ACM MSRR is a combination of such individual homogeneous units assembled manually. The ACM-R2 is an improvement to ACM MSRR and has capabilities of forming 3D structures. The ACM-R2 MSRR is equipped with pitch and yaw motors in the joint unit between units for proving 2 DOF. The ACM-R3 is designed using custom frame body and wheels for providing robust support in formation of 3D structures and also facilitating manual assembly of robots with ±90° offsets with respect to each other.

Brown Jr. et al. prototyped a two-sided tracked vehicle called Millibot [38] capable of forming 2D structures for applications like movement in uneven terrains, stair climbing, and so forth. The Millibot MSRR is approximately an elliptical structure robot capable of self-docking using male and female connectors via latching mechanism actuated by SMA and is shown in Figure 7(a). The male connectors are installed in the front on a lifter capable of lifting objects vertically with the help of harmonic drives. Amoeba-I [39, 40] is another tracked MSRR with self-mobility proposed by Liu et al. for forming 3D structures. Each unit is a tracked elliptical structure capable of moving itself and is equipped with pitch joint on one side and yaw joint on the other. The robots when manually connected using physical links provide various DOF as shown in Figure 7(b). The amoeba-I MSRR locomotion combinations are numerous depending on the orientation of link between the modules as well as actuation of corresponding joints. Li et al. developed improvised version of Millibot, JL 1 [41] and JL 2 [42], in terms of DOF by providing yaw and pitch control mechanism to each bot and also gear based docking mechanism at the cost of weight of robot. The major difference between JL-1 and JL-2 is earlier employed latching mechanism for docking and later employed gripper for docking. The gripper on JL-2 can also be utilized as manipulator arm for holding objects in the environment. Lyder et al. published Thor [43] MSRR made up of modular blocks. The blocks are analytically developed motors, gears, right angle joints, gears, and wheels that can be utilized for forming various single robotic structures similar to Lego structures. The blocks can be assembled in various configurations due to symmetry in block designs and Thor is a robot build with a gripper using such blocks. Thor robot is equipped with wheels for mobility and gripper to dock with neighboring modules and so forth and hence making it a MSRR.

Yim designed Polypod [44] MSRR that falls under chain structure category and with capabilities of forming 3D structures. Polypod consists of two types of modules: segments and nodes. The nodes are rigid modules in cubical structure with single connector on each face providing six connectors from batteries. The segments are formed using 10-bar linkages providing two degree of freedom to the system and are capable of expanding or contracting in length as well as inclining towards left and right. The segments and nodes together facilitate formation of complex structures in 3D as shown in Figure 8.

The Polypod is actuated using small DC motor and position sensors are used for measuring angles of the linkages. The control architecture is implemented in three levels with highest level deciding the behavioral modes, the middle level executing the behavioral mode, and the lower level translating the commands to actuator joint space. The connection plates between the modules also facilitate the electrical connectivity for power and communications. In spite of absence of wheels the system is capable of movements like snake, caterpillar, rolling track turning, the moon walk dance, and so forth.

Castano et al. designed CONRO [45] MSSR to form structures like snakes or hexapods in 3D. Each module in CONRO consists of three segments:

(a) Passive connector.

(b) Body.

(c) Active connector.

Two servo motors with rotation axis in orthogonal orientation are attached to the body as represented in Figure 9. The pitch motor is connected between the active connector and body. The yaw motor is connected to body and the passive connector. The docking mechanism and communication are handled using the feedback from IR transceivers present on the faces of active and passive connectors. The SMA equipped

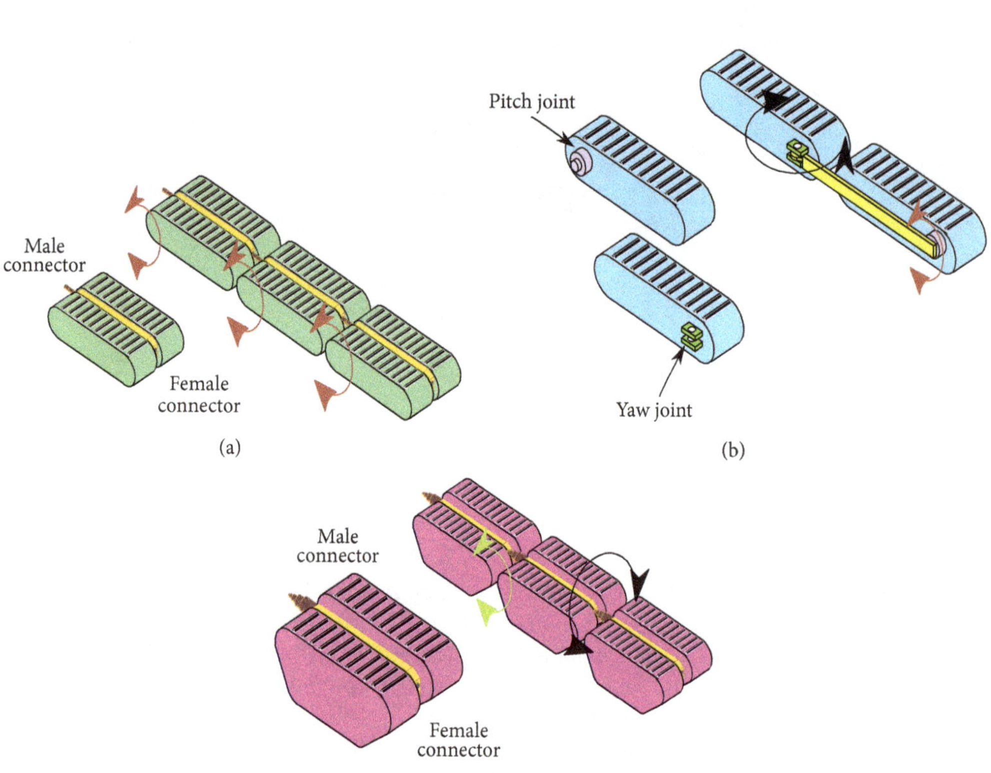

FIGURE 7: Chain structured mobile MSRR hardware models and structures: (a) Millibot, (b) Amoeba, and (c) JL-1.

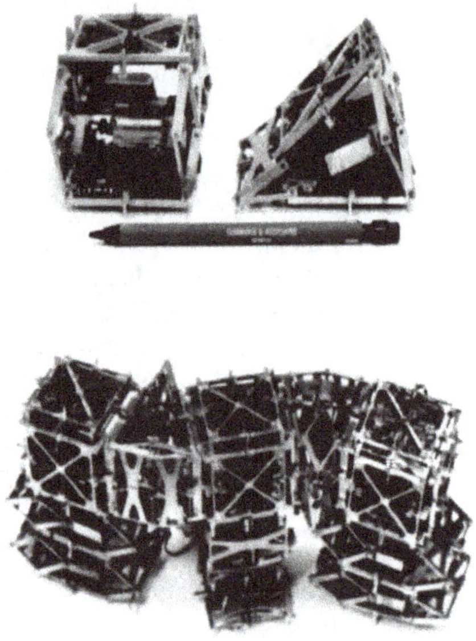

FIGURE 8: Polypod MSRR (source: http://robotics.stanford.edu/users/mark/photos.html).

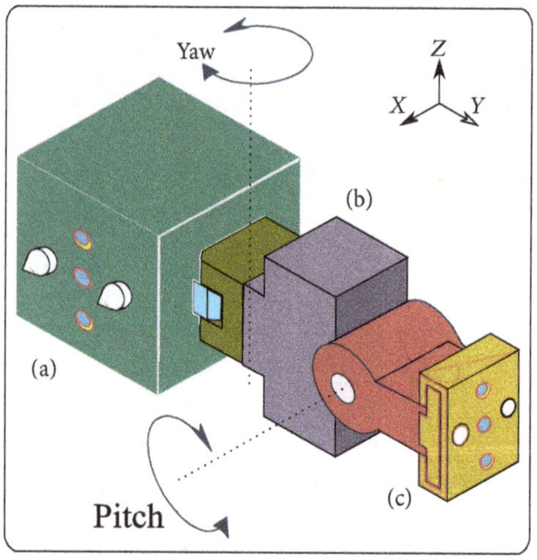

FIGURE 9: Model diagram of CONRO MSRR.

locking system present in passive connector latches the modules together after successful docking. A hormone based centralized and decentralized control for coordinate movements in modular robots was researched on CONRO robots in [46, 47]. Further research on docking and alignment issues in CONRO robot modules are addressed in detail in [48].

ModReD [49] MSRR proposed by Dasgupta et al. is similar to CONRO MSRR with modification in DOF. The ModRED robot consists of 3 cuboid blocks with 2 pitch motors: one at first block and the other at last block. A prismatic motor is placed along with pitch motor in last block for elongation of bond in horizontal plane between the center and last block. A roll motor is placed in center block for rotating front block with respect to center block. The first and last blocks are equipped with brackets as connectors with grooves and pins in structure of square along with a solenoid controlled mechanism for latching.

Polybot [50, 51] MSRR is a chain structure inspired robotic design capable of forming 3D structures. The Polypod is a cubic structure prototyped in three major versions: G1, G2, and G3. The G1 version of Polybot is a quick prototype with connection plates on front and back faces of 5 cm cube. The connection plates orientation with respect to each other can be changed with DC motor mounted outside the cube whose axis of rotation is normal to the side faces. The G1 prototype has no mechanism for latching and unlatching and hence docking is done manually. Since the connection plates are equipped with grooved pins and holes symmetrically, it is possible to dock two polybot G1 modules back to back even with 90° offsets. The Polybot G2 is similar to G1 and additionally equipped with electromechanical latches and SMA controlled by software. The docking mechanism is guided by IR transceivers mounted on face plate and the robot is shown in Figure 10. The Polybot G3 are miniaturized modules with dimensions around $50 * 50 * 50\,mm^3$. The externally visible DC motor in G1 and G2 version is made internally by changing the mechanism to dc pancake motor with harmonic gear along with active braking feature.

Transmote [52] module design is similar to Polybot with major difference in latching mechanism and number of connection surfaces. The front side face of transmote is equipped with a conical structure used for docking with female socket present at back of robot. The transmote facilitates twist and lock mechanism controlled by a servo motor for docking between robots. The transmote MSRR has connection provision on one side face along with front and back faces providing more stability to few 3D structures. The GZ-I MSRR robotic module proposed in [53] is similar to transmote with three connector faces and slightly different physical construction. The GZ-I modules are not equipped with docking sensors, actuators, and so forth and hence are assembled manually.

The YaMor [54] robot is a semicylindrical box structured robot capable of forming 2D chain structures. A triple beam in shape "⊔" is connected to side faces of semicylindrical box at free ends of beams. Each robot module has one DOF and the system does not support autonomous docking. The velcros placed on beams, side faces, and back of the robots are used for docking with neighboring modules manually. The YaMor robot is a complete integrated solution with wireless communication capabilities and FPGA for reconfigurable computation purposes.

2.3. Hybrid Structured Systems. Mondada et al. developed a completely integrated autonomous robot called S-BOT [55–62] capable of forming lattice structures in 2D and chain structures in 3D and hence a hybrid category robot. The robot is a cylindrical structured track robot designed for research in swarm robotics. The robots are capable of localization and navigation in uneven terrains. The robots employ gripper mechanism for docking with a ring covering the periphery of robot. Since the ring is present around the periphery, the docking can be done almost from every direction. The optical sensors present in gripper modules form a closed control loop for providing feedback on docking process. The S-Bot employs the same features of modular robots such as modularity and reconfiguration.

FIGURE 10: Polybot G2 MSRR (source: http://elek3ronik.blogspot.in/2007/05/chain-reconfiguration-robot.html).

M^3 MSRR researched by Kutzer et al. is capable of forming 3D chain and lattice structures along with mobility features and is developed in two versions: M^3 [63] and M^3 express [64]. The modules are "L" shaped models with two wheels on parallel sides of long beam and one omnidirectional wheel on outside face of short beam parallel to surface and perpendicular to common rotational axis of two other wheels as shown in Figure 11. The wheels play dual role, enabling mobility and connection plates for docking. The M^3 module is equipped with two hooks on wheels separated by 180°. The units are latched together when wheels of two modules come face to face with an offset of 180° or 360°. The custom designed slip rings aid robots with docking as well as mobility using same wheels. In the M^3 express module each wheel is equipped with two magnets at the ends of diameter, a yoke and four locking pins. The Yorks are connected to servo motors in a sliding mechanism for activating a slip disk with metallic screws. The disk is normally separated due to internal springs and the actuation of servo motor mounts the slip disk into wheels bringing metallic screws on to face of wheels at the ends of other diameters for docking.

Imobot [65, 66] is another mobile hybrid MSRR prototyped by Harry et al. The iMobot MSRR is cuboid structured formed from assembly of two semicylindrical modules as shown in Figure 12. The side faces of iMobot are equipped chamfered flat sheets capable of rotating continuously and hence providing mobile abilities to robot. The semicylindrical modules are capable of rotating 180° along their axis independently. The four rotation mechanisms together aid iMobot to mimic movements such as crawling, rolling, and standing along with lattice and chain structures. The iMobot

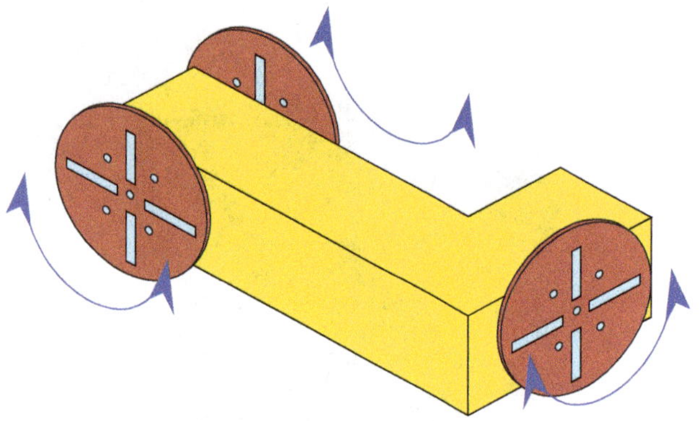

Figure 11: Model diagram of M^3 MSRR.

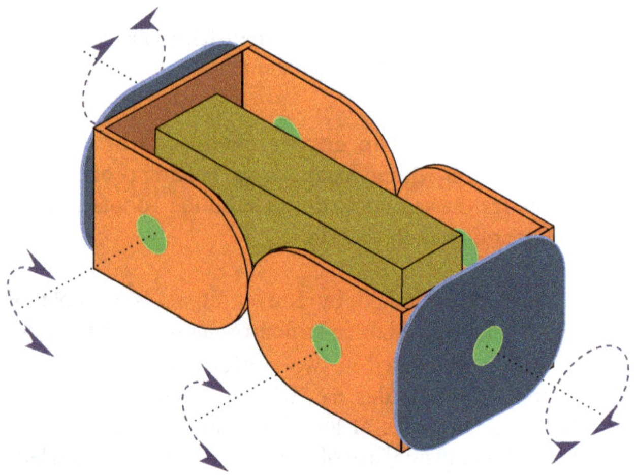

Figure 12: Model diagram of iMobot MSRR.

modules can assemble manually all the sides, hence forming various complex structures required for numerous real-time applications.

The SMORES [67] MSRR design is similar to iMobot consisting of a single semicylindrical cubic structure on which three of four side faces of cube are equipped with circular discs. Two circular discs on parallel faces play dual role in movement and docking and third for rolling neighboring modules after docking. Another internal motor provides pitch movement abilities to system by lifting third wheel. The locomotion is designed using orthogonally placed gears. Each face is equipped with four magnets with the same polarity magnets occupying alternate positions and hence at a time eight magnets participate in a docking when the connection plates face each other with an offset of 90° or 270°. The docking keys selector present internally can extend through the center of all faces creating necessary gap for undocking.

Trimobot [68] is a fully integrated mobile category hexagonal MSRR capable of forming lattice structures in 2D and chain structures in 3D. The robot is equipped internally with three omnidirectional wheels on alternate sides of hexagonal structures for movement in 2D plane. The sides of trimobot are fixed with 5 passive connection faces and an active connection face on outside. A pitch joint is embedded with active connector face on one side of hexagonal structure to facilitate lifting of modules in vertical plane and hence forming chain structures in 3D. The active connector face is also equipped with camera for docking purposes. The docking is enabled using four hooks present on active connector face and are controlled using rotation mechanism. The hooks are activated during docking when the passive and active connector faces of various modules face each other.

M-Tran is a hybrid configuration modular robot capable of forming 3D structures in both lattice and chain configurations and has three versions: M-Tran I [69, 70], M-Tran II [71, 72], and M-Tran III [73]. M-Tran robotics system consists of active and passive modules in the semicylindrical structures and a link permanently fixed in active unit as shown in Figure 13. The active and passive modules and links are equipped with four permanent magnets in a square structure on outside faces providing three connection surfaces on each module and two connection surfaces on link. The passive units can be coupled at the back of active units in two different angular orientations, 0° and 360° and 90° and 270°, due to the alignment of magnets. The connection surfaces were also designed to aid electrical connectivity between the modules. The servo motors present in active unit enabled the rotation of link and the connection is established between units after a link present on active units enters the passive unit. The latching process is controlled by SMA coils by extending or retracting the magnets in passive units docked with magnets in link. The M-Tran II latches/unlatches link with passive part at 89% more efficiency in relation to M-Tran I with a trade-off observed in time has improved torques and hardware used for sensing and control purposes. The M-Tran III is an improvised design in relation to previous versions. The latching/unlatching between link and passive part is replaced with hooks controlled by motor and hence providing more stable connection.

Superbot module proposed in [74, 75] is formed by permanently bonding two semicylindrical cells using link similar to iMobot MSRR. The cells are capable of rotating 180° along their individual axis and also can roll with respect to bond binding them. The superbot MSRR has connectors on all faces making 6 connectors in total available for each superbot module. The rotating bond and two cells together provide 3 DOF to each superbot module: 180° yaw, 180° pitch, and 270° roll. The superbot is capable of forming both lattice and chained structures and hence making it a hybrid category robot. The CKbot [76] MSRR design proposed by Yim et al. is similar to SMORES MSRR with reduction in self-mobility and rolling capabilities in individual units. The CKbot MSRR have autodocking/undocking features enabled by magnetic faces and also via screws if manual assembly is required and so forth. The CKbot MSRR is designed to test the self-healing capabilities of robotic system with the aid of vision after sudden events such as explosions.

Zykov et al. developed Molecubes [77], a cubic structure based hybrid category MSRR. The cube is assembly of two

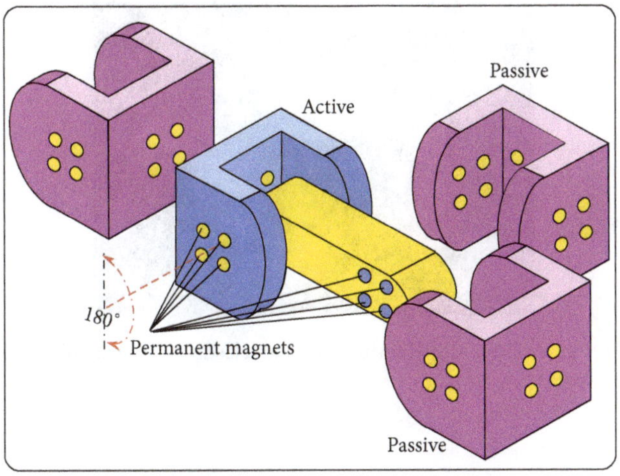

FIGURE 13: Model diagram of M-Tran MSRR.

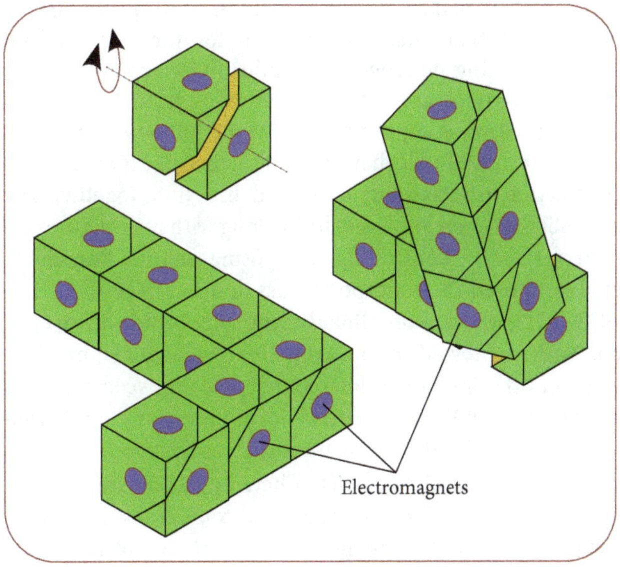

FIGURE 14: Model diagram and structures of Molecubes MSRR.

parts made by splitting cubic structure of 10 cm along the plane perpendicular to a longest diagonal as shown in Figure 14. One-half of the cube can be rotated with respect to another in multiples of 120° with the help of internal servo motor coupled with worm gear. The system is capable of forming both chained and lattice structures. The permanent pole magnets present around the center on faces facilitates coupling and the polarity of electromagnets at center can be utilized for severing or strengthening the bonds.

The UBot [78–80] MSRR system consists of cubic structured cells capable of rotating in discrete steps along longest diagonal similar to Molecubes. The internal faces are chamfered for facilitating rotation. The UBot robotic cells are categorized into active and passive modules with active modules providing four active connection interfaces and passive modules providing four passive connection interfaces. The active and passive modules have the same outer structures and rotation mechanisms. The hooks present on active connection interfaces enable firm docking with passive connectors. The active and passive modules are latched using hook and sliding mechanism guided by position sensors for forming lattice and chain structures in 3D making UBot a hybrid category robot.

Roombots [81, 82] MSRR is another hybrid architecture designed to from chained and lattice structure in 3D. Each roombot robot has two cells of spherical structure bonded together and each cell is a combination of two half-spheres mounted on each other along faces as shown in Figure 15. The locomotion is facilitated by three gear motors: one at the bond between cells and one is present in each cell for rotating other half spheres. Each roombot robot can be equipped with 10 active connections from neighboring modules to 1 active connection from a half sphere and 8 passive connections. The connection mechanism between various roombots is implemented with mechanical latches for holding neighboring modules at the holes present on surfaces.

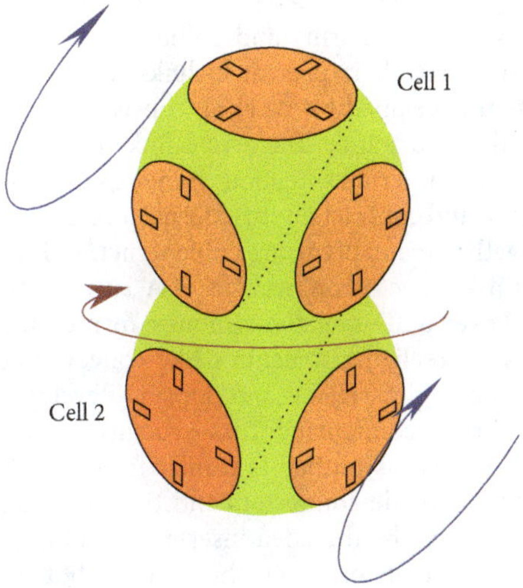

FIGURE 15: Model diagram of roombot MSRR.

Soldercubes [83, 84] developed by Neubert et al. is a hybrid category MSRR with shape of an individual Soldercube similar to a cell in dual-cell structure of roombots. The six genderless connector faces of each cell facilitate docking between modules and coordinate movement. The connector faces are custom made symmetrically designed PCB boards with soldering contacts. The contacts on connector faces can melt upon transmission of current at low temperatures, hence making a bond between modules for forming structures along with mechanical and electrical connections. The bond can be broken using the same mechanism of melting the contacts. The Soldercubes module has embedded mechanism for rotation of single connector face providing single DOF to the module but facilitating various DOF after docking with similar modules.

2.4. Truss Structured Systems. Hamlin et al. prototyped a truss based MSRR-tetrobot [85, 86] for forming random structures using heterogeneous units: links and Joints. The links in tetrobot are cylindrical rods of fixed length and reconfiguration is supported only at the joints. A three-axis concentric multilink spherical joint capable of expansion and contraction in 3D is designed to hold three links together. The assembly between joints and links along with reconfiguration is performed by controlling joints using motors. Ramchurn et al. proposed the conceptual truss design MSRR-ORTHOBOT [87] with telescopic links having split toroids at two ends and with one toroid connected to link via revolute joint. The split-toroid joint aids in interconnectivity between modules providing 2 DOF rotation. The locomotion of coordinated system is simulated for structures such as hexapod.

Odin [88] MSRR consists of heterogeneous units: Cubic Closed Packed (CCP) joints and telescopic links along with capabilities to form structures in 3D as shown in Figure 16. The CCP has twelve female connector sockets each with internal female PCB connector. The telescopic links are extendable cylindrical structures with flexible connectors on both ends equipped with male PCB connectors. The modules are not capable of autonomous docking and are fitted manually. The joints act as power sharing and communication interfaces between the controllers present in links. The Morpho [89] truss system developed by Yu et al. consists of active links, passive links, and joints. The active links can expand and contract due to internal actuation of motors and the passive links expand and contract due to external forces. The links are joined together manually using cubic structured interfacing unit with a connector on each face. A surface membrane is covered over a 3D-skeleton structure formed using links and joints for realizing structures like conveyor belts with adapting topologies. Hjelle and Lipson developed Hinge MSRR [90] for reconfiguring truss structures. The design of truss system used as testbed is similar to odin MSRR. The joints have 18 female connectors and the struts are passive cylinders fastened by threaded inserts. Instead of providing locomotion in struts or joints, the Hinge robot maneuvers from one strut to another till it reaches destination and rotates the struts with help of servos by firmly holding them and hence reconfiguring the structure.

A concept of shape-shifting materials [91] was introduced by Amend and Lipson for programmable structures. The system consists of links and nodes like general truss systems. The links are beams of granular material instead of static metal structures. The nodes are connectors between the beams supporting transfer of granular materials from one beam to another. The nodes are capable of jamming the movement and hence modifying the stiffness of beams for changing structures. Galloway et al. developed a reconfigurable truss system called factory floor [92] to demonstrate the idea of autoassembly of truss structured systems. The CKBots equipped with a manipulator is used for assembly of custom structures by placing various elements together. The joints in factor floor MSRR are cubical structures with passive connectors on each face and the struts are hollow cuboid rods with grippers at both ends for docking. The pressing action performed by manipulator at the center of strut creates a couple force internally leading to opening of grippers.

FIGURE 16: Odin MSRR (http://modular.tek.sdu.dk/index.php?page=robots).

3. Free-Form Structured Systems

Tokashiki et al. [93] prototyped a MSRR capable of forming free-form structures in 2D. The cylindrical structured MSRR (referred as Transform henceforth) is equipped with gear on top and bottom of the cylinder actuated by motors as shown in Figure 17. The robots are also equipped with 6 pole magnets around the periphery for providing bonding between the robots by attraction. The robots can move around when the gears of neighboring modules are locked with each other with magnets maintaining the structural integrity of the system. Goldstein et al. developed cylindrical structured MSRR named Claytronics [94–96] of diameter 44 mm for demonstrating various structures in 2D. The periphery of cylindrical structure is equipped with 24 spherical electromagnets in two rings present one below the other. The robots by themselves are immobile and require support of neighboring robots for forming structures as well as locomotion (on friction less surfaces). The modules have point contacts due to the shape of electromagnets and hence can implement various structures at much faster pace compared to other latched and rotating structures. Slime [97, 98] is another cylindrical design capable of forming free-form structure similar to Claytronics MSRR. The slime MSRR is equipped with 6 solenoids each controlling a 60° section of 360° periphery. Each cylinder section is equipped with a velcro to make contact with the neighboring robots. The spring action regulated by pneumatic air cylinders can extend and retract the cylinder sections for making and breaking the bond between robots. An extra solenoid placed downwards controls the position of a friction plate with respect to ground for increasing/decreasing friction during attachment/detachment process. The mini form factor MSRR, Catoms [99], is another cylindrical structure utilizing electrostatic forces for locomotion. The Catoms MSRR consists of a cylindrical wafer of 1 mm diameter and electrode strips placed vertically around the periphery of cylinder. The electrodes are sourced such that every alternate electrode holds charges of opposite polarities. The stability of structures is maintained by static fields and locomotion mechanism

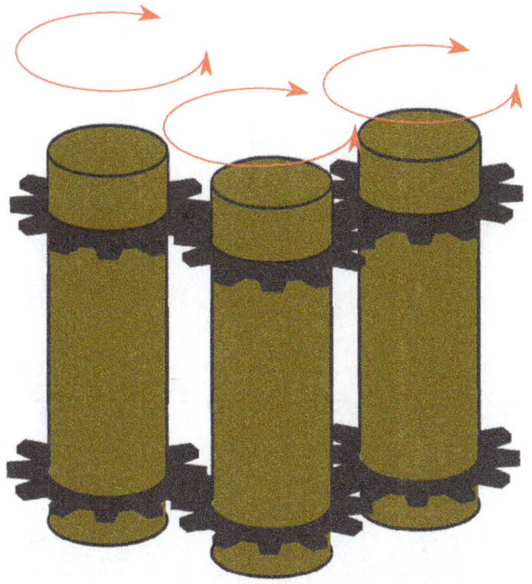

FIGURE 17: Architecture diagram of Transform MSRR.

is controlled by changing the polarities of electrodes on modules.

A micro form factor scratch drive MSRR, MEMS [100, 101], was developed by Donald et al. for forming free-form structures. The module consists of an arm and a scratch drive forming an "L" shaped structure whose structures are controlled by the voltages applied to module. The long beam acts as scratch drive for turning and the short beam in the structure is used for movements. The pulsating voltages applied to the system from bottom surface create various structures in arm and scratch drive with different frictional effects contributing to the movement. The authors explored various control algorithms and movement strategies for aligning the robots in a structure required using pulsating voltages.

The MSRR modules summarized so far are designed in various shapes such as squares and triangle for 2D scenarios and cube, cuboid, cylinder, and so forth for 3D scenarios so that the modules can have maximum contact surfaces for docking with neighbors while providing stability to coordinated structure as they adopt in the environment. The interfacing mechanism between individual MSRR modules for docking also plays vital role in restructuring the systems. Many researchers advocated the alignment issues raised during relocation of modules and stability of the interfaces for handling the increasing loads. Numerous sensor and actuator assemblies coupled with precise docking algorithms are attempted for autonomous docking of MSRR modules. The paradigm adopted for connection interfaces can be listed as male, female, active, and passive interfaces. The active connection interfaces are generally constructed using mechanical/electrical actuation assemblies for docking and the same are absent in passive connection interfaces. The passive connection interfaces still contribute to docking due to presence of passive materials like permanent magnets, sockets for screws, velcros, and so forth. The active and passive terminology is widely applied for genderless docking mechanisms and gender based docking designs differentiate between interfaces using male and female connection faces.

Table 2 provides a broad comparison of various MSRR robotic designs explained in the previous sections. The comparison is listed as per the categories mentioned in Figure 1. Since the shape generally defines the robustness of structures and the number of actuators along with type of actuator defines the parameters such as form factor and power consumption, the details of actuators and structures are also listed. The connection faces, structure, and number of connection faces on each MSRR module aid in identifying the probable structures possible when visualized in association with the shape of a robotic module. Since the connection faces are implemented using wide range of technologies, various jargons are adopted for categorizing them. The number of connection faces column in Table 2 lists details of a single robotic module in a MSRR design and is separated into two subcolumns: active and passive types for providing better visualization while interpreting locomotion capabilities. The entries in connection faces column for robotic module are listed as male (M), female (F), and dual role (DL, active and passive interfaces present on the same face). In the case of presence of heterogeneous modules in MSRR designs, the listed number is total of the active and passive interfaces present on heterogeneous units.

4. Conclusion

Research in MSRR has also extended to development of Robotic development environments, communication protocols (wired and wireless), middleware development [106, 107], human machine interface improvement, and so forth which are generally coupled with MSRR robotic modules providing a complete platform for rapid research in MSRR for development of algorithms, prototype validation, and so forth. The details on such platforms and protocols are not within scope of this paper and hence are not summarized.

In this paper a summary of various modular self-reconfigurable robotic structures is provided in terms of form factor, mobility, structural capabilities, and reconfiguration strategies. Research in MSRR as can be visualized from the summary is a deeply creative process employing technologies from mechanics and electronics and also requires deep understanding in merits/demerits of various sensor and actuation technologies. The research involves intensive prototyping and many MSRR models developed in past research with limited autonomous capabilities can be researched again due to availability of miniaturized sensor and actuator assemblies. This paper intends to provide preliminary study for prospective researchers by providing various innovations, strategies, and technologies employed in MSRR research.

Conflicts of Interest

The authors declare that they have no conflicts of interest.

TABLE 2: Comparison of MSRR designs.

Struct.	Category: Reconf.	Category: Form ftr.	Locom.	Robot	Shape	DOF	Coupling: Interface	Coupling: Actuator	Number of Conn. faces: Activ./M	Number of Conn. faces: Passv./F	Ref.
Lattice	Detrmn.	Mini	Coord.	Microunit I	Square	2D	Latch	SMA	2F	2M	[15]
				Microunit II	Cubical	3D	Latch	SMA	3F	3M	[16]
		Macro	Coord.	Metamorphic I	Hexagon	2D	Lock	DC motor	3M	3F	[6–8]
				Metamorphic II	Square	2D	Lock	DC motor	2M	2F	[7]
				Fracta	Triangular	2D	Elect. mag.	Current	3M	3F	[9]
				Vertical	Cubical	2D	Key & lock	—	0	4	[17]
				3D unit	Cubical	3D	Hooks	DC motor	6	0	[11, 12]
				Molecule	Cuboid	3D	Elect. mag.	Current	0	6	[10, 102]
				Crystalline	Cuboid	2D	Key & lock	DC motor	2M	2F	[18]
				I-cubes	Cubical	3D	Key & lock	Servo motor	2M	6F	[13, 14]
				Telecubes	Cubical	3D	Swtc. mgnts.	SMA	6 Dl. faces		[19]
				ATRON	Octagonal	3D	Hooks	DC motor	4M	4F	[30, 31]
				EM-cube	Cubical	2D	Elect. mag.	Current	2	2	[20]
				M-blocks 2D	Cubical	2D	Perm. mag.	—	0	6	[21]
				M-blocks 3D	Cubical	3D	Perm. mag.	—	0	6	[22]
	Stocst.	Mini	Mobile	Petro	Tetrahedral	3D	Latch	Manual	0	4	[32]
			Extern.	Pebbles	Cubical	2D	El-pm. mag.	Current	4	0	[24]
				Miche	Cubical	3D	Rot. pr. mg.	Motor	3	3	[23]
		Macro	Extern.	Stochastic 2D	Cubical	2D	Elect. mag.	Current	4	0	[25]
				Stochastic 3D	Cubical	3D	Elect. mag.	Current	6	0	[26]
				Prog. parts	Triangular	2D	Perm. mag.	DC motor	3	0	[27]
				X-bot	Cubical	2D	Perm. mag.	SMA	4	0	[28, 29]
Chain	Detrmn.	Macro	Mobile	CEBOT I	Cuboid	3D	Latch	SMA	1M	1F	[33]
				CEBOT II	Cuboid	3D	Latch	DC motor	1M	1F	[34]
				ACM-R1	Rectangle	2D	Latch	Manual	0	2	[35]
				ACM-R2	Rectangle	3D	Latch	Manual	0	2	[36]
				ACM-R3	Rectangle	3D	Latch	Manual	0	2	[37]
				Millibot	Elliptical	2D	Latch	SMA	1M	1F	[38]
				UNI-Rover	Cylind. arm	3D	Latch	Servo motor	1	1	[103]
				Amoeba - I	Cuboid	3D	Latch	Manual	0	2	[39, 40]
				JL-1	Trapezoid	3D	Latch	DC motor	1M	1F	[41]
				JL-2	Trapezoid	3D	Gripper	DC motor	1M	1F	[42]
				Thor	Cuboid. arm	3D	Gripper	DC motor	1	0	[43]
				Steering	Cuboid	2D	Latch	—	1M	1F	[104]
			Co-ord.	Polypod	Cubical	3D	Latch	Manual	0	6	[44]
				CONRO	Cuboid	3D	Latch	SMA	1M	3F	[45–48]
				Polybot I	Cubical	3D	Latch	Manual	1M	1F	[50]
				Polybot II	Cubical	3D	Latch	SMA	1M	1F	[51]
				Polybot III	Cubical	3D	Latch	SMA	1M	1F	[51]
				Yamor	Semicylnd.	3D	Velcros	Manual	0	4	[54]
				GZ-1	Cubic	3D	Latch	Manual	0	4	[53]
				Transmote	Cubic	3D	Key & lock	Manual	1M	2F	[52]
				ModRED	Cuboid	3D	Latch	Solenoid	2 Dl. faces		[49]

Table 2: Continued.

Struct.	Category: Reconf.	Category: Form ftr.	Locom.	Robot	Shape	DOF	Coupling: Interface	Coupling: Actuator	Number of Conn. faces: Activ./M	Number of Conn. faces: Passv./F	Ref.
Hybrid	Determn.	Macro	Mobile	S-BOT	Cylindrical	3D	Gripper	Motor	1	1	[55–59]
				M^3	L shaped	3D	Hooks	DC motor	3 DL. conn.		[63]
				M^3 express	L shaped	3D	Latch	SMA, servo	3 DL. conn.		[64]
				iMobot	Cuboidal	3D	Latch	Manual	0	6	[65]
				SMORES	Cubical	3D	Perm. mag	DC motor	3	1	[67]
				Trimobot	Hexagonal	3D	Hooks	DC motors	1M	6F	[68]
			Co-ord.	M-Tran I	Semicylnd.	3D	Perm. mag	SMA	6	2	[69, 105]
				M-Tran II	Semicylnd.	3D	Perm. mag	SMA	6	2	[71, 72]
				M-Tran III	Semicylnd.	3D	Hooks	DC motor	6	2	[73]
				Superbot	Cuboid	3D	Latch	Manual	0	6	[74, 75]
				Molecubes	Cubical	3D	Elect. mag.	Current	6	0	[77]
				CKBot	Cubical	3D	Perm. magnets	Manual	0	4	[76]
				UBot	Cubical	3D	Hooks	DC motor	2	2	[78–80]
				Roombots	Cuboidal	3D	Latch	Manual	0–10	0–10	[81]
				Neurobot	Cubical	3D	Latch	DC motor	1M	1F	[5]
				Soldercubes	Cubical	3D	Binder mat.	Current	6	0	[83, 84]
Truss	Determn.	Macro	Co-ord.	Tetrobot	Cylindrical	3D	Spherical jnt.	Manual	2M	3F	[85, 86]
				ORTHO-BOT	Linact	3D	Split toroid	Manual	0	2	[87]
				Odin	Cylindrical	3D	CCP jnt.	Manual	2M	12F	[88]
				Morpho	Cubical	3D	Cubic jnt.	Manual	1	6	[89]
				Shape shift.	Amorphous	3D	3-way pipe	Manual	2M	3F	[91]
				Hinge	Cylindrical	3D	18-axis node	Manual	2M	18F	[90]
				Factory Flr.	Cylindrical	3D	Grippers	Couples	6M	2F	[92]
Free Fr.	Determn.	Micro	Coord.	MEMS	L shaped	2D	Alignment	Voltage	—	—	[100, 101]
		Mini	Coord.	Claytronics	Cylindrical	2D	Electr. mag.	Current	24	0	[94–96]
			Coord.	Catom	Cylindrical	2D	Electrodes	Voltage	8	0	[99]
		Macro	Coord.	Slimebot	Cylindrical	2D	Velcros	Pnu. air cyld.	0	6	[97, 98]
			Coord.	Transform.	Cylindrical	2D	Perm. mag.	—	6	2	[93]

References

[1] J. Baca, P. Pagala, C. Rossi, and M. Ferre, "Modular robot systems towards the execution of cooperative tasks in large facilities," *Robotics and Autonomous Systems*, vol. 66, pp. 159–174, 2015.

[2] M. Yim, W.-M. Shen, B. Salemi et al., "Modular self-reconfigurable robot systems [Grand challenges of robotics]," *IEEE Robotics and Automation Magazine*, vol. 14, no. 1, pp. 43–52, 2007.

[3] M. Yim, P. White, M. Park, and J. Sastra, "Modular self-reconfigurable robots," in *Encyclopedia of Complexity and Systems*, pp. 19–33, Springer, New York, NY, USA, 2009.

[4] K. Gilpin and D. Rus, "Modular Robot Systems," *IEEE Robotics & Automation Magazine*, vol. 17, no. 3, pp. 38–55, 2010.

[5] P. Moubarak and P. Ben-Tzvi, "Modular and reconfigurable mobile robotics," *Robotics and Autonomous Systems*, vol. 60, no. 12, pp. 1648–1663, 2012.

[6] G. S. Chirikjian, "Kinematics of a metamorphic robotic system," in *Proceedings of the IEEE International Conference on Robotics and Automation*, pp. 449–455, IEEE, San Diego, Calif, USA, May 1994.

[7] A. P. Chirikjian, D. Stein, C.-J. Chiang, and Grigory, "Design and Implementation of metamorphic robots," in *Proceedings of the Design Engineering Technical Conference and Computers in Engineering Conference*, pp. 1–10, ASME, Irvine, Calif, USA, 1996.

[8] A. Pamecha, I. Ebert-Uphoff, and G. S. Chirikjian, "Useful metrics for modular robot motion planning," *IEEE Transactions on Robotics and Automation*, vol. 13, no. 4, pp. 531–545, 1997.

[9] S. Murata, H. Kurokawa, and S. Kokaji, "Self-assembling machine," in *Proceedings of the IEEE International Conference on Robotics and Automation*, pp. 441–448, San Diego, Calif, USA, 1994.

[10] D. Rus, "Self-reconfiguring robots," *IEEE Intelligent Systems*, vol. 13, no. 4, pp. 2–4, 1998.

[11] H. Kurokawa, S. Murata, E. Yoshida, K. Tomita, and S. Kokaji, "A 3-D self-reconfigurable structure and experiments," in *Proceedings of the IEEE/RSJ International Conference on Intelligent Robots and Systems. Innovations in Theory, Practice and Applications*, vol. 2, pp. 860–865, Victoria, Canada, 1998.

[12] S. Murata, H. Kurokawa, E. Yoshida, K. Tomita, and S. Kokaji, "A 3-D self-reconfigurable structure," in *Proceedings of the IEEE International Conference on Robotics and Automation*, vol. 1, pp. 432–439, IEEE, Leuven, Belgium, May 1998.

[13] C. Ünsal, H. Kiliççöte, and P. K. Khosla, "Modular self-reconfigurable bipartite robotic system: implementation and motion planning," *Autonomous Robots*, vol. 10, no. 1, pp. 23–40, 2001.

[14] C. Unsal and P. Khosla, "Mechatronic design of a modular self-reconfiguring robotic system," in *Proceedings of the Millennium Conference on IEEE International Conference on Robotics and Automation (ICRA '00)*, vol. 2, pp. 1742–1747, IEEE, April 2000, Symposia Proceedings Cat. No. 00CH37065.

[15] E. Yoshida, S. Murata, S. Kokaji, K. Tomita, and H. Kurokawa, "Micro self-reconfigurable robotic system using shape memory alloy," in *Proceedings of the Symposium on Distributed Autonomous Robotic Systems (DARS '00)*, vol. 4, pp. 145–154, Springer, Tokyo, Japan, 2000.

[16] E. Yoshida, S. Murata, S. Kokaji, A. Kamimura, K. Tomita, and H. Kurokawa, "Get back in shape! [SMA self-reconfigurable microrobots]," *IEEE Robotics & Automation Magazine*, vol. 9, no. 4, pp. 54–60, 2002.

[17] K. Hosokawa, T. Tsujimori, T. Fujii et al., "Self-organizing collective robots with morphogenesis in a vertical plane," in *Proceedings of the IEEE International Conference on Robotics and Automation*, pp. 2858–2863, May 1998.

[18] D. Rus and M. Vona, "A physical implementation of the self-reconfiguring crystalline robot," in *Proceedings of the IEEE International Conference on Robotics and Automation (ICRA '00)*, vol. 2, pp. 1726–1733, San Francisco, Calif, USA, April 2000.

[19] J. Suh, S. Homans, and M. Yim, "Telecubes: mechanical design of a module for self-reconfigurable robotics," in *Proceedings of the IEEE International Conference on Robotics and Automation*, pp. 4095–4101, IEEE, Washington, DC, USA, May 2002.

[20] B. K. An, "EM-Cube: cube-shaped, self-reconfigurable robots sliding on structure surfaces," in *Proceedings of the IEEE International Conference on Robotics and Automation (ICRA '08)*, pp. 3149–3155, Pasadena, Calif, USA, May 2008.

[21] J. W. Romanishin, K. Gilpin, and D. Rus, "M-blocks: momentum-driven, magnetic modular robots," in *Proceedings of the 26th IEEE/RSJ International Conference on Intelligent Robots and Systems (IROS '13)*, pp. 4288–4295, IEEE, Tokyo, Japan, November 2013.

[22] J. W. Romanishin, K. Gilpin, S. Claici, and D. Rus, "3D M-Blocks: self-reconfiguring robots capable of locomotion via pivoting in three dimensions," in *Proceedings of the IEEE International Conference on Robotics and Automation (ICRA '15)*, pp. 1925–1932, Seattle, Wash, USA, May 2015.

[23] K. Gilpin, K. Kotay, and D. Rus, "Miche: modular shape formation by self-dissasembly," in *Proceedings of the IEEE International Conference on Robotics and Automation (ICRA '07)*, pp. 2241–2247, IEEE, Roma, Italy, April 2007.

[24] K. Gilpin, A. Knaian, and D. Rus, "Robot pebbles: one centimeter modules for programmable matter through self-disassembly," in *Proceedings of the IEEE International Conference on Robotics and Automation (ICRA '10)*, pp. 2485–2492, May 2010.

[25] P. White, K. Kopanski, and H. Lipson, "Stochastic self-reconfigurable cellular robotics," in *Proceedings of the IEEE International Conference on Robotics and Automation (ICRA '04)*, vol. 3, pp. 2888–2893, New Orleans, La, USA, April 2004.

[26] P. White, V. Zykov, J. Bongard, and H. Lipson, "Three dimensional stochastic reconfiguration of modular robots," in *Proceedings of the International Conference on Robotics: Science and Systems (RSS '05)*, pp. 161–168, June 2005.

[27] J. Bishop, S. Burden, E. Klavins et al., "Programmable parts: a demonstration of the grammatical approach to self-organization," in *Proceedings of the IEEE/RSJ International Conference on Intelligent Robots and Systems*, pp. 3684–3691, IEEE, Alberta, Canada, August 2005.

[28] P. J. White and M. Yim, "Scalable modular self-reconfigurable robots using external actuation," in *Proceedings of the IEEE/RSJ International Conference on Intelligent Robots and Systems (IROS '07)*, pp. 2773–2778, November 2007.

[29] P. J. White and M. Yim, "Reliable external actuation for extending reachable robotic modular self-reconfiguration," in *Proceedings of the 11th International Symposium on Experimental Robotics*, pp. 13–23, Athens, Greece, 2009.

[30] M. W. Jørgensen, E. H. Østergaard, and H. H. Lund, "Modular ATRON: modules for a self-reconfigurable robot," in *Proceedings of the IEEE/RSJ International Conference on Intelligent

Robots and Systems (IROS '04), pp. 2068–2073, IEEE, Sendai, Japan, October 2004.

[31] U. P. Schultz, M. Bordignon, and K. Stoy, "Robust and reversible self-reconfiguration," in *Proceedings of the IEEE/RSJ International Conference on Intelligent Robots and Systems (IROS '09)*, pp. 5287–5294, October 2009.

[32] B. Salem, "PetRo: development of a modular pet robot," in *Proceedings of the 23rd IEEE International Symposium on Robot and Human Interactive Communication*, pp. 483–488, IEEE, August 2014.

[33] T. Fukuda and S. Nakagawa, "Dynamically reconfigurable robotic system," in *Proceedings of the IEEE International Conference on Robotics and Automation*, pp. 1581–1586, IEEE Computer Society Press, Philadelphia, Pa, USA, 1988.

[34] T. Fukuda and S. Nakagawa, "Method of autonomous approach, docking and detaching between cells for dynamically reconfigurable robotic system CEBOT," *JSME International Journal. Series 3: Vibration, Control Engineering, Engineering for Industry*, vol. 33, no. 2, pp. 263–268, 1990, http://ci.nii.ac.jp/naid/130003964367/en/.

[35] G. Endo, K. Togawa, and S. Hirose, "Study on self-contained and terrain adaptive active cord mechanism," in *Proceedings of the IEEE/RSJ International Conference on Intelligent Robots and Systems. Human and Environment Friendly Robots with High Intelligence and Emotional Quotients (Cat. No.99CH36289)*, vol. 3, no. 1, pp. 1399–1405, Kyongju, South Korea, 1999.

[36] K. Togawa, M. Mori, and S. Hirose, "Study on three-dimensional active cord mechanism: development of ACM-R2," in *Proceedings of the IEEE/RSJ International Conference on Intelligent Robots and Systems (IROS '00)*, vol. 3, pp. 2242–2247, November 2000.

[37] M. Mori and S. Hirose, "Development of active cord mechanism ACM-R3 with agile 3D mobility," in *Proceedings of the IEEE/RSJ International Conference on Intelligent Robots and Systems*, vol. 3, pp. 1552–1557, IEEE, Maui, Hawaii, USA, 2001.

[38] H. B. Brown Jr., J. M. Vande Weghe, C. A. Bererton, and P. K. Khosla, "Millibot trains for enhanced mobility," *IEEE/ASME Transactions on Mechatronics*, vol. 7, no. 4, pp. 452–461, 2002.

[39] L. Jinguo, M. Shugen, L. Zhenli, W. Yuechao, L. Bin, and W. Jing, "Design and experiment of a novel link-type shape shifting modular robot series," in *Proceedings of the IEEE International Conference on Robotics and Biomimetics (ROBIO '05)*, pp. 318–323, July 2005.

[40] B. Li, S. Ma, J. Liu, M. Wang, T. Liu, and Y. Wang, "AMOEBA-I: a shape-shifting modular robot for urban search and rescue," *Advanced Robotics*, vol. 23, no. 9, pp. 1057–1083, 2009.

[41] D. Li, H. Fu, and W. Wang, "Ultrasonic based autonomous docking on plane for mobile robot," in *Proceedings of the IEEE International Conference on Automation and Logistics (ICAL '08)*, pp. 1396–1401, IEEE, Qingdao, China, September 2008.

[42] W. Wang, W. Yu, and H. Zhang, "JL-2: a mobile multi-robot system with docking and manipulating capabilities," *International Journal of Advanced Robotic Systems*, vol. 7, no. 1, pp. 9–18, 2010.

[43] A. Lyder, R. Franco, M. Garcia, and K. Stoy, "Genderless connection mechanism for modular robots introducing torque transmission between modules," in *Proceedings of the ICRA 2010 Workshop on Modular Robots: The State of the Art*, pp. 77–82, 2010.

[44] M. Yim, "New locomotion gaits," in *Proceedings of the IEEE International Conference on Robotics and Automation*, pp. 2508–2514, IEEE Computer Society Press, San Diego, Calif, USA, 1994.

[45] A. Castano, R. Chokkalingam, and P. Will, "Autonomous and self-sufficient CONRO Modules for Recon gurable robots," in *Distributed Autonomous Robotic Systems 4*, chapter 5, pp. 155–164, Springer, Tokyo, Japan, 2000.

[46] W.-M. Shen, B. Salemi, and P. Will, "Hormone-inspired adaptive communication and distributed control for CONRO self-reconfigurable robots," *IEEE Transactions on Robotics and Automation*, vol. 18, no. 5, pp. 700–712, 2002.

[47] W.-M. Shen, Y. Lu, and P. Will, "Hormone-based control for self-reconfigurable robots," in *Proceedings of the 4th International Conference on Autonomous Agents (AGENTS '00)*, vol. 00, pp. 1–8, 2000, http://portal.acm.org/citation.cfm?doid=336595.336602.

[48] M. Rubenstein, K. Payne, P. Will, and W.-M. Shen, "Docking among independent and autonomous CONRO self-reconfigurable robots," in *Proceedings of the IEEE International Conference on Robotics and Automation (ICRA '04)*, vol. 3, pp. 2877–2882, May 2004.

[49] P. Dasgupta, J. Baca, A. Dutta, S. M. G. Hossain, and C. Nelson, "Mechanical design and computational aspects for locomotion and reconfiguration of the ModRED Modular Robot," in *Proceedings of the 12th International Conference on Autonomous Agents and Multiagent Systems (AAMAS '13)*, pp. 1359–1360, ACM, Saint Paul, Minn, USA, May 2013.

[50] M. Yim, D. Duff, and K. Roufas, "PolyBot: a modular reconfigurable robot," in *Proceedings of the IEEE International Conference on Robotics and Automation (ICRA '00)*, vol. 1, pp. 514–520, San Francisco, Calif, USA, 2000.

[51] M. Yim, Y. Zhang, K. Roufas, D. Duff, and C. Eldershaw, "Connecting and disconnecting for chain self-reconfiguration with polybot," *IEEE/ASME Transactions on Mechatronics*, vol. 7, no. 4, pp. 442–451, 2002.

[52] G. Qiao, G. Song, J. Zhang, H. Sun, W. Wang, and A. Song, "Design of transmote: a modular self-reconfigurable robot with versatile transformation capabilities," in *Proceedings of the IEEE International Conference on Robotics and Biomimetics (ROBIO '12)*, pp. 1331–1336, December 2012.

[53] Y. Li, H. Zhang, and S. Chen, "A four-legged robot based on GZ-I modules," in *Proceedings of the IEEE International Conference on Robotics and Biomimetics (ROBIO '08)*, pp. 921–926, Bangkok, Thailand, February 2009.

[54] R. Moeckel, C. Jaquier, K. Drapel, E. Dittrich, A. Upegui, and A. J. Ijspeert, "Exploring adaptive locomotion with YaMoR, a novel autonomous modular robot with Bluetooth interface," *Industrial Robot*, vol. 33, no. 4, pp. 285–290, 2006.

[55] F. Mondada, G. C. Pettinaro, A. Guignard et al., "Swarm-bot: a new distributed robotic concept," *Autonomous Robots*, vol. 17, no. 2-3, pp. 193–221, 2004.

[56] M. Dorigo, "SWARM-BOT: an experiment in swarm robotics," in *Proceedings of the IEEE Swarm Intelligence Symposium (SIS '05)*, pp. 192–200, IEEE, Pasadena, Calif, USA, June 2005.

[57] E. Tuci, R. Gross, V. Trianni, F. Mondada, M. Bonani, and M. Dorigo, "Cooperation through self-assembly in multi-robot systems," *ACM Transactions on Autonomous and Adaptive Systems*, vol. 1, no. 2, pp. 115–150, 2006.

[58] R. Gro, M. Bonani, F. Mondada, and M. Dorigo, "Autonomous self-assembly in swarm-bots," *IEEE Transactions on Robotics*, vol. 22, no. 6, pp. 1115–1130, 2006.

[59] R. Groß, E. Tuci, M. Dorigo, M. Bonani, and F. Mondada, "Object transport by modular robots that self-assemble," in *Proceedings of the IEEE International Conference on Robotics and Automation (ICRA '06)*, pp. 2558–2564, May 2006.

[60] V. Trianni, S. Nolfi, and M. Dorigo, "Cooperative hole avoidance in a swarm-bot," *Robotics and Autonomous Systems*, vol. 54, no. 2, pp. 97–103, 2006.

[61] R. O'Grady, R. Groß, A. L. Christensen, F. Mondada, M. Bonani, and M. Dorigo, "Performance benefits of self-assembly in a swarm-bot," in *Proceedings of the IEEE/RSJ International Conference on Intelligent Robots and Systems (IROS '07)*, pp. 2381–2387, November 2007.

[62] G. Fu, A. Menciassi, and P. Dario, "Development of a genderless and fail-safe connection system for autonomous modular robots," in *Proceedings of the IEEE International Conference on Robotics and Biomimetics (ROBIO '11)*, pp. 877–882, IEEE, Karon Beach, Thailand, December 2011.

[63] M. D. M. Kutzer, M. S. Moses, C. Y. Brown, D. H. Scheidt, G. S. Chirikjian, and M. Armand, "Design of a new independently-mobile reconfigurable modular robot," in *Proceedings of the IEEE International Conference on Robotics and Automation (ICRA '10)*, pp. 2758–2764, May 2010.

[64] K. C. Wolfe, M. S. Moses, M. D. M. Kutzer, and G. S. Chirikjian, "M^3Express: a low-cost independently-mobile reconfigurable modular robot," in *Proceedings of the IEEE International Conference on Robotics and Automation*, pp. 2704–2710, IEEE, St. Paul, Minn, USA, May 2012.

[65] G. G. Ryland and H. H. Cheng, "Design of iMobot, an intelligent reconfigurable mobile robot with novel locomotion," in *Proceedings of the IEEE International Conference on Robotics and Automation (ICRA '10)*, pp. 60–65, May 2010.

[66] D. Ko and H. H. Cheng, "Programming reconfigurable modular robots," in *Proceedings of the 8th IEEE/ASME International Conference on Mechatronic and Embedded Systems and Applications (MESA '12)*, pp. 160–165, July 2012.

[67] J. Davey, N. Kwok, and M. Yim, "Emulating self-reconfigurable robots—design of the SMORES system," in *Proceedings of the 25th IEEE/RSJ International Conference on Robotics and Intelligent Systems (IROS '12)*, pp. 4464–4469, October 2012.

[68] Y. Zhang, G. Song, S. Liu, G. Qiao, J. Zhang, and H. Sun, "A modular self-reconfigurable robot with enhanced locomotion performances: design, modeling, simulations, and experiments," *Journal of Intelligent and Robotic Systems: Theory and Applications*, vol. 81, no. 3-4, pp. 377–393, 2016.

[69] S. Murata, E. Yoshida, A. Kamimura, H. Kurokawa, K. Tomita, and S. Kokaji, "M-TRAN: self-reconfigurable modular robotic system," *IEEE/ASME Transactions on Mechatronics*, vol. 7, no. 4, pp. 431–441, 2002.

[70] E. Yoshida, S. Murata, A. Kamimura, K. Tomita, H. Kurokawa, and S. Kokaji, "A motion planning method for a self-reconfigurable modular robot," in *Proceedings of the RSJ/IEEE International Conference on Intelligent Robots and Systems*, vol. 1, pp. 590–597, Maui, Hawaii, USA, 2000.

[71] H. Kurokawa, A. Kamimura, E. Yoshida, K. Tomita, S. Kokaji, and S. Murata, "M-TRAN II: metamorphosis from a four-legged walker to a caterpillar," in *Proceedings of the IEEE/RSJ International Conference on Intelligent Robots and Systems (IROS '03)*, vol. 3, pp. 2454–2459, Las Vegas, Nev, USA, October 2003.

[72] A. Kamimura, H. Kurokawa, E. Yoshida, K. Tomita, S. Kokaji, and S. Murata, "Distributed adaptive locomotion by a modular robotic system, M-TRAN II," in *Proceedings of the IEEE/RSJ International Conference on Intelligent Robots and Systems (IROS '04)*, vol. 3, pp. 2370–2377, IEEE, Septemper-October 2004, IEEE Cat. No. 04CH37566.

[73] H. Kurokawa, K. Tomita, A. Kamimura, S. Kokaji, T. Hasuo, and S. Murata, "Distributed self-reconfiguration of M-TRAN III modular robotic system," *International Journal of Robotics Research*, vol. 27, no. 3-4, pp. 373–386, 2008.

[74] B. Salemi, M. Moll, and W.-M. Shen, "SUPERBOT: a deployable, multi-functional, and modular self-reconfigurable robotic system," in *Proceedings of the IEEE/RSJ International Conference on Intelligent Robots and Systems (IROS '06)*, pp. 3636–3641, October 2006.

[75] W.-M. Shen, M. Krivokon, H. Chiu, J. Everist, M. Rubenstein, and J. Venkatesh, "Multimode locomotion via SuperBot robots," in *Proceedings of the IEEE International Conference on Robotics and Automation (ICRA '06)*, pp. 2552–2557, Orlando, Fla, USA, May 2006.

[76] M. Yim, B. Shirmohammadi, J. Sastra, M. Park, M. Dugan, and C. J. Taylor, "Towards robotic self-reassembly after explosion," in *Proceedings of the IEEE/RSJ International Conference on Intelligent Robots and Systems (IROS '07)*, pp. 2767–2772, IEEE, San Diego, Calif, USA, November 2007.

[77] V. Zykov, E. Mytilinaios, M. Desnoyer, and H. Lipson, "Evolved and designed self-reproducing modular robotics," *IEEE Transactions on Robotics*, vol. 23, no. 2, pp. 308–319, 2007.

[78] S. T. S. Tang, Y. Z. Y. Zhu, J. Z. J. Zhao, and X. C. X. Cui, "The UBot modules for self-reconfigurable robot," in *Proceedings of the ASME/IFToMM International Conference on Reconfigurable Mechanisms and Robots (ReMAR '09)*, pp. 529–535, June 2009.

[79] Y. Zhu, J. Zhao, X. Cui et al., "Design and implementation of UBot: a modular self-reconfigurable robot," in *Proceedings of the 10th IEEE International Conference on Mechatronics and Automation (ICMA '13)*, pp. 1217–1222, IEEE, Takamatsu, Japan, August 2013.

[80] Y. Zhu, H. Jin, X. Zhang, J. Yin, P. Liu, and J. Zhao, "A multi-sensory autonomous docking approach for a self-reconfigurable robot without mechanical guidance," *International Journal of Advanced Robotic Systems*, vol. 11, 2014.

[81] A. Sproewitz, A. Billard, P. Dillenbourg, and A. J. Ijspeert, "Roombots-mechanical design of self-reconfiguring modular robots for adaptive furniture," in *Proceedings of the IEEE International Conference on Robotics and Automation (ICRA '09)*, pp. 4259–4264, IEEE, Kobe, Japan, May 2009.

[82] A. Sproewitz, P. Laprade, S. Bonardi et al., "Roombots—towards decentralized reconfiguration with self-reconfiguring modular robotic metamodules," in *Proceedings of the 23rd IEEE/RSJ International Conference on Intelligent Robots and Systems (IROS '10)*, pp. 1126–1132, October 2010.

[83] J. Neubert, A. Rost, and H. Lipson, "Self-soldering connectors for modular robots," *IEEE Transactions on Robotics*, pp. 1–14, 2014.

[84] J. Neubert and H. Lipson, "Soldercubes: a self-soldering self-reconfiguring modular robot system," *Autonomous Robots*, vol. 40, no. 1, pp. 139–158, 2016.

[85] G. J. Hamlin and A. C. Sanderson, "Tetrobot: a modular system for hyper-redundant parallel robotics," in *Proceedings of the IEEE International Conference on Robotics and Automation 1995*, May 1995.

[86] G. J. Hamlin and A. C. Sanderson, "TETROBOT: a modular approach to parallel robotics," *IEEE Robotics and Automation Magazine*, vol. 4, no. 1, p. 42, 1997.

[87] V. Ramchurn, R. C. Richardson, and P. Nutter, "ORTHO-BOT: a modular reconfigurable space robot concept," in *Climbing and Walking Robots*, pp. 659–666, Springer, Berlin, Germany, 2006.

[88] A. Lyder, R. F. M. Garcia, and K. Stoy, "Mechanical design of Odin, an extendable heterogeneous deformable modular

robot," in *Proceedings of the IEEE/RSJ International Conference on Intelligent Robots and Systems (IROS '08)*, pp. 883–888, IEEE, Nice, France, September 2008.

[89] C.-H. Yu, K. Haller, D. Ingber, and R. Nagpal, "Morpho: a self-deformable modular robot inspired by cellular structure," in *Proceedings of the IEEE/RSJ International Conference on Intelligent Robots and Systems (IROS '08)*, pp. 3571–3578, September 2008.

[90] D. Hjelle and H. Lipson, "A robotically reconfigurable truss," in *Proceedings of the ASME/IFToMM International Conference on Reconfigurable Mechanisms and Robots (ReMAR '09)*, pp. 73–78, June 2009.

[91] J. R. Amend and H. Lipson, "Shape-shifting materials for programmable structures," in *Proceedings of the in International Conference on Ubiquitous Computing: Workshop on Architectural Robotics*, September-October 2009.

[92] K. C. Galloway, R. Jois, and M. Yim, "Factory floor: a robotically reconfigurable construction platform," in *Proceedings of the IEEE International Conference on Robotics and Automation (ICRA '10)*, pp. 2467–2472, IEEE, May 2010.

[93] H. Tokashiki, H. Amagai, S. Endo, K. Yamada, and J. Kelly, "Development of a transformable mobile robot composed of homogeneous gear-type units," in *Proceedings of the IEEE/RSJ International Conference on Intelligent Robots and Systems (IROS '03)*, vol. 2, pp. 1602–1607, October 2003.

[94] S. Goldstein and T. Mowry, "Claytronics: an instance of programmable matter," in *Proceedings of the Wild and Crazy Ideas Session of ASPLOS*, vol. 17, Boston, Mass, USA, October 2004.

[95] S. C. Goldstein, J. D. Campbell, and T. C. Mowry, "Programmable matter," *Computer*, vol. 38, no. 6, pp. 99–101, 2005.

[96] J. Campbell, P. Pillai, and S. C. Goldstein, "The robot is the tether: active, adaptive power routing for modular robots with unary inter-robot connectors," in *Proceedings of the IEEE IRS/RSJ International Conference on Intelligent Robots and Systems (IROS '05)*, pp. 2960–2967, August 2005.

[97] M. Shimizu, A. Ishiguro, and T. Kawakatsu, "A modular robot that exploits a spontaneous connectivity control mechanism," in *Proceedings of the IEEE IRS/RSJ International Conference on Intelligent Robots and Systems (IROS '05)*, pp. 2658–2663, Edmonton, Canada, August 2005.

[98] M. Shimizu, T. Mori, and A. Ishiguro, "A development of a modular robot that enables adaptive reconfiguration," in *Proceedings of the IEEE/RSJ International Conference on Intelligent Robots and Systems (IROS '06)*, pp. 174–179, Beijing, China, October 2006.

[99] M. E. Karagozler, S. C. Goldstein, and J. R. Reid, "Stress-driven MEMS assembly + electrostatic forces = 1mm diameter robot," in *Proceedings of the IEEE/RSJ International Conference on Intelligent Robots and Systems (IROS '09)*, pp. 2763–2769, IEEE, St. Louis, Mo, USA, October 2009.

[100] B. R. Donald, C. G. Levey, C. G. McGray, I. Paprotny, and D. Rus, "An untethered, electrostatic, globally controllable MEMS micro-robot," *Journal of Microelectromechanical Systems*, vol. 15, no. 1, pp. 1–15, 2006.

[101] B. R. Donald, C. G. Levey, and I. Paprotny, "Planar microassembly by parallel actuation of MEMS microrobots," *Journal of Microelectromechanical Systems*, vol. 17, no. 4, pp. 789–808, 2008.

[102] K. Kotay, D. Rus, M. Vona, and C. McGray, "The self-reconfiguring robotic molecule," in *Proceedings of the IEEE International Conference on Robotics and Automation 1998*, pp. 424–431, May 1998.

[103] A. Kawakami, A. Torii, K. Motomura, and S. Hirose, "SMC rover: planetary rover with transformable wheels," in *Experimental Robotics VIII*, vol. 5 of *Springer Tracts in Advanced Robotics*, pp. 498–506, Springer, Berlin, Germany, 2003.

[104] M. Delrobaei and K. A. McIsaac, "Design and steering control of a center-articulated mobile robot module," *Journal of Robotics*, vol. 2011, Article ID 621879, 14 pages, 2011.

[105] A. Kamimura, S. Murata, E. Yoshida, H. Kurokawa, K. Tomita, and S. Kokaji, "Self-reconfigurable modular robot—experiments on reconfiguration and locomotion," in *Proceedings of the IEEE/RSJ International Conference on Intelligent Robots and Systems*, pp. 606–612, IEEE, Maui, Hawaii, USA, November 2001.

[106] M. Mamei and F. Zambonelli, "Programming modular robots with the TOTA middleware," in *Proceedings of the 4th International Conference on Engineering Self-organising Systems (ESOA '06)*, pp. 99–114, Springer, Hakodate, Japan, 2007, http://dl.acm.org/citation.cfm?id=1763581.1763591.

[107] W. Hongxing, L. Shiyi, Z. Ying, Y. Liang, and W. Tianmiao, "A middleware based control architecture for modular robot systems," in *Proceedings of the IEEE/ASME International Conference on Mechtronic and Embedded Systems and Applications (MESA '08)*, pp. 327–332, Beijing, China, October 2008.

Permissions

All chapters in this book were first published in JR, by Hindawi Publishing Corporation; hereby published with permission under the Creative Commons Attribution License or equivalent. Every chapter published in this book has been scrutinized by our experts. Their significance has been extensively debated. The topics covered herein carry significant findings which will fuel the growth of the discipline. They may even be implemented as practical applications or may be referred to as a beginning point for another development.

The contributors of this book come from diverse backgrounds, making this book a truly international effort. This book will bring forth new frontiers with its revolutionizing research information and detailed analysis of the nascent developments around the world.

We would like to thank all the contributing authors for lending their expertise to make the book truly unique. They have played a crucial role in the development of this book. Without their invaluable contributions this book wouldn't have been possible. They have made vital efforts to compile up to date information on the varied aspects of this subject to make this book a valuable addition to the collection of many professionals and students.

This book was conceptualized with the vision of imparting up-to-date information and advanced data in this field. To ensure the same, a matchless editorial board was set up. Every individual on the board went through rigorous rounds of assessment to prove their worth. After which they invested a large part of their time researching and compiling the most relevant data for our readers.

The editorial board has been involved in producing this book since its inception. They have spent rigorous hours researching and exploring the diverse topics which have resulted in the successful publishing of this book. They have passed on their knowledge of decades through this book. To expedite this challenging task, the publisher supported the team at every step. A small team of assistant editors was also appointed to further simplify the editing procedure and attain best results for the readers.

Apart from the editorial board, the designing team has also invested a significant amount of their time in understanding the subject and creating the most relevant covers. They scrutinized every image to scout for the most suitable representation of the subject and create an appropriate cover for the book.

The publishing team has been an ardent support to the editorial, designing and production team. Their endless efforts to recruit the best for this project, has resulted in the accomplishment of this book. They are a veteran in the field of academics and their pool of knowledge is as vast as their experience in printing. Their expertise and guidance has proved useful at every step. Their uncompromising quality standards have made this book an exceptional effort. Their encouragement from time to time has been an inspiration for everyone.

The publisher and the editorial board hope that this book will prove to be a valuable piece of knowledge for researchers, students, practitioners and scholars across the globe.

List of Contributors

Zhongjian Wu and Chengming Luo
College of IOT Engineering, Hohai University, Changzhou 213022, China

Xinnan Fan and Jianjun Ni
Jiangsu Universities and Colleges Key Laboratory of Special Robot Technology, Hohai University, Changzhou 213022, China

Haiyan An
Tianjin Key Laboratory for Advanced Mechatronic System Design and Intelligent Control, School of Mechanical Engineering, Tianjin University of Technology, Tianjin 300384, China

Bin Li, Shoujun Wang and Weimin Ge
Tianjin Key Laboratory for Advanced Mechatronic System Design and Intelligent Control, School of Mechanical Engineering, Tianjin University of Technology, Tianjin 300384, China
National Demonstration Center for Experimental Mechanical and Electrical Engineering Education (Tianjin University of Technology), China

Min Lin and Tuanjie Li
School of Electro-Mechanical Engineering, Xidian University, Xi'an 710071, China

Zhifei Ji
College of Mechanical and Energy Engineering, Jimei University, Xiamen 361021, China

Nga Thi-Thuy Vu, Nam Phuong Tran and Nam Hoai Nguyen
Hanoi University of Science and Technology, Vietnam

Satwinder Singh and Ekta Singla
Mechanical Engineering Department, IIT Ropar, Rupnagar 140001, India

Ruiqing Mao
Xuzhou Institute of Technology, Xuzhou, Jiangsu 221111, China

Xiliang Ma
Xuzhou Institute of Technology, Xuzhou, Jiangsu 221111, China
School of Mechanical and Electrical Engineering, China University of Mining and Technology, Xuzhou, Jiangsu 221116, China

Benjamin Vedder and Jonny Vinter
Department of Electronics, RISE Research Institutes of Sweden, Sweden

Magnus Jonsson
School of Information Technology, Halmstad University, Sweden

Qing Chang
School of Mechanical Engineering, Tianjin University of Commerce, Tianjin, 300134, China

Fanghua Mei
School of Automation Science and Electrical Engineering, Beijing University of Aeronautics and Astronautics, Beijing 100191, China

Navid Negahbani and Enrico Fiorz
Dipartimento di Meccanica, Politecnico di Milano, Campus Bovisa Sud, Via La Masa 1, 20156 Milano, Italy

Hermes Giberti
Dipartimento di Ingegneria Industriale e dell'Informazione, Università degli Studi di Pavia, Via A. Ferrata 5, 27100 Pavia, Italy

Arthur Seibel, Stefan Schulz and Josef Schlattmann
Workgroup on System Technologies and Engineering Design Methodology, Hamburg University of Technology, 21073 Hamburg, Germany

Toma Morisawa, Kotaro Hayashi and Ikuo Mizuuchi
Department of Mechanical Systems Engineering, Tokyo University of Agriculture & Technology, 2-24-16 Nakacho, Koganei, Tokyo 184-8588, Japan

Michiel Aernouts, Ben Bellekens and Maarten Weyn
IMEC, IDLab, Faculty of Applied Engineering, University of Antwerp, Groenenborgerlaan 171, 2000 Antwerp, Belgium

Thumeera R. Wanasinghe, George K. I. Mann and Raymond G. Gosine
IS Lab, Faculty of Engineering and Applied Science, Memorial University of Newfoundland, St. John's, NL, Canada A1B 3X5

Xuelei Wang and Bin Zhang
College of Engineering, China Agricultural University, Beijing 100083, China

Dongjie Zhao
College of Engineering, China Agricultural University, Beijing 100083, China
School of Mechanical and Automobile Engineering, Liaocheng University, Liaocheng 252059, China

Ying Zhao
School of Mechanical and Automobile Engineering, Liaocheng University, Liaocheng 252059, China

Pengchao Zhang
Key Laboratory of Industrial Automation of Shaanxi Province, Shaanxi University of Technology, Hanzhong, Shaanxi 723000, China

YuKang Jia, Zhicheng Wu and Yanyan Xu
School of Information Science and Technology, Beijing Forestry University, No. 35 Qinghuadong Road, Haidian District, Beijing 100083, China

Dengfeng Ke
National Laboratory of Pattern Recognition, Institute of Automation, Chinese Academy of Sciences, No. 95 Zhongguancundong Road, Haidian District, Beijing 100190, China

Kaile Su
College of Information Science and Technology, Jinan University, No. 601, West Huangpu Avenue, Guangzhou, Guangdong 510632, China

S. Sankhar Reddy Chennareddy, Anita Agrawal and Anupama Karuppiah
EEE Department, NH-17B, BITS Pilani KK Birla Goa Campus, Goa 403726, India

Bilal M. Yousuf
Electronics & Power Engineering, PN Engineering College, National University of Sciences and Technology, Karachi, Pakistan

Asim Mehdi, Abdul Saboor Khan, Aqib Noor and Arslan Ali
Electrical Department, Fast-National University of Computing and Emerging Sciences, Karachi, Pakistan

Thilina H. Weerakkody, Thilina Dulantha Lalitharatne and R. A. R. C. Gopura
Bionics Laboratory, Department of Mechanical Engineering, University of Moratuwa, Katubedda, Sri Lanka

Index

A

Acoustic Localization System, 1

Adaptive Neuro-fuzzy Inference Systems (ANFIS), 39

Adaptive-robust Control, 86-87, 89, 91, 93, 95-99

Angle Of Arrival, 1, 128

Ant Colony Optimization, 115

Autonomous Underwater Vehicles, 1, 12, 14

B

Backstepping Method, 170

Ball-screw-driven Transmission, 86

C

Cebot, 184, 188, 196, 199

Central Pattern Generators, 75, 83, 85

Centroid, 159, 161-163, 165-166

Coal Mine Rescue Robot, 63

Coal Mine Robot, 57-61, 63

Connectionist Temporal Classification, 176, 182

D

Degrees Of Freedom, 15-17, 30, 33, 35, 46-47, 50-52, 55, 185

Differential Explosion, 170, 175

Direct Kinematics Problem, 101-102, 106, 108-109

Disobedient Robots, 110

Disturbance Observer, 170-171, 175

Dynamic Surface Adaptive Robust Control Method, 170

E

Electromagnetic Radiation, 1

Excavator Arm, 39-40, 43, 45

F

Flight Simulator, 26

Forward Kinematics, 109

Fuzzy Systems, 13, 39

G

Genetic Algorithm, 39, 45, 109

Global Navigation Satellite Systems, 1

Global Positioning System, 141

Graphical User Interface, 66, 73

H

Heterogeneous Robots, 110-112, 116-117, 126-127

Hexaglide Robot, 87, 89

Hexapod Robot, 75, 77-79, 82-85, 99-100

Hidden Markov Model, 176

I

Inertial Measurement Unit, 73

Inertial Measurement Units, 1

Intuitionistic Fuzzy Sets, 112, 126

K

Kinematic Redundancy, 46-47

Kinematics, 15-17, 19, 23-27, 31, 33, 35, 37-38, 87-89, 91, 93, 96, 101-102, 106, 108-109, 198

L

Large-scale Direct Slam, 129, 139

Linear Actuators' Orientations, 108

Long Short-termmemory, 176

M

Manipulator Robot, 39

Mechanism Position Analysis, 17

Minimum-enclosing-circle (MEC), 159, 163

Mobile Spray Robotic System, 159

Modular Self-reconfigurable Robots, 183

Multirobot System, 45, 110

Multirobot Task Allocation System, 110

N

Neural Network, 39-40, 43, 45, 55, 75, 78-80, 83, 85, 109, 115, 176-179, 181-182

Node Energy Consumption, 57-58

Node Energymodel, 58

O

Obstacle Avoidance, 39-40, 48, 55, 65, 73, 125

Overall Jacobian Matrix, 19

P

Parallel Kinematic Machines, 86, 100

Parallel Kinematic Mechanism, 25

Particle Swarm Optimization, 1, 14

Polyvinylidene Fluoride Patches, 27

Printed Circuit Board, 66, 73

Q

Q-learning Algorithm, 58-60

Qlsorp Algorithm, 60-63

R
Real-time Kinematic Satellite Navigation, 65, 73
Recurrent Neural Networks, 176, 181-182
Robotics, 8, 58, 60, 62-64, 66, 68, 70-72, 74, 118, 120, 122, 124, 126-127, 130, 132, 134, 136, 138-140, 142, 146, 148, 150, 152, 154, 180, 182-186, 188, 190-192, 194, 196
Roll-pitch Orientation, 101-102, 108

S
Single-robot Tasks, 111
Singular Configuration, 19, 31
Slam Algorithm, 128-132, 134-135, 138
Speech Recognition, 181-182

T
Target Spray Robotic System, 159-161, 163, 165, 167, 169
Task Space Locations, 47, 51
Tensegrity Systems, 26-27, 37
Tensegrity Walking Robot, 27
Time Difference Of Arrival, 1
Tracking Controller, 170-171

U
Unconventional Robotic Parameters, 46-48, 51, 53, 55
Unmanned Marine Vehicle, 170

V
Velocity, 1, 7-8, 15, 19, 23, 35, 40, 45, 73, 75-80, 82, 88, 91, 95, 115-118, 120-121, 149, 167, 170-171

W
Wireless Sensor Networks, 57-58, 60, 62-64

www.ingramcontent.com/pod-product-compliance
Lightning Source LLC
Chambersburg PA
CBHW080704200326
41458CB00013B/4956
* 9 7 8 1 6 3 2 3 8 7 2 3 3 *